Advanced Research on Hydraulic Engineering and Hydrological Modelling

Advanced Research on Hydraulic Engineering and Hydrological Modelling

Guest Editors

Shiqiang Wu
Ang Gao

MDPI

Basel • Beijing • Wuhan • Barcelona • Belgrade • Novi Sad • Cluj • Manchester

Guest Editors

Shiqiang Wu
Hydraulic Engineering
Department
Nanjing Hydraulic Research
Institute
Nanjing
China

Ang Gao
Hydraulic Engineering
Department
Nanjing Hydraulic Research
Institute
Nanjing
China

Editorial Office
MDPI AG
Grosspeteranlage 5
4052 Basel, Switzerland

This is a reprint of the Special Issue, published open access by the journal *Water* (ISSN 2073-4441), freely accessible at: https://www.mdpi.com/journal/water/special_issues/ZRUO2681VH.

For citation purposes, cite each article independently as indicated on the article page online and as indicated below:

Lastname, A.A.; Lastname, B.B. Article Title. *Journal Name* **Year**, *Volume Number*, Page Range.

ISBN 978-3-7258-6073-9 (Hbk)
ISBN 978-3-7258-6074-6 (PDF)
https://doi.org/10.3390/books978-3-7258-6074-6

Cover image courtesy of Ang Gao

Contents

About the Editors

Shiqiang Wu

Shiqiang Wu holds a doctorate in engineering and serves as a doctoral supervisor. His research focuses on river–lake water system connectivity and water security, urban water environment improvement, high-dam flood discharge and energy dissipation, and atomization. He has been recognized through several national and provincial talent programs, including the National Hundred and Thousand Talents Program, the Ministry of Water Resources' "5151 Talent Program," and Jiangsu Province's "333 High-level Talent Training Program" (second level). He has been honored as an "Outstanding Young Expert with Significant Contributions" and receives the Special Allowance from the State Council. He has led the completion of more than 200 research projects, including national science and technology initiatives, major water conservancy projects, 863 Program projects, key R&D programs, and major engineering research projects. He is a recipient of three National Science and Technology Awards and over ten provincial/ministerial-level science and technology awards. He has published more than 180 papers, authored or co-authored 18 monographs and standards, and holds over 40 national invention patents.

Ang Gao

Ang Gao, Ph.D., specializes in engineering and environmental hydraulics. He has led research projects funded by the National Natural Science Foundation of China, a sub-project of the National Key R&D Program, and central-level research funds. He has also conducted over 20 hydraulic research and consulting projects for major infrastructures, including the Baihetan Hydropower Station and the Xudapu Nuclear Power Plant. He is recognized as a Young Scientific and Technological Talent of Jiangsu Province and an International Talent of the Ministry of Water Resources, and is registered as a science and technology consulting expert in Jiangsu and Zhejiang. He also serves as a training instructor for the IAEA. He has received 10 provincial/ministerial awards, published 36 papers, holds 32 invention patents, contributed to 2 industry standards, and developed 5 advanced practical technologies. He serves as a Guest Editor for *Water*, a Young Editorial Board member for *South-to-North Water Transfers and Water Science & Technology* and *Water Resources Development Research*, and a reviewer for multiple SCI-indexed journals.

Preface

This Reprint aims to offer a comprehensive and multidisciplinary overview of recent advancements in hydraulic engineering and hydrological modeling. The collected works focus on fundamental principles, governing mechanisms, and technological innovations driving hydraulic processes, while also addressing challenges in the design and management of water infrastructure. By integrating theoretical insights with engineering practice, this volume seeks to advance both scientific understanding and practical applications in hydraulics and hydrology. The contributions cover a broad scope, reflecting the diversity of contemporary research in the field. Topics can generally be categorized as follows: (1) high-dam hydraulics; (2) channel hydraulics; (3) cylinder flow; (4) air–water two-phase flow; and (5) hydrological modeling. Collectively, these studies underscore the interdependence of hydraulic and hydrological processes across various spatial and temporal scales and emphasize the need for interdisciplinary approaches to tackle water-related challenges. This Reprint also highlights the value of coupling numerical modeling with experimental studies, as their integration can yield more robust insights into hydraulic performance and infrastructure safety. The motivation behind this compilation stems from the growing global demand for water security, sustainable reservoir management, and resilient hydraulic structures amid evolving environmental and social conditions. Dams and reservoirs play a crucial role in flood control, energy generation, irrigation, and ecological regulation. Ensuring their long-term safety and efficiency demands continuous innovation, rigorous scientific analysis, and reliable engineering solutions. By synthesizing methodological advances and applied research, the contributions in this Reprint serve as a timely resource for current practice and future development. This publication is intended for a wide audience: researchers and graduate students will discover new theoretical insights and methodologies; engineers and designers will find practical applications; and practitioners, project managers, and policymakers will benefit from case-based findings that support evidence-based decision-making and sustainable water management strategies. In summary, the studies presented herein provide valuable perspectives on hydraulic and hydrological sciences and point to emerging trends in the design, operation, and management of dams and reservoirs. It is our hope that this Reprint will inspire further research, foster interdisciplinary collaboration, and contribute to the safe and sustainable development of water infrastructures worldwide.

Shiqiang Wu and Ang Gao
Guest Editors

water

Editorial

Advanced Research on Hydraulic Engineering and Hydrological Modelling

Ang Gao * and Shiqiang Wu

Nanjing Hydraulic Research Institute, Nanjing 210029, China; sqwu@nhri.cn
* Correspondence: agao@nhri.cn

1. Introduction to the Special Issue

Hydraulics and hydrology are ancient disciplines that play a crucial role in ensuring water security, safeguarding water environments, and maintaining water ecology. In recent years, the construction of high dams, such as the Baihetan Project, has not only resolved numerous problems but also generated a wealth of practical theories, advanced techniques, and valuable practical experiences in the fields of hydraulics and hydrology [1,2]. Physical models, numerical models, and prototype measurements are traditional research methods [3–5]. However, with the rapid advancement of emerging technologies like information technology [6] and big data [7], the integration of these technologies with hydraulics and hydrology has emerged as a prominent research topic in the current academic landscape. It is essential to summarize and disseminate the above-mentioned related research findings to further promote the development of this area of interest.

Consequently, this Special Issue of *Water* was launched. Since the call for papers was announced, twelve original articles and one review article have been accepted for publication following a rigorous peer-review process (Contributions 1–13). These papers can be categorized into several domains: high dam hydraulics, channel hydraulics, cylinder flow, air–water two-phase flow, and hydrological modeling. To provide a better understanding of this Special Issue, we hereby summarize the key highlights of the published papers.

2. Overview of the Contributions to This Special Issue

Regarding hydraulic research on high dams, Zheng et al. (Contribution 1) investigated the emergency discharge capacity of ultra-high dams under extreme events such as floods and earthquakes by reviewing dams over 200 m worldwide. The study revealed inadequate discharge capacity, poor structural safety at high heads, excessive hoisting requirements, and complex sealing technologies. To overcome these challenges, strategies such as multi-stage gates and siphon layouts were proposed, emphasizing enhanced gate pressure resistance, intelligent control, and integrated risk management to ensure continuous safe drawdown and strengthen disaster resilience. High-altitude hydraulic projects face increased cavitation risk due to reduced atmospheric pressure, and conventional aeration designs perform poorly under such conditions. Guo et al. (Contribution 2) conducted decompression chamber tests simulating various pressures and flow rates to analyze air discharge, cavity characteristics, and air–water concentration. Results showed that lower pressure significantly reduces aeration and air–water content. A correction formula was developed to accurately predict aeration performance under low pressure, providing a theoretical basis for ventilation design in high-altitude hydraulic structures. Cavitation is a major safety concern in high-head spillways. Dong et al. (Contribution 3) addressed

 https://doi.org/10.3390/w17172562

the severe backflow and ventilation failure of conventional aerators in mild-slope chutes under low Froude numbers by proposing a curved aerator with locally increased slope to reduce impact angle. Shape parameters were derived and validated through numerical simulations. Results show that appropriately lowering the jet impact point eliminates backflow, significantly outperforms traditional designs, and effectively resolves cavitation protection challenges under mild slopes and high discharge conditions. Overtopping and seepage are the primary causes of tailings dam failures. Gao et al. (Contribution 4) conducted a 1:100 physical model test and combined numerical simulations to investigate the breaching process and downstream hazard evolution. Results show that failure progresses through seepage stabilization and flow-slide development stages before full crest breach. Measures such as controlling the phreatic line, adding drainage systems, and improving flood regulation were proposed to enhance dam safety and provide technical guidance for emergency management.

Regarding channel hydraulics, Smith et al. (Contribution 5) used the River2D model combined with field data to simulate velocity and shear stress distributions in a shallow river reach under ice-covered and open-water conditions. The results show that ice cover reduces overall flow velocity but enhances local recirculation and alters high shear stress distribution, while narrowing the main channel and increasing water depth, which may affect fish habitats. This study provides an important basis for understanding winter hydraulic characteristics and ecological impacts. China's water resources are unevenly distributed between the north and south, making inter-basin water transfer projects highly demanding and risky. Zhang et al. (Contribution 6) addressed the issues of complex channel response and insufficient scheduling safety in such projects by developing a hydraulic model for the Yangtze–Huaihe Water Transfer Project. They proposed a safe regulation range based on the earliest and latest adjustment times and identified flow variation as the key factor through the Sobol sensitivity analysis, further deriving a power function-based rapid prediction formula. Results indicate high prediction accuracy, enabling the replacement of traditional models, improving scheduling efficiency, and ensuring safe project operation. Climate change has increased flash flood risks in steep-slope watersheds, while existing hydrodynamic models often neglect canopy interception and litter storage, reducing simulation accuracy. Ámon et al. (Contribution 7) integrated these processes into the HEC-RAS model to improve flash flood simulation in steep basins and compared two approaches for representing litter: as rainfall loss and as a high-porosity soil layer. Results show that litter significantly reduces surface runoff and peak discharge. This study provides a scientific basis for small watershed flood modeling and risk management.

In terms of cylinder flow research, Wang et al. (Contribution 8) developed a coupled internal wave–cylinder–topography model using large eddy simulation to examine how topography influences cylinder loading. The results show that irregular topography significantly alters the direction of horizontal forces, enhances shallow water effects, and greatly increases reverse forces on the lower section, while higher internal wave amplitudes intensify flow strength and vortex activity. This study reveals the hydrodynamic response of structures under complex terrain, providing valuable guidance for the safe design of hydraulic structures. Most studies on cylinder flow neglect boundary effects and lack the systematic analysis of how confinement influences flow structures. Xu et al. (Contribution 9) used two-dimensional numerical simulations to examine the impact of boundary confinement on cylinder flow characteristics. The results show that confinement does not alter the four fundamental flow regimes but reduces vortex size, shifts vortex centers toward the boundary, and significantly increases drag and vortex shedding frequency when $G/D < 3.5$.

Prediction formulas for drag and shedding frequency were proposed, providing reference for the design and vibration control of nearshore structures.

In terms of high-speed air–water two-phase flow research, Luo et al. (contribution 10) conducted fluent-based numerical simulations to analyze concentric double-nozzle non-submerged cavitation jets, focusing on the influence of nozzle geometry and pressure parameters on the flow field. Results indicate that contraction and expansion sections significantly enhance cavitation intensity, with the expansion section enlarging the cavitation zone, and smaller contraction angles yielding better performance; moderately increasing throat diameter and nozzle pressure further improves impact pressure. The study proposes an optimized nozzle design method, providing theoretical guidance for large-scale surface strengthening applications. Composite cavitation nozzles can significantly enhance cavitation performance, yet the influence of key structural parameters and effective optimization strategies remain unclear. Huang et al. (Contribution 11) employed large-eddy simulation combined with the response surface method to optimize nozzle structural parameters. The results indicate that the optimized design substantially increased the vapor volume fraction compared with conventional nozzles, with model prediction errors of only 2.04% and impact force test errors ranging from 1.5% to 2.7%. This study introduces a systematic optimization approach, providing theoretical and technical support for cavitation jet nozzle design.

In terms of hydrological modeling research, Jin et al. (Contribution 12) proposed an improved vertical exchange approach based on node water balance, integrating vertical flow computation into the 1D hydraulic equation with iterative solutions to resolve unreasonable backflow calculations and instability in conventional models. The results indicate that the improved method achieves coupling results highly consistent with the ICM model while significantly enhancing computational efficiency, providing a high-accuracy, low-subjectivity solution for urban flood simulation and drainage system optimization. The Jakarta government plans to construct a coastal reservoir at the river mouth to alleviate water shortages and flood risks, but feasibility assessment methods tailored to Indonesian conditions are lacking. Soekarno et al. (Contribution 13) proposed a pre-feasibility analysis framework for coastal reservoirs, incorporating hydrological calculations, operational simulations, and flood and sediment assessments. The results show that the reservoir provides limited flood regulation, lowering upstream water levels only in the early stages. This study offers a scientific basis for planning inter-basin water supply and flood control projects in Indonesia.

3. Conclusions

This Special Issue features multidisciplinary scholarly works that aim to provide an in-depth comprehension of the scientific principles and mechanisms, encompassing critical technologies, challenges, and concepts, in the fields of hydraulic engineering and hydrological modelling. The guest editors anticipate that the papers published in this Special Issue will appeal to researchers, designers, and practitioners involved in the design and management of dams and reservoirs and will also assist in identifying future research directions. The research findings and methodologies presented in this Special Issue, including hydraulic research on high dams, channel hydraulics, cylinder flow, air–water two-phase flow, and hydrological modelling, hold great research significance. These technological contributions can aid relevant scholars and project managers in analyzing and managing the safety of the major structures in reservoir dams.

Funding: This research was funded by National Natural Science Foundation of China (grant no. 52209096, 52179073); special funds for basic scientific research operations of central-level public welfare research institutes (grant no. Y124017, Y124001); and Jiangsu Province Youth Science and Technology Talent Support Program (grant no. JSTJ-2024-182).

Data Availability Statement: Not applicable.

Acknowledgments: As Guest Editor of this Special Issue, I would like to express my deep appreciation to all the authors whose valuable work was published in this Issue and who have thus contributed to the success of this edition.

Conflicts of Interest: The authors declare no conflicts of interest.

List of Contributions:

1. Zheng, X.; Zhang, L.; Yang, J.; Du, S.; Wu, S.; Luo, S. Technical Challenges of Safety Emergency Drawdown for High Dam and Large Reservoir Project. *Water* **2023**, *15*, 1538. https://doi.org/10.3390/w15081538.
2. Guo, Y.; Zhang, L.; Yu, L.; Luo, S.; Liu, C.; Liu, Y. The Impact of Decreased Atmospheric Pressure on Forced Aeration of Discharged Flow. *Water* **2024**, *16*, 353. https://doi.org/10.3390/w16020353.
3. Dong, Y.; Li, G.; Liu, S.; Li, S.; Li, P.; Wei, Y. A Study on the Shape of Parabolic Aeration Facilities with Local Steepness in Slow Slope Chutes. *Water* **2024**, *16*, 1574. https://doi.org/10.3390/w16111574.
4. Gao, Z.; Liu, J.; He, W.; Lu, B.; Wang, M.; Tang, Z. Study of a Tailings Dam Failure Pattern and Post-Failure Effects under Flooding Conditions. *Water* **2024**, *16*, 68. https://doi.org/10.3390/w16010068.
5. Smith, K.; Cockburn, J.M.H.; Villard, P.V. Rivers under Ice: Evaluating Simulated Morphodynamics through a Riffle-Pool Sequence. *Water* **2023**, *15*, 1604. https://doi.org/10.3390/w15081604.
6. Zhang, N.; Ren, H.; Lin, F. Study on the Hydraulic Response of an Open-Channel Water Transmission Project after Flow Switching. *Water* **2023**, *15*, 3201. https://doi.org/10.3390/w15183201.
7. Ámon, G.; Bene, K.; Ray, R.; Gribovszki, Z.; Kalicz, P. Improving Flash Flood Hydrodynamic Simulations by Integrating Leaf Litter and Interception Processes in Steep-Sloped Natural Watersheds. *Water* **2024**, *16*, 750. https://doi.org/10.3390/w16050750.
8. Wang, Y.; Xu, M.; Wang, L.; Shi, S.; Zhang, C.; Wu, X.; Wang, H.; Xiong, X.; Wang, C. Numerical Investigation of the Stress on a Cylinder Exerted by a Stratified Current Flowing on Uneven Ground. *Water* **2023**, *15*, 1598. https://doi.org/10.3390/w15081598.
9. Xu, Z.; Wu, S.; Wu, X.; Xue, W.; Wang, F.; Gao, A.; Zhang, W. Analysis of Flow Characteristics around a Square Cylinder with Boundary Constraint. *Water* **2023**, *15*, 1507. https://doi.org/10.3390/w15081507.
10. Luo, Y.; Zang, J.; Zheng, H. Flow Field and Gas Field Distribution of Non-Submerged Cavitation Water Jet Based on Dual-Nozzle with Concentric Configuration. *Water* **2023**, *15*, 2904. https://doi.org/10.3390/w15162904.
11. Huang, G.; Qiu, C.; Song, M.; Qu, W.; Zhuang, Y.; Chen, K.; Huang, K.; Gao, J.; Hao, J.; Hao, H. Optimization of Composite Cavitation Nozzle Parameters Based on the Response Surface Methodology. *Water* **2024**, *16*, 850. https://doi.org/10.3390/w16060850.
12. Jin, X.; Mu, Y. Hydrodynamic Simulation of Urban Waterlogging Based on an Improved Vertical Flow Exchange Method. *Water* **2024**, *16*, 1563. https://doi.org/10.3390/w16111563.
13. Soekarno, I.; Adityawan, M.B.; Sandi, C.; Amatullah, S.I.; Farid, M.; Suryadi, Y.; Yakti, B.P.; Chrysanti, A.; Kuntoro, A.A.; Widyaningtias; et al. A Methodology for Water Resource Management and the Planning of the Coastal Reservoir in Indonesia. *Water* **2024**, *16*, 344. https://doi.org/10.3390/w16020344.

References

1. Zhang, J.M. Hydraulics of high-speed flows: Recent achievements and future outlook. *J. Hydroelectr. Eng.* **2021**, *40*, 1–18.
2. Liu, Z.J.; Guo, S.L.; Xu, X.F.; Xu, S.C.; Cheng, J.Q. Application of Copula functions in hydrology and water resources: A state-of-the-art review. *Adv. Water Sci.* **2021**, *32*, 148–159.
3. Savage, B.M.; Johnson, M.C. Flow Over Ogee Spillway: Physical And Numerical Model Case Study. *J. Hydraul. Eng.* **2001**, *127*, 640–649. [CrossRef]
4. Duan, W.G.; Hou, D.M.; Wang, C.H.; Hu, H.; Tang, X.F. Hydraulic prototype observation and analysis of Three Gorges Dam discharge structures. *J. Hydraul. Eng.* **2019**, *50*, 1339–1349.
5. Zeng, Y.J.; Yan, G.H.; Shi, X.R.; Sun, Y.X.; Zhu, C.Y. Hydraulic and Structural Dynamic Safety Prototype Observation and Analysis of a High-head Spillway Tunnel. *J. Water Resour. Archit. Eng.* **2020**, *18*, 111–116.
6. Chen, J.H.; Wang, C.H.; Guo, S.Y. Application analysis of digital technology in high dam and large reservoir survey. *China Water Resour.* **2022**, 62–64.
7. Jiang, Y.Z.; Ye, Y.T.; Zhao, H.L.; Liang, L.L.; Cao, Y.; Gu, J.J. Research status and prospects on water conservancy big data. *J. Hydroelectr. Eng.* **2020**, *39*, 1–32.

water

MDPI

Article

A Study on the Shape of Parabolic Aeration Facilities with Local Steepness in Slow Slope Chutes

Yuping Dong [1], Guodong Li [1,*], Shaobin Liu [2], Shanshan Li [1], Pengfeng Li [2] and Yong Wei [2]

1 State Key Laboratory of Eco-Hydraulics in Northwest Arid Region, Xi'an University of Technology, Xi'an 710048, China; dongyup0119@163.com (Y.D.); shanshanli@xaut.edu.cn (S.L.)
2 Northwest Engineering Corporation Limited, Xi'an 710065, China; liushaob@nwh.cn (S.L.); lipengfeng1994@163.com (P.L.); wyandch@163.com (Y.W.)
* Correspondence: gdli2008@xaut.edu.cn

Abstract: For flood discharge structures with high water heads, aeration facilities are usually installed in engineering to promote water flow aeration and prevent cavitation damage to the overflow surface. Actual engineering has shown that as the slope of the discharge channel bottom decreases or water level changes lead to a decrease in the Froude number, the cavity morphology after conventional aeration facilities or allotype aerators is poor. This article proposes a curved aeration facility scheme based on the idea of locally increasing the bottom slope to reduce the impact angle, which is formed by the convex parabolic bottom plate and concave parabolic bottom plate. The convex parabolic bottom plate is tangent to a flat bottom plate behind the offset, and the concave parabolic bottom plate is tangent to the downstream. The jet landing point is controlled at the junction of the convex parabolic bottom plate and the concave parabolic bottom plate, and the lower jet trajectory is in line with the parabolic bottom plate. The corresponding parabolic bottom plate calculation formulas were theoretically derived, and the design method of the shape parameters of the aeration facility was provided. Through specific engineering case studies, it was found that: (1) As the Z_{AC}/Z_{AG} value increases, point C becomes closer to point G, the slope of the water tongue landing point C becomes steeper, and the cavity is less likely to return water. (2) When the position of the water tongue landing point is 0.5–0.8 times the height of the water tongue impact point, there is almost no water accumulation in the calculated cavity. At this time, the platform length $L_{AB} = 0.5L_{AF}$, the convex parabolic section length L_{BC} = (0.45–0.6) L_{AG}, the concave parabolic section length L_{CD} = (0.43–0.11) L_{AG}, the convex parabolic section calculation formula is z (x) = $-A_1x_2$ (A_1 = 0.0059–0.00564), and the concave parabolic section calculation formula is $A_2x_2 - B_2x_2$ (A_2 = 0.003347–0.01927).This solved the problem of aeration and corrosion reduction under small bottom slope, large-unit discharge, and low Froude number engineering conditions.

Keywords: parabolic aerated facilities; no backwater; small bottom slope; impact angle; the lower jet trajectory

1. Introduction

In high head flood discharge structures, the water flow velocity is high. Low-pressure areas can easily form in the contraction part of the inlet section of the flood discharge tunnel, the side wall behind the gate groove, the turning section of the tunnel, the top of the overflow weir, the uneven part of the discharge groove, and the energy dissipation ridge. When the pressure in these areas drops below the saturated vapor pressure of the water, the gas core in the water will quickly expand and form bubbles. These bubbles will flow with the water flow and quickly collapse if the pressure rises again [1]. The collapse process of bubbles will generate great impact force and high temperature, causing serious damage to the walls of flood discharge structures, namely cavitation damage [2]. Aeration and erosion reduction measures are usually adopted to prevent cavitation damage on the flow surface [3,4]. At present, aeration facilities can be divided into conventional aeration facilities and small bottom slope aeration facilities [5]. A conventional aeration facility consists of several transverse and vertical grooves around the chute walls that are connected to an air supply. Air is entrained into the water flow through the air cavity neighboring the water flow surface owing to the difference in atmospheric pressure. The cavity jet length, air entrainment coefficient, and air diffusion evolution near the wall are the main parameters used to describe the aerator efficiency [6]. For example, Pfister and Hager's [7,8] investigation focuses on the flow structure and the air transport downstream of chute aerators: Minimum F_r values for efficient air entrainment by aerators are $F_r = 6$ for offsets, $F_r = 5$ for deflector aerators with the deflector angle = 6°, and $F_r = 4$ for the deflector angle = 11°. Steep deflectors are more efficient than those with small values of the deflector angle. Aerators with a floor slope of 30° on steep chutes operate more efficiently than those on flat chutes. Therefore, as the bottom plate angle of the discharge channel decreases, the efficiency of conventional aeration facilities significantly decreases. This is mainly due to the serious return water in the cavity under the action of small bottom slopes, which blocks the ventilation holes and leads to the failure of aeration facilities [9]. In recent years, domestic scholars have proposed some small bottom slope aeration facilities based on engineering practice, namely irregular aerators. These special-shaped aerators have the same structural parameters as conventional aerators. By controlling the lateral distribution of water tongue landing points and reducing the impact angle between the water tongue and the bottom plate behind the aerator, the shape of the aerator is changed to some extent, eliminating cavity backflow. For example, Pang et al. [10] proposed a U-shaped groove aerator; Sun et al. [11] proposed a concave aerator in conjunction with the Xiao wan spillway tunnel; Wang et al. [12] proposed a V-shaped groove aerator; Deng et al. [6] proposed a wedge-shaped aerator arranged at the end of the pressurized inlet of the small bottom slope open channel. When the bottom slope of the discharge channel is small, the application range of these irregular aerators is small, and cavity water accumulation is still inevitable. According to research results, the key factor causing cavity water accumulation is the impact angle between the jet water tongue and the bottom plate. When the impact angle decreases to a certain extent, the water accumulation disappears. The critical impact angle given by Zhang Liheng et al. [9] is 9°. Based on this, some scholars have proposed "aerator + local steep slope" [13], such as Sun Zhenxing, Wang Fangfang, and others [14] proposed a "planar V-shaped ridge + local steep slope" aeration facility to solve the problem of aeration and erosion reduction in flood discharge cavities under conditions of gentle bottom slopes and large water level fluctuations; Liu et al. [15] proposed a "bottom plate bending aerator" by fitting the local bottom plate curve with the lower jet trajectory. The aerated facility has a continuous and simple two-dimensional structure, which not only effectively suppresses cavity backflow, but also has a good flow state. However, due to the lack of specific design methods for shape parameters, the aerated facility still needs to continuously adjust the local bottom plate shape behind the offset in specific engineering applications, which is time-consuming and laborious. Based on the research results of Liu et al. [15], this article proposes a curved aeration facility scheme based on the idea of locally increasing the bottom slope to reduce the impact angle of the jet flow. The parabolic curve of the bottom plate is theoretically derived, and a design method for the shape parameters of the aeration facility is provided. The reliability and rationality of this design method are verified through engineering examples. Through engineering examples, it is verified that the parabolic aerated facility proposed in this article effectively eliminates the problem of cavity backflow and solves the problem of cavity backflow in small bottom slope aerated facilities.

2. The Principle of the Double Parabolic Aerator Shape Design

The impact angle of the jet flow behind offset is the most critical factor causing cavity backflow. When the impact angle decreases to a certain extent, the accumulated water in the cavity disappears [16]. Based on this idea, as shown in Figure 1, this study transforms the local bottom slope behind the offset into a body structure composed of the flat bottom plate L_{AB}, the convex parabolic bottom plate L_{BC}, and the concave parabolic bottom plate L_{CD}. The lower edge OC of the jet flow impacts on the connection point C of the convex bottom plate and the concave parabolic bottom plate, where the local slope is the largest. By designing the parabolic curves reasonably, the lower jet trajectory fits to the bottom plate, effectively reducing the jet impact angle, eliminating cavity backflow, and ensuring sufficient aeration flow.

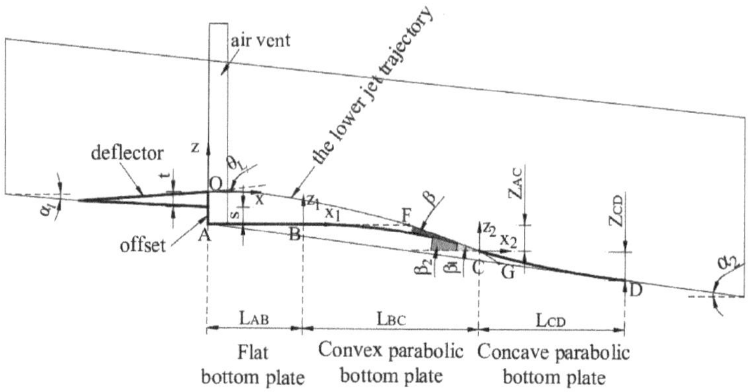

Figure 1. Definition sketch with notation.

Establish three Cartesian coordinate systems, xOz, x_1Bz_1, and x_2Cz_2, with the vertex O, B, and C as the origin, as shown in Figure 1. In the figure, the deflector height is t and the deflector angle is γ, and the jet take-off angle is θ_L. The bottom slopes of the upstream and downstream discharge channels are, respectively, $\alpha_1$and α_2. β_1 is the angle between the jet landing point C and the x-axis, β_2 is the tangent angle of the convex parabolic the bottom plate at point C, β is the impact angle between the lower jet trajectory and the convex parabolic bottom plate, and $\beta = \beta_1 - \beta_2$. The endpoint A at the bottom of the offset intersects with the lower jet trajectory at point F, and B is the point on the segment AF. Create a convex parabolic segment BC that is tangent to line segment AF at point B, tangent to the lower jet trajectory at point C, and then create a concave parabolic segment CD that is tangent to the convex parabolic segment at point C, and tangent to the downstream bottom plate at point D. This is the parabolic aerator design scheme.

Next, determine the relevant design parameters for the jet landing point C that fits the body shape.

3. The Lower Jet Trajectory

To construct the relationship between the lower edge of the jet water tongue behind the ridge and the local bottom plate, it is necessary to understand the trajectory of the jet water tongue movement.

Jet trajectories are typically described by the physically based point-mass parabola versus the take-off conditions, thereby assuming constant density and neglecting jet disintegration and aerodynamic interaction. However, the effective take-off angle of a jet trajectory is not identical with the boundary angle of the terminal structure due to pressure rearrangement. Therefore, take-off angles are usually derived from prototype observation or from physical model tests, thereby assuming a parabolic trajectory geometry and fitting the take-off angles, depending on the deflector's geometry and the hydraulic conditions [17].

In the case of Froude number $F_r < 7$, under the condition of a small bottom slope, the jet distance is short, the influence of aeration resistance is small, and the effect of gravity is significant, so only the correction of the jet take-off angle needs to be made [18]. According to mass point dynamics, the lower jet trajectory equation reads:

$$z(x) = x\tan\theta_L - \frac{x^2}{2F_r^2 h\cos^2\theta_L} \tag{1}$$

Here, the approach flow Froude number is F_r; the approach flow depth is h; the jet take-off angle is θ_L.

There have been many achievements in the study of the take-off angles of the lower jet trajectory, and significant achievements have been made, such as Chanson [19], Steiner et al. [20], Rutschmann and Hager [21], Wu and Ruan [22], Pfister [23], and Long Qiang et al. [24]. The shape of the parabolic aerator is achieved by designing the reasonable convex concave parabolic curve to achieve the lower jet trajectory which fits the bottom plate. The accuracy of the take-off angle of the lower jet trajectory directly affects the shape of the cavity, and also is a key parameter for the successful design of parabolic aerated facilities. In recent years, turbulence numerical simulation has been widely used in hydraulic engineering. Deng and Xu [25], Luo and Diao [26], and Gao et al. [27] simulated the aeration flow in tunnels and spillways. The results show that the numerical simulation results are in good agreement with the experimental data, which shows that turbulence numerical simulation is feasible for simulating the hydraulic characteristics of the Large Discharge Spillway Tunnel. Based on this, the numerical simulation method is used to calculate the take-off angle of the lower jet trajectory.

4. Calculation Formulas of the Parabolic Bottom Plate

As shown in Figure 1, the bottom plate structure behind the offset is divided into three parts: the first part is the platform bottom plate, the second part is the convex parabolic bottom plate, and the third part is the concave parabolic bottom plate. Through theoretical analysis, the impact point G is a key parameter for constructing the relationship between the lower edge of the jet water tongue and the local bottom plate. Therefore, the relationship between the platform section and the horizontal jet distance is established, and the relationship between the lower edge of the water tongue and the convex concave bottom plate is established. The specific process is as follows:

4.1. The Length of the Platform Bottom Plate

Put Z = $-\Delta$ into Equation (1), from which can be obtained the intersection point of the lower jet trajectory and the platform extension line:

$$L_{AF} = 0.5F_r^2 h\sin 2\theta_L\left(1 \pm \sqrt{1 + \frac{2\Delta}{F_r^2 h\sin^2\theta_L}}\right) \tag{2}$$

In the formula: Δ is the total height, m; if $\theta_L > 0$, take "+", if $\theta_L < 0$, take "$-$".
Therefore, the length of the platform bottom plate is written as:

$$L_{AB} = kL_{AF} \tag{3}$$

where k is the coefficient to be determined, and $0.45 \le k < 0.50$.

4.2. Calculation Formulas of the Convex Parabolic Bottom Plate

In the coordinate system x_1Bz_1, the formula for calculating the convex parabolic bottom plate is:

$$z(x_1) = a_1x_1^2 + b_1x_1 + c_1 \tag{4}$$

In the formula, a_1, b_1, and c_1 are undetermined coefficients.
When $x_1 = 0$ and $z_1 = 0$, $c_1 = 0$.

Taking the derivative of calculating Equation (4) yields the following Equations (5) and (6):

$$\left.\frac{dz_1}{dx_1}\right|_{x_1=0} = 2a_1x_1 + b^1 = 0 \tag{5}$$

$$\left.\frac{dz_1}{dx_1}\right|_{x_1=L_{BC}} = 2a_1L_{BC} + b_1 = -\tan\beta_2 \tag{6}$$

The calculation formula of the convex parabolic the bottom plate is:

$$z(x_1) = \frac{-\tan\beta_2}{2L_{BC}}x_1^2 \tag{7}$$

In the formula: $L_{AF} < L_{BC} + L_{AB} < L_{AG}$.

$$\beta_2 = \beta_1 - \beta$$

In the formula: β is the actual impact angle between the lower jet trajectory and the bottom plate; if $\beta = 0$, it indicates that the lower jet trajectory is completely tangent to the bottom plate. β_1 is obtained from Equation (8); based on $\beta_2 = \beta_1 - \beta$, β_2 can be obtained.

$$\tan\beta_1 = -\left.\frac{dz}{dx}\right|_{x=L_{AC}} \tag{8}$$

According to Equation (7), the length of L_{BC} depends crucially on the vertical distance Z_{AC} from the point C to the platform bottom plate. The value of Z_{AC} is closely related to the elevation of the original impact point G, as the elevation of the jet landing point C will not be lower than the original impact point G. Therefore,

$$Z_{AC} = nZ_{AG} \tag{9}$$

In the formula, $n \leq 1.0$ is used to ensure that the point C position is as far away as possible and the elevation is low, and n is taken as high as possible.

After the Z_{AC} value is determined, the L_{BC} value can be determined.

4.3. Calculation Formula of the Concave Parabolic Bottom Plate

In the second coordinate system x_2Cz_2, the formula of the concave parabolic the bottom plate is obtained as follows:

$$z(x_2) = \frac{-\tan\alpha_2 + \tan\beta_2}{2L_{CD}}x_2^2 - \tan\beta_2 x_2 \tag{10}$$

Here: $L_{CD} > L_{AG} - L_{AC}$.

The calculation formulas (3), (7), (9), and (10) are the general equations for parabolic bottom plates for an aerated facility. This equation is applicable to conventional two-dimensional aerated facilities.

5. Engineering Application

The ecological drainage (emptying) tunnel on the left bank of the particular hydropower station has a total length of approximately 1260 m. The outlet of the pressure tunnel is connected to the working gate chamber, and the working gate chamber is connected to the non-pressure tunnel section. The longitudinal slope of the non-pressure section bottom plate is i = 0.1, and the verified flood level is H = 3894.18 m, Q = 983 m^3/s.

5.1. The Shape Design Based on This Paper's Approach

The original first aerator in the non-pressure section of the ecological drainage tunnel was located at the 0 + 920.20 m section of the gate chamber. Due to the gentle slope of the bottom of the spillway, the stability of the cavity of conventional aeration facilities is poor, and is prone to the formation of water accumulation. Water accumulation will reduce the length of the cavity, causing a decrease in ventilation volume. In severe cases, it can block the ventilation holes and cause poor air intake. Aeration facilities have instead become artificially convex bodies, forming cavitation sources and causing cavitation and erosion damage. This article adopts a combination of "the small deflector + the offset + parabolic bottom plate" aerators, which is a parabolic aerator. The small deflector height is 0.3 m, with a slope of 1:10, and the offset height is 0.7 m. The reason for using a combination of "the small deflector + the offset + parabolic bottom plate" aerators is that the small deflector + the offset causes little disturbance to the original water surface and easily forms a stable cavity. The parabolic bottom plate can reduce the impact angle between the jet water tongue and the original bottom plate, effectively eliminating cavity backflow. This combination setting fully utilizes the advantages of various body structures, overcomes their shortcomings, and achieves the goals of aerating and reducing corrosion.

Given the hydraulic parameters under the verified water level, use the parabolic bottom plate calculation formula in the third subsection of the text to design the parameters of each part of the bottom plate. In the design process, not only should the convex parabolic bottom plate be tangent to the platform bottom plate, but also the concave parabolic section should be tangent to the downstream bottom plate. Therefore, iterative calculation methods should be used in the calculation and design process.

5.1.1. Known Conditions

Known hydraulic parameters in front of the deflector: the water depth h = 6.1 m, Froude number $F_r = 4.163$, total height of the aerator $\Delta = 1.0$ m, the small deflector height is 0.3 m, with a slope of 1:10, and the offset height is 0.7 m, upstream slope $i_1 = 0$, downstream bottom slope $i_2 = 0.1$; numerical simulation is used to simulate the combination of "the small deflector + the offset" combination aerator and calculate the take-off angle $\theta_L = 4\,°$of the lower jet trajectory.

5.1.2. The Control Parameters Calculation

① L_{AF} length

Calculate the position of the intersection point between the lower jet trajectory and the platform extension line using Equation (2):

$L_{AF} = 23.611$ m;

② L_{AB} length

According to Equation (3), let k = 0.5; then, the length of the platform segment is:

$L_{AB} = 11.806$ m;

③ L_{AG} length

Calculate the jet distance L_{AG} when the jet flow impacts the original bottom plate, $L_{AG} = 40.20$ m;

④ Z_{AG} height

Calculate the height of the impact point G where the jet flow impacts the original bottom plate relative to the platform section: $Z_{AG} = 4.02$ m.

5.1.3. The Design Parameters Calculation

① Calculate $\boldsymbol{Z_{AC}}$

After determining the $\boldsymbol{Z_{AC}}$ value using Equations (3) and (7), a reasonable body shape can be obtained. According to Equation (9), let n = 0.3, 0.4, 0.5, 0.6, 0.7, 0.8, 0.9, and calculate the $\boldsymbol{Z_{AC}}$ values listed in Table 1, represented by $M_{0.3}$, $M_{0.4}$, $M_{0.5}$;

Table 1. Type design based on this paper's approach.

Design Shape	Convex Parabolic Bottom Plates				Concave Parabolic Bottom Plates			$\beta_2/°$
	Z_{AC} /m	L_{AB}/m	L_{BC} /m	Calculation Formulas	L_{CD} /m	Calculation Formulas	Z_{CD} /m	
$M_{0.3}$	1.206	11.806	13.861	$-0.00628x^2$	37.292	$0.000994x^2 - 0.1741x$	5.108	9.876
$M_{0.4}$	1.608	11.806	16.305	$-0.00605x^2$	25.140	$0.001936x^2 - 0.1972x$	3.736	11.163
$M_{0.5}$	2.010	11.806	18.461	$-0.00590x^2$	17.601	$0.003347x^2 - 0.2177x$	2.796	12.289
$M_{0.6}$	2.412	11.806	20.416	$-0.00579x^2$	12.163	$0.005608x^2 - 0.2363x$	2.045	13.301
$M_{0.7}$	2.814	11.806	22.214	$-0.00571x^2$	7.909	$0.009704x^2 - 0.2534x$	1.397	14.225
$M_{0.8}$	3.216	11.806	23.887	$-0.00564x^2$	4.394	$0.019277x^2 - 0.26926x$	0.811	15.078
$M_{0.9}$	3.618	11.806	25.460	$-0.00558x^2$	1.380	$0.066795x^2 - 0.28420x$	0.265	15.873

② Calculate L_{BC} and β_2

According to $L_{AF} - L_{AB} < L_{BC} < L_{AG} - L_{AB}$, within this range, in descending order, β_1 is calculated by initially assuming the L_{BC} value, substituting it into Equation (8);

③ Calculation formula of the convex parabolic segment

Substitute the β_2 value and the L_{BC} value into Equation (7) to obtain the calculation formula for the convex parabolic bottom plate;

④ Calculate L_{CD}

Within the range of $L_{CD} > L_{AG} - L_{AC}$, in ascending order, the L_{CD} value that is inputted into Equation (10) to calculate Z'_{CD} is preliminarily assumed;

⑤ Calculation formula of the concave parabolic section

Take β_2, L_{BC}, α_2. Substitute into Equation (10) to obtain the calculation formula for the concave parabolic bottom plate.

5.2. The Influence of the Jet Landing Point C Position on the Bottom Plate Shape Parameters

List the parameters of each part of the bottom plate obtained above in Table 1, and plot the relationship between Z_{AC}/Z_{AG} and the relative length L_{BC}/L_{AG}, the relative length L_{CD}/L_{AG}, and the relative total length L_{AD}/L_{AG}, as shown in Figure 2. As the Z_{AC}/Z_{AG} value increases, the relative length L_{BC}/L_{AG} gradually increases, the relative length L_{CD}/L_{AG} significantly decreases, and the relative total length L_{AD}/L_{AG} gradually decreases. This indicates that the larger the Z_{AC}/Z_{AG} value, the closer point C is to point G, the steeper the slope of the jet landing point C, and the less likely the cavity is to backflow.

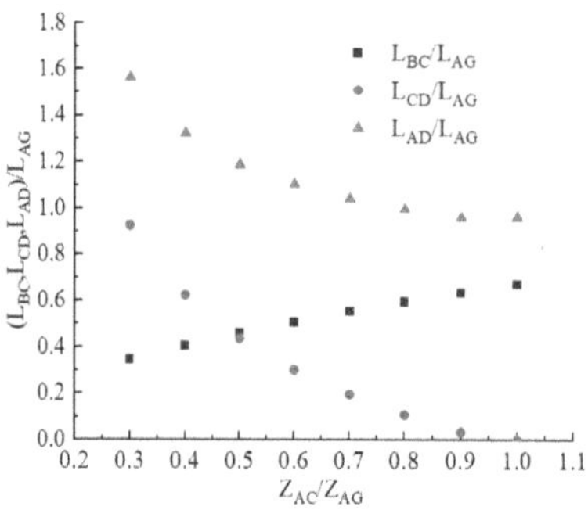

Figure 2. Relationship between Z_{AC}/Z_{AG} and the relative lengths of each part of the bottom plate.

5.3. Numerical Simulation

To visually observe the shape of the cavity, turbulent numerical simulation was used to simulate and calculate these seven types of body shapes.

5.3.1. Computational Model

The numerical simulation calculation area includes the upstream reservoir, pressure tunnel section, gate chamber section, and non-pressure section, with a total length of about 1260 m, as shown in Figure 3. The RNG k-ε turbulence model is suitable for complex flows such as strongly turbulent water flow, jet collision, and separated flow. It considers the anisotropy of turbulence and corrects turbulence viscosity, which can repair complex turbulence simulations to a certain extent, and especially for situations with large flow amplitudes, the RNG k-ε model has more advantages, so this article adopts the anisotropic RNG k-ε turbulence model [25]. A coupled solution was obtained using the PISO algorithm, using unsteady flow to approximate constant flow for iterative calculation.

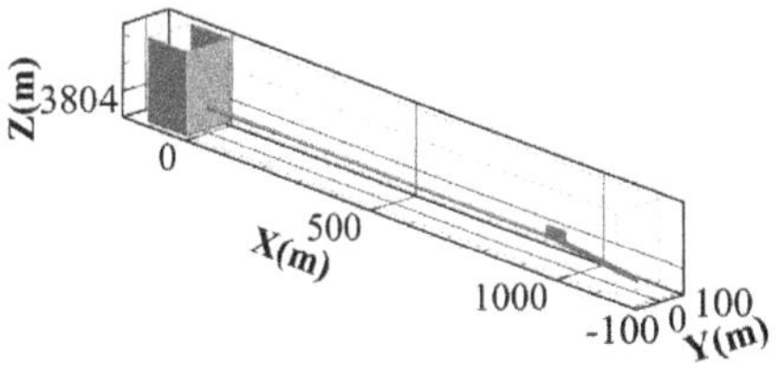

Figure 3. Calculation Model.

In order to accurately observe the shape of the cavity, the aerator section is the focus of simulation calculations. Due to its clear interface tracking, conservation, interface reconstruction techniques, wide applicability, and ease of implementation, the VOF method is particularly suitable for capturing air/water interfaces in simulations [26,27]. In this paper, the VOF [28–30] method is used to calculate the water/air interface. The basic principle of the VOF method is to define a fluid volume function F in each grid unit in the computational domain. When F = 1, it means that the grid unit is completely filled with a certain fluid phase. When F = 0, it indicates that there is no fluid phase in the grid cell. When 0 < F < 1, it indicates that the grid cell is filled with at least two fluid phases, and the cell is referred to as the interface between the fluids. The value of F can be obtained by solving Equation (11), which needs to satisfy the following transport equation:

$$\frac{\partial F}{\partial t} + \frac{\partial uF}{\partial x} + \frac{\partial vF}{\partial y} = 0 \tag{11}$$

In the formula, F is the ratio of the fluid volume within a certain unit to the total volume of that unit.

5.3.2. Grid Division and Boundary Conditions

Due to the fast convergence speed, high accuracy, and high computational efficiency of structured grids, the overall model of ecological drainage tunnels adopts structured grid division.

However, the grid size has a direct impact on the accuracy of capturing convective features. If the grid is too large, it may not be possible to capture the subtle structural changes at the lower edge of the jet water tongue behind the ridge; if the grid is too small, although it can improve accuracy, it will increase computation time and resource consumption. In order to reduce time and improve calculation speed, only the aeration facility section is encrypted. The grid size of this area along the X and Z directions is 0.1 m, and the grid size in the Y direction is 0.1 m, in order to obtain a clearer cavity shape. The connecting calculation areas are set as gradient grids, and the total number of grids is kept below 1.2 million; Figure 4 shows the local grid division of the aerator section. In the case of a known flow rate, the inlet boundary condition of the upstream reservoir area adopts a pressure inlet boundary, setting a characteristic water level for the verification flood level, corresponding to a flow rate of 983 m^3/s. The outlet adopts a pressure outlet, and the ventilation port and the top surface of the reservoir are in contact with the atmosphere.

The pressure inlet is given at standard atmospheric pressure, and the wall surface adopts a no-slip boundary condition. The viscous bottom layer is treated using the wall function method. The constant discrimination condition for numerical simulation calculation is that the error between the inlet flow rate and the outlet flow rate does not exceed ±5%.

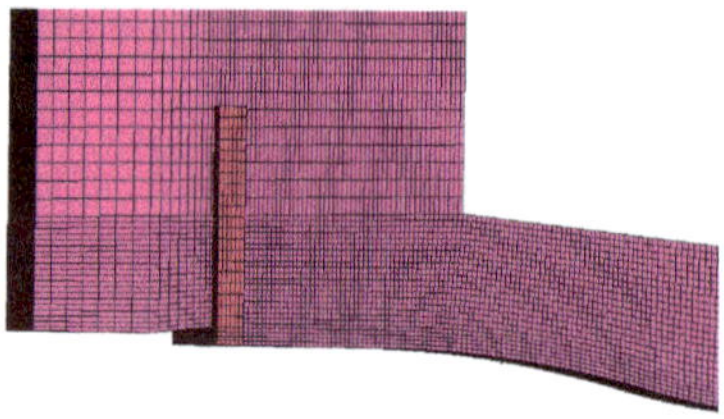

Figure 4. Local grid division of the aerator section.

5.3.3. Numerical Simulation Calculation Results

Calculate Cavity Shape

The numerical simulation first calculates "the small deflector + the offset" combination aerator, and the calculated cavity shape is shown in Figure 5. Then, seven reasonable shapes designed in Sections 4.1 and 4.2 are simulated and calculated. The calculated cavity shape of the longitudinal profile of the discharge channel bottom plate and the bottom plate are shown in Figure 6. (1) Due to the dual effects of small bottom slope and gravity, traditional aerators have severe cavity backflow and poor flow state. The seven reasonable types of aerators calculated using the design method in this article have stable cavities and better flow state than traditional aerating facilities. When $Z_{AC}/Z_{AG} \leq 0.5$, there is a small amount of water accumulation in the cavity, and when $Z_{AC}/Z_{AG} \geq 0.5$, there is almost no water accumulation in the cavity. (2) As the Z_{AC}/Z_{AG} value increases, the maximum net cavity length shows a trend of first increasing and then gradually decreasing. (3) The maximum net cavity length on the bottom plate of the discharge channel shows a shape of "convexity on the middle and concavity on both sides", which is mainly due to the fact that in the two-dimensional continuous aeration facility the middle jet flow has a high velocity, while the jet flow on both sides has a low velocity.

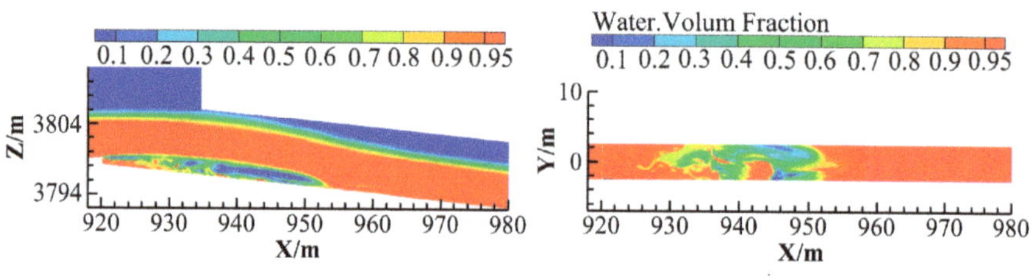

Figure 5. Calculated cavity shape of "the small deflector + the offset" combination aerator.

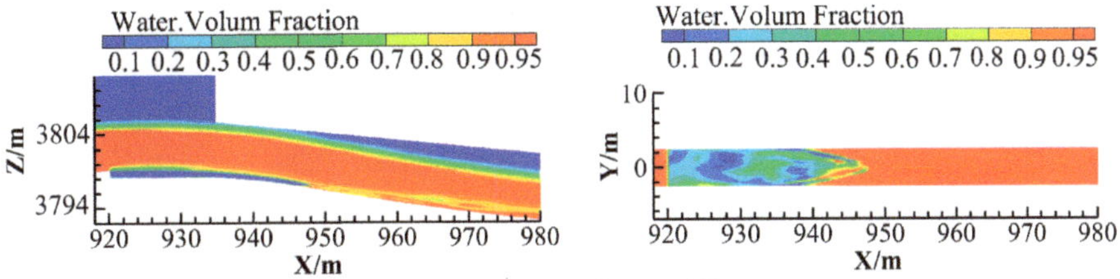

$Z_{AC} = 0.3Z_{AG}$

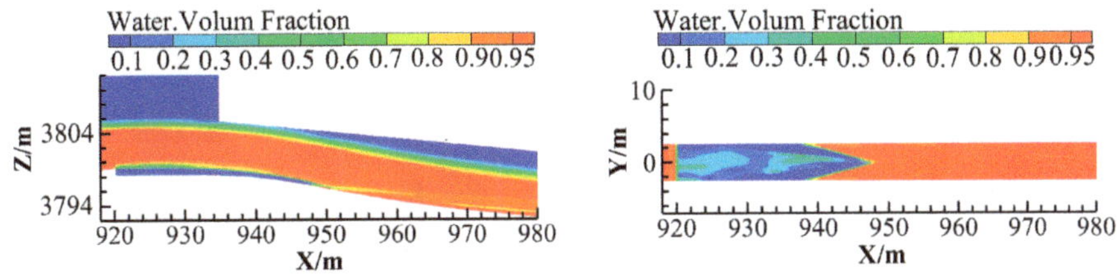

$Z_{AC} = 0.4Z_{AG}$

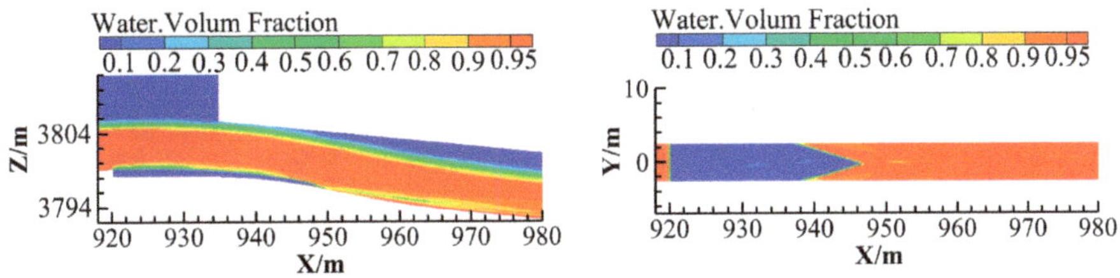

$Z_{AC} = 0.5Z_{AG}$

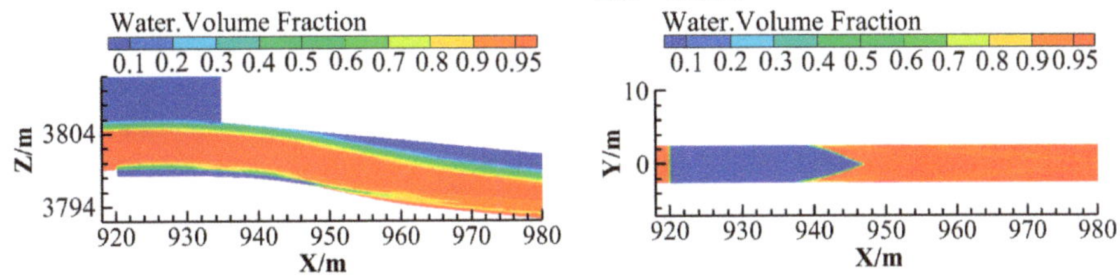

$Z_{AC} = 0.6Z_{AG}$

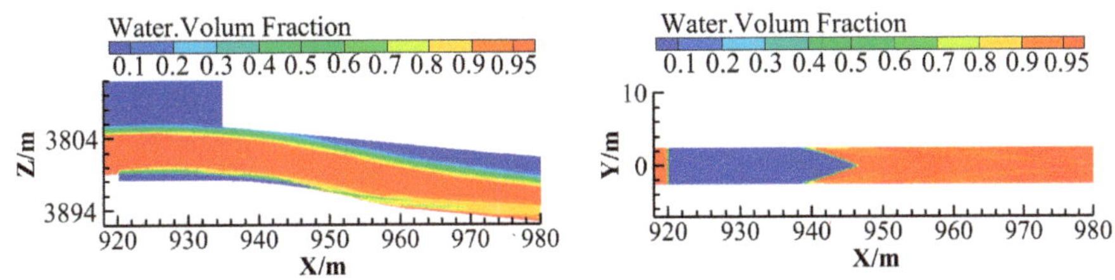

$Z_{AC} = 0.7Z_{AG}$

Figure 6. *Cont.*

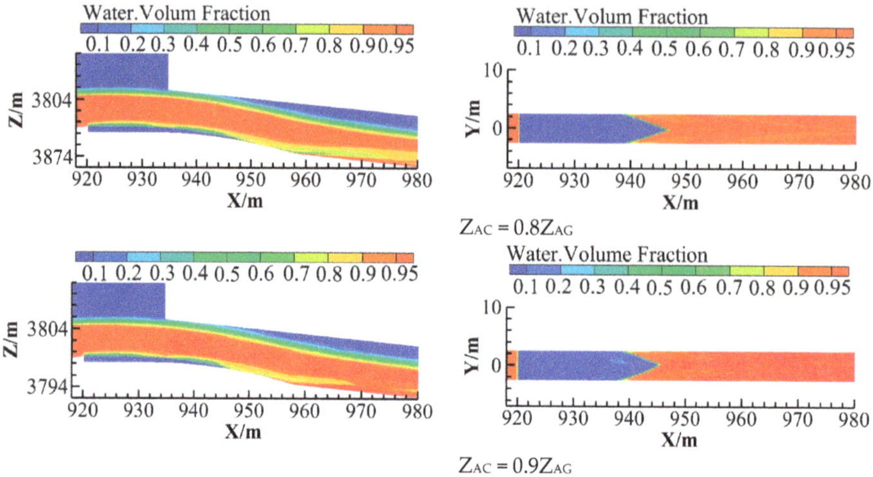

Figure 6. Calculated cavity shape behind the offset.

Distribution Characteristics of Bottom Plate Pressure

Time-average pressure distribution of the local bottom plates is shown in Figure 7 and the calculated value of time-average pressure at the centerline of the bottom plate is shown in Figure 8. Under different Z_{AC}/Z_{AG} values, we plot the relationship between relative convex parabolic segment L_{AC}/L_{AG} and relative impact force peak position $L_{A\text{-}IFP}/L_{AG}$, and the relationship between relative total length L_{AD}/L_{AG} and relative centrifugal force peak position $L_{A\text{-}CFP}/L_{AG}$ is shown in Figure 9. (1) The pressure change in the platform bottom plate is not significant, with a range of -2 KPa $< p <$ 0 KPa. When $Z_{AC}/Z_{AG} = 0.6$, the negative pressure is minimum, and when $Z_{AC}/Z_{AG} = 0.9$, the negative pressure is maximum. (2) The pressure distribution in the convex concave parabolic section shows a "bimodal" pattern. The maximum pressure value generated when the jet flow impacts the convex parabolic bottom plate is the impact force peak, and the maximum pressure value generated when the centrifugal force acts on the concave parabolic bottom plate is the centrifugal force peak. At the same Z_{AC}/Z_{AG} value, the pressure significantly increases to the impact force peak, gradually decreases and fluctuates within a certain range, then gradually increases to the centrifugal force peak and significantly decreases and tends to stabilize. (3) As the Z_{AC}/Z_{AG} value increases, the impact force peak first increases and then decreases, and the maximum net cavity length also increases first and then decreases, while the centrifugal force peak gradually decreases. Possible causes of this phenomenon are as follows: One reason is that as the Z_{AC}/Z_{AG} value increases, the concave parabolic section gradually decreases and rapid stream, resulting in the flow unable to discharge smoothly, causing the upstream jet flow to impact the bottom plate in advance, and the impact force peak and the centrifugal force peak to gradually coincide. Secondly, although the take-off angle of the lower jet trajectory is accurately calculated in the process of body shape design, the Froude number Fr calculation uses the average velocity in front of the aerator, resulting in a difference between the designed point C and the impact force peak position. Overall, the position of the impact force peak fluctuates back and forth within a certain range of the point C.

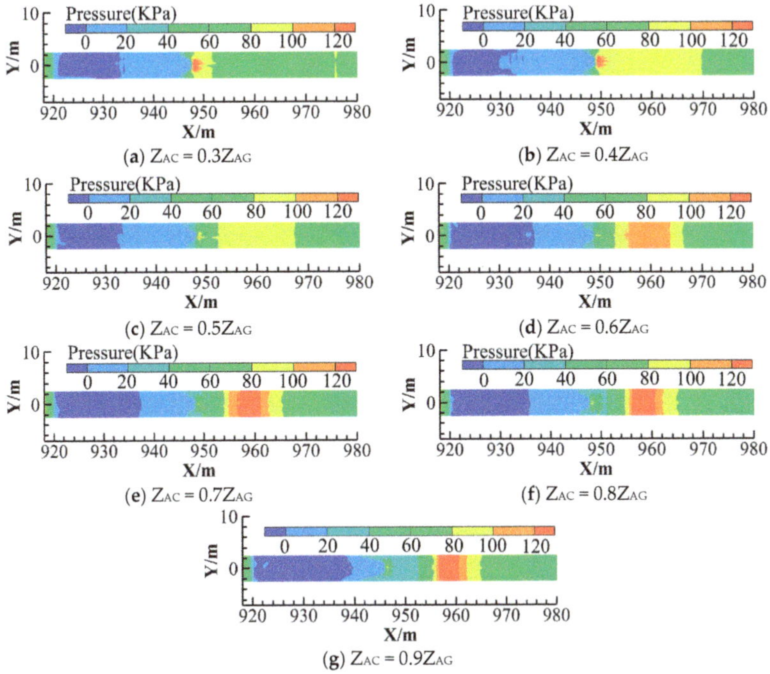

Figure 7. The time-average pressure distribution of local bottom plates under different Z_{AC}/Z_{AG} values.

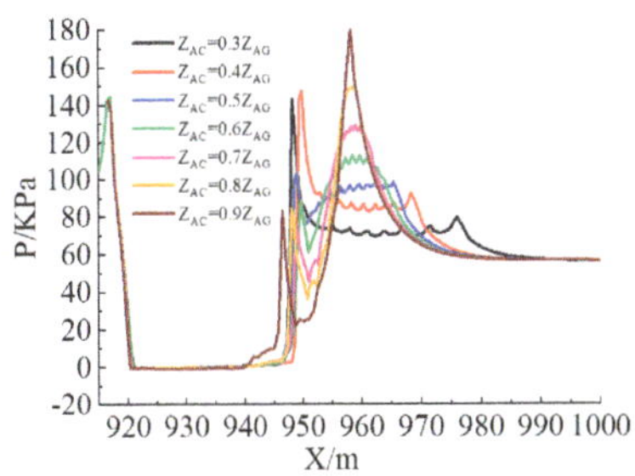

Figure 8. The time-average pressure distribution of local bottom plates centerline under different Z_{AC}/Z_{AG} values.

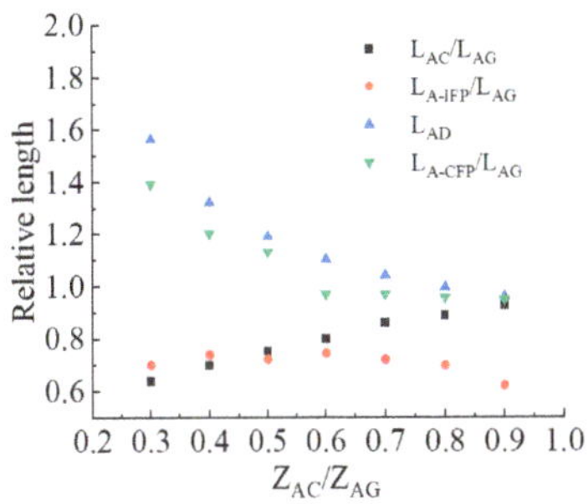

Figure 9. The relationship between point C, point D, and bimodal relative position under different Z_{AC}/Z_{AG} values.

But regardless of the reason, when $Z_{AC}/Z_{AG} \geq 0.5$, the calculated net cavity can still maintain a certain length and almost no water accumulation after the aerator.

6. Discussion

Through the above research, it has been found that under small bottom slopes and low Froude number engineering conditions, the "aerator + local steep slope" aerated facility can effectively suppress or reduce cavity backflow, and obtain a relatively stable aerated cavity. However, whether it is a special-shaped aerator or a combination of "aerator + local steep slope" aerator, when the bottom plate of the discharge channel decreases or the Froude number decreases, the accumulated water in the cavity cannot be eliminated due to the dual effects of gravity and small bottom slope, indicating a relatively weak adaptability. The double parabolic aerator is designed for engineering conditions such as small low slopes or flat bottom slopes. The basic principle of the design is to locally increase the bottom slope behind the offset, reduce the impact angle between the jet water tongue and the bottom plate, and keep the lower edge of the jet water tongue as close as possible to the local bottom plate. Locally increasing the bottom slope only changes the shape of the local bottom plate and does not change the size of the original bottom slope, which is independent of the size of the original bottom slope. It is suitable for a wider and stronger range of bottom slopes. In addition, both the irregular aerator and the "aerator + local steep slope" measures are determined through repeated engineering tests, which is time-consuming and laborious. The reason for this is that there is no design method with body shape parameters for reference, and it all depends on engineering experience. The double parabolic aerated facility has its own design method for body shape parameters, which can be used as a preselection scheme for designing the optimal body shape parameters of the aerated facility, saving time and effort.

7. Conclusions

To solve the problem of aeration and erosion reduction in discharge structures under conditions of small bottom slope, low Froude number, and large single-width flow rate, this paper proposes a new double parabolic aeration facility based on the concept of local steepening. The convex parabolic section is tangent to the flat section behind the ridge, and the concave parabolic section is tangent to the downstream bottom plate. The water tongue landing point is controlled at the junction of the convex concave parabolic curve, and the lower edge of the water tongue is in contact with the parabolic bottom plate. The calculation formula for the parabolic bottom plate has been theoretically derived, and a design method for the shape parameters of the aeration facility has been provided. Through engineering examples, it has been found that the larger the Z_{AC}/Z_{AG} value, the closer the C point is to the G point, and the less likely the cavity behind the offset is to retain water. When the water tongue landing point is 0.5–0.8 times the height of the water tongue impact point, there is almost no water accumulation in the calculated cavity. Therefore, the larger the Z_{AC}/Z_{AG} value, the better, under the condition of meeting the downstream connection conditions of the project. The research results of this article make up for the lack of experience in the design of aerated structures, and promote the quantification of aerated corrosion reduction in discharge structures, which has important engineering value. In addition, to increase the volume of the aerated cavity, systematic large-scale model experiments will be carried out in subsequent work to optimize the structural parameters of the double parabolic aerated facility and provide a design method for the optimal aerated facility body parameters.

Author Contributions: Conceptualization, G.L. and Y.D.; methodology, G.L. and Y.D.; validation, G.L. and Y.D.; formal analysis, Y.D.; investigation, Y.D.; resources, S.L. (Shaobin Liu); data curation, Y.D.; writing—original draft preparation, Y.D.; writing—review and editing, Y.D.; visualization, S.L. (Shanshan Li); supervision, P.L.; project administration, Y.W.; funding acquisition, G.L. All authors have read and agreed to the published version of the manuscript.

Funding: This study was funded by National Natural Science Foundation of China (Grant No. 52079107), National Natural Science Foundation Youth Program (Grant No. 52309105).

Data Availability Statement: The original contributions presented in the study are included in the article, further inquiries can be directed to the corresponding author.

Conflicts of Interest: Author Guodong Li was employed by State Key Laboratory of Eco-Hydraulics in Northwest Arid Region, Xi'an University of Technology. The remaining authors declare that the research was conducted in the absence of any commercial or financial relationships that could be construed as a potential conflict of interest.

References

1. Xu, W.L.; Zhang, Y.L.; Luo, J.; Arong Zhang, Q. The impact of particles on the collapse characteristics of cavitation bubbles. *Ocean. Eng.* **2017**, *131*, 15–24. [CrossRef]
2. Zhang, H.; Hu, H. Simulation of cavitation collapse process based on lattice Boltzmann method. *J. Huazhong Univ. Sci. Tech.* **2024**, *52*, 42–48. (In Chinese)
3. Zhang, J.M. Hydraulics of high-speed flows: Recent achievements and future Outlook. *J. Hydroelectr. Eng.* **2021**, *40*, 1–16. (In Chinese)
4. Bai, R.; Zhang, F.; Liu, S.; Wang, W. Air concentration and bubble characteristics downstream of a chute aerator. *Int. J. Multiph. Flow* **2016**, *87*, 156–166. [CrossRef]
5. Shi, Q. *High-Velocity Aerated Flow;* China Water Resources and Hydropower Press: Beijing, China, 2007.
6. Wei, W.R.; Deng, J. Wedge Aerator at the Bottom Outlet in Flat Tunnels. *J. Hydraul. Eng.* **2023**, *149*, 04022037. [CrossRef]
7. Pfister, M.; Hager, W.H. Chute aerators. I: Air transport characteristics. *Hydraul. Eng.* **2010**, *136*, 352–359. [CrossRef]
8. Pfister, M.; Hager, W.H. Chute aerators. II: Hydraulic design. *Hydraul. Eng.* **2010**, *136*, 360–367. [CrossRef]
9. Zang, L.H.; Xu, W.L. Experimental investigation on the backwater behind aeration facility. *J. Sci. Technol. Eng.* **2006**, *6*, 2387–2389. (In Chinese)
10. Pan, C.J.; Yuan, Y.Z. The study on the layout of aerators in free-flow spillway tunnel with inlet raised and the mild slope. *J. Hydraul. Eng.* **1993**, *6*, 61–66. (In Chinese)
11. Sun, S.K.; Liu, H.T.; Wang, X.S. Optimistic study on the layout of aeration facilities under mild gradient. *J. Water Resour. Hydropower Eng.* **2004**, *35*, 26–29. (In Chinese)
12. Wang, H.Y.; Dang, G.Q.; Yang, Q. Experimental study on V-type aerator for spillway tunnel with inlet raised. *J. Hydraul. Eng.* **2005**, *36*, 1371–1374. (In Chinese)
13. Gao, A.; Wu, S.; Wang, F. Research progress of aeration cavitation reduction technology and aeration devices. *Adv. Sci. Technol. Water Resour.* **2019**, *39*, 86–94. (In Chinese)
14. Sun, Z.; Wang, F.; Sun, C.; Feng, Y.; Fan, G.; Wu, S. Experimental analysis of hydraulic characteristics of aerators in spillway tunnels with mild slope. *J. Hydroelectr. Eng.* **2023**, *42*, 57–66. (In Chinese)
15. Liu, S.B. Study on the Configuration of Aeration Facility on Small Bottom Slope. Master's Thesis, Xi'an University of Technology, Xi'an, China, 2013. (In Chinese)
16. Wang, H.Y.; Dai, G.Q.; Liu, C.; Yang, Q. Discussion of aerator bottom cavity backflow of discharge structures. *Chin. J. Hydro Dyn.* **2009**, *24*, 425–431. (In Chinese)
17. Pfister, M.; Hager, W.H. Deflector-generated jets. *J. Hydraul. Res.* **2009**, *47*, 466–475. [CrossRef]
18. Yang, Y.S.; Yang, Y.Q.; Shuang, Q.H. The hydraulic and aeration characteristics of low froude number over a step aerator. *J. Hydraul. Eng.* **2000**, *2*, 27–31. (In Chinese)
19. Chanson, H. Predicting the filling of ventilated cavities behind spillway aerators. *J. Hydraul. Res.* **1995**, *33*, 361–372. [CrossRef]
20. Steiner, R.; Heller, V.; Hager, W.H.; Minor, H.-E. De-flector ski jump hydraulics. *J. Hydraul. Eng. ASCE* **2008**, *134*, 562–571. [CrossRef]
21. Rutschmann, P.; Hager, W.H. Air entrainment by spillway aerators. *J. Hydraul. Eng.* **1990**, *116*, 765–782. [CrossRef]
22. Wu, J.H.; Ruan, S.P. Cavity length below chute aerators. *J. Sci. China Technol. Sci.* **2008**, *51*, 170–178. (In Chinese) [CrossRef]
23. Pfister, M. Jet impact angel on chute downstream of aerator. In Proceedings of the 4th IAHR International Symposium on Hydraulic Structures, Porto, Portugal, 12–13 January 2012.
24. Long, L.C.; Liu, C.; Deng, J.; Wei, W.R.; Ran, Y.B. Study on takeoff angle characteristics of continuous flip bucket's ski-jump nappe. *J. China Inst. Water Resour. Hydropower Res.* **2023**, *21*, 83–91. (In Chinese)
25. Deng, J.; Xu, W.L.; Lei, J.; Diao, M.J. Numerical simulation of hydraulic characteristics of high spillway tunnel. *J. Hydraul. Eng.* **2005**, *36*, 1209–1218. (In Chinese)
26. Luo, Y.Q.; Diao, M.J.; He, D.M.; Bai, S.X. Three-dimensional numerical simulation analysis of air entrainment and erosion reduction of high dam open flow spillway tunnel. *Adv. Water Sci.* **2012**, *23*, 110–116. (In Chinese)
27. Gao, X.P.; Jia, L.F.; Song, H.F.; Cui, G.H. Three-dimensional numerical simulation of aerated water flow behind the aerated sill of a spillway. *J. Hydroelectr. Eng.* **2014**, *33*, 90–96. (In Chinese)
28. Xu, J.; Wu, J.; Yu, P.; Fei, M. Fin performance of 3D aerator devices with backward lateral deflectors. *J. Hydrodyn.* **2020**, *32*, 410–413. [CrossRef]

29. Teng, P.; Yang, J. CFD Modeling of Two-phase Flow of a Spillway Chute Aerator of Large Width. *J. Appl. Water Eng. Res.* **2016**, *4*, 163–177. [CrossRef]
30. Teng, P.; Yang, J.; Pfister, M. Studies of Two-phase Flow at a Chute Aerator with Experiments and CFD Modelling. *Model. Simul. Eng. J.* **2016**, *2016*, 4729128. [CrossRef]

water

MDPI

Article

Hydrodynamic Simulation of Urban Waterlogging Based on an Improved Vertical Flow Exchange Method

Xi Jin * and Yan Mu

School of Civil Engineering and Architecture, Wuhan University of Technology, Wuhan 430070, China; muyan3342@163.com

* Correspondence: jinxi@whut.edu.cn

Abstract: In the 1D–2D coupled simulation of urban waterlogging, the calculation process of vertical flow exchange is independent from the 1D hydraulic calculation, resulting in a failure to consider the node head and pipe flow during the exchange flow calculation, which may lead to irrational results and further affect the stability of the model calculation. However, setting an upper limit for the exchange flow may introduce excessive subjective factors into the simulation process. In this study, a vertical flow exchange method based on the water balance of nodes is proposed. When a node is in an overloaded state, the calculation of vertical flow exchange at the node is integrated into the 1D hydraulic simulation process, thus taking into consideration the influence of the node head and pipe flow when calculating vertical flow exchange. Additionally, the iterative solution method used in the 1D hydraulic model ensures numerical harmony between the vertical flow exchange, node head and pipe flow, thus ensuring the stability of the coupled calculation. For the non-overloaded nodes, the calculation of the vertical flow exchange was conducted using a variable-head orifice discharge formula, enabling the consideration of changes in the surface water depth during the calculation of the node backflow. Using the InfoWorks ICM model as a benchmark, a comparative analysis of case simulation results demonstrated that the improved vertical flow exchange method was able to accurately and stably simulate the process of vertical flow exchange. When used with the improved vertical exchange method, the coupled model gave simulation results that closely matched those of the benchmark model.

Keywords: urban waterlogging; coupled model; SWMM; vertical flow exchange

1. Introduction

In recent years, with the increasing frequency of extreme rainfall disasters, more cities have implemented planning and construction for drainage and flood control. In this process, urban hydrodynamic models have been widely applied [1–4]. Early hydrodynamic models focused on simulating surface runoff processes and hydrodynamic processes in underground drainage systems, but they were unable to simulate surface waterlogging [5,6]. In order to address this deficiency, Hsu M. H. (2000) used the SWMM (Storm Water Management Model) as a one-dimensional (1D) model coupled with a two-dimensional (2D) diffusive overland flow model, in order to achieve a coupled urban waterlogging simulation [7]. However, in this method, the flow exchange between models was unidirectional. When the drainage network was overloaded, water flowed from the drainage system to the ground surface; when the drainage system had sufficient discharge capacity, surface water on the ground could not flow back to the drainage system. In further research, a series of coupled models capable of bidirectional exchange between an underground drainage system and surface water were developed [8–10]. Although coupled models have the advantages of detailed results and adeptness in handling simulations of the bidirectional exchange of water flows, their computational processes are complex and their efficiency is low. In order to achieve a balance between simulation accuracy and computational efficiency, some researchers explored waterlogging simulation methods that could avoid solving hydrodynamic equations yet still maintain good accuracy, such as with flood analysis algorithms based on GIS [11], an enhanced inundation model for flood calculations [12], and some simple models of cellular automata [13]. However, these methods can only determine the extent of waterlogging and water depth, but are unable to simulate the hydrodynamic processes of waterlogging, so their application scope is limited.

Currently, the coupled 1D–2D model is the main method for simulating urban waterlogging. Vertical flow exchange is a key segment in the coupled process, and it significantly influences the accuracy of the simulation results and the water balance between the 1D and 2D models [14–16]. The current calculation method for vertical flow exchange mainly uses the head difference between the water levels of nodes in the 1D model (H_{1D}) and the surface water levels of the 2D model (H_{2D}) as the driving head, and orifice or weir flow formulas are utilized to calculate the flow exchange between the 1D and 2D models [17].

The typical calculation formula for node overflow is as follows [2]:

$$Q_{of} = c_o A_{mh} \sqrt{2g(H_{1D} - H_{2D})} \qquad \text{if } H_{1D} > H_{2D} \tag{1}$$

where Q_{of} is the calculated overflow value (m^3/s); c_o is the orifice discharge coefficient; A_{mh} is the manhole area of nodes (m^2); g is the gravitational acceleration (m/s^2); H_{1D} is the water level of the 1D model's nodes (m); and H_{2D} is the water level of the 2D model (m).

The typical calculation formula for the node backflow is as follows [2]:

$$Q_{bf} = \begin{cases} c_w W H_{2D} \sqrt{2gH_{2D}} & \text{if } H_{1D} \leq Z_{2D} < H_{2D} \text{ and } Q_{bf} < Q_{bm} \\ c_o A_{mh} \sqrt{2g(H_{1D} - H_{2D})} & \text{if } Z_{2D} \leq H_{1D} < H_{2D} \text{ and } Q_{bf} < Q_{bm} \\ Q_{bm} & \text{if } Q_{bf} \geq Q_{bm} \end{cases} \tag{2}$$

where Q_{bf} is the calculated value of the backflow (m^3/s), and in order to achieve a stable simulation, the value of Q_{bf} cannot exceed the maximum backflow rate of Q_{bm}; c_w is the weir discharge coefficient; W is the perimeter of the node or the width of the rainwater outlet (m); and Z_{2D} is the ground elevation where the node is located (m).

When calculating node backflow or intercepted discharge by inlets, in order to consider the water interactions between the surface and drainage system more accurately, water energy upstream of the node or inlet (sum of the head difference and velocity head) is used as the driving head for Formula (2) [18]. Furthermore, some studies have discussed the relationship between the discharge coefficient and the surface flow pattern, and have concluded that the orifice and weir discharge coefficient have a significant correlation with the Froude number of surface flow [18,19]. The vertical water exchange method takes into consideration that velocity head and flow pattern are often used for detailed 1D models of drainage systems, including the full drainage structure composed of conduits, manholes, gullies and inlets. These are coupled with 2D models in a fully distributed way [20,21], where the runoff volumes are estimated and applied directly to the elements of the 2D model of the overland surface, and exchange with the 1D model through inlets. However, due to limitations in data availability and computational power, the Fully Distributed models are generally suitable for small-scale cases. For medium to large-scale models, the Semi-Distributed Urban Stormwater models remain the preferred choice. In Semi-Distributed models, conceptual, empirical or physical-based methods transform runoff routing into inflow hydrographs, which are then applied to the selected computational nodes of the drainage system. Not every inlet is modeled, but they are clustered to computational ones. In this case, the discharge coefficient has been decoupled from specific inlets and has become a subjectively determined parameter that needs to be calibrated using actual measurement data.

The determination of discharge coefficients for flow exchange equations over a range of hydraulic conditions and inlet types has a significant impact on the simulation results. Recent studies related to water exchange through manholes or inlets using the weir and orifice-type equations have been implemented using physical model experiments [22–24] and numerical simulations [25]. The values of the discharge coefficients are influenced by the inlet geometry and the approaching flow characteristics (flow depth and Froude number). Rubinato et al. [19] studied water exchange of scaled circular manholes under subcritical flow with a Froude number of 0.151–0.691; the results revealed that under subcritical flow conditions, water exchange is sensitive to hydraulic heads within interaction nodes and surface water. Hence, the uncertainty related to head loss in hydraulic structures, surface roughness and other parameters in urban hydrology may have significant implications to water exchange and should be considered in the selection of weir/orifice formulas and the determination of discharge coefficients. Under supercritical flow conditions, the flow velocity parameter has to be considered while determining the discharge coefficients. Cosco et al.'s study [18] highlighted the strong relationship between discharge coefficient ranges and the upstream Froude number under supercritical flow, and formulated power functions in order to express the weir/orifice discharge coefficients as a function of Froude number. Ali Zaiter et al. (2024) gave a comprehensive review about "equations and methodologies of inlet drainage system discharge coefficients" [26]. Ali's review summarized orifice and weir discharge coefficient ranges of rectangular inlets, grated circular inlets and circular non-grated inlets under subcritical flow and supercritical flow, respectively, and concluded that there is no uniform pattern or trend to determine the discharge coefficient of each inlet as it may vary significantly with the change in Froude number, inlet geometry, grate geometry, grate orientation, flow velocity and water depth. Therefore, weir and orifice discharge coefficients have to be calibrated and adjusted experimentally, taking into consideration the crucial factors that affect the inlet discharge capacity [27]. Russo et al. (2015) analyzed extreme floods in Barcelona using the 1D–2D coupled model created by InfoWorks ICM, and gave a detailed description of how to calculate and calibrate the discharge coefficients of manholes and storm inlets [28].

The calculation of flow exchange through orifice or weir formulas is not only complex, but also involves significant uncertainty in the selection of discharge coefficients. Moreover, the introduction of an artificially set value of Q_{bm} adds subjectivity and uncertainty to the simulation of coupled models. The fundamental reason for the instability in the coupling process when not using a limitation for Q_{bm} is that the calculation of the backflow is independent of the 1D model. Thus, interactions among variables such as the vertical flow exchange, node head and conduit flow are neglected, which may lead to irrational backflow values. When an irrational value is involved in the coupling process, computational instability may occur. Fan Y. (2017) suggested determining the Q_{bm} based on the simulation stability of the model and the actual situation of the node [29]. According to this method, the appropriate value of Q_{bm} can only be obtained through trial and error during the simulation process. In order to implement a stable coupled simulation without artificially set values, Peng G. (2022) calculated the backflow while considering the free space of a manhole and connected conduits [30]; however, this was a compromise solution and the interactions among the vertical flow exchange, node head and conduit flow were still neglected.

This study introduces the principle of node inflow and outflow balance into the calculation process for vertical flow exchange, and the calculation of vertical flow exchange is integrated into the 1D model in order to consider interactions among the vertical flow exchange, node head and conduit flow so that harmonious results can be obtained. This improved method simplifies the calculation process for vertical flow exchange and enhances the stability of the coupled process.

2. Coupled 1D–2D Hydrodynamic Model

2.1. Hydrodynamic Models

The SWMM is a hydrologic–hydraulic water quality simulation model that utilizes dynamic modeling techniques [31]. As an open-source model, it is widely used in the field of urban stormwater simulation. The SWMM is suitable for the 1D hydrodynamic simulation of drainage systems in urban areas. In this study, the SWMM is used as a 1D model for the hydrological simulation of sub-catchments and the hydrodynamic calculation of drainage systems.

Conservative shallow water equations are used as the governing equations in the 2D model, and they are solved using a self-developed simulation module with the Godunov scheme for the finite volume method [32]. Conservative shallow water equations are adopted as the governing equations in the 2D model. The governing equations can be written in matrix form, as shown in Formulas (3) and (4).

$$\frac{\partial U}{\partial t} + \frac{\partial F}{\partial x} + \frac{\partial G}{\partial y} = S_f + S_b \quad (3)$$

where U, F and G represent the conservation vector, the flux vectors in the x and y directions, respectively, and S_b and S_f are the source terms representing the bed slope and friction effects, respectively. In addition, t, x, and y represent time and the coordinate axes of a Cartesian coordinate system. The specific forms of U, F, G, S_b and S_f are shown in Formula (4).

$$U = \begin{bmatrix} h \\ hu \\ hv \end{bmatrix}, F = \begin{bmatrix} hu \\ hu^2 + \frac{1}{2}gh^2 \\ huv \end{bmatrix}, G = \begin{bmatrix} hv \\ huv \\ hv^2 + \frac{1}{2}gh^2 \end{bmatrix}, S_b = \begin{bmatrix} 0 \\ -gh\frac{\partial z_b}{\partial x} \\ -gh\frac{\partial z_b}{\partial y} \end{bmatrix}, S_f = \begin{bmatrix} 0 \\ -C_f u\sqrt{u^2+v^2} \\ -C_f v\sqrt{u^2+v^2} \end{bmatrix} \quad (4)$$

where h represents water depth (m); u and v represent the flow velocities in the x and y directions, respectively (m/s); z_b represents the absolute elevation of the ground surface (m); and $C_f = \frac{gn^2}{h^{1/3}}$, in which n is the Manning coefficient. Additionally, water level η is used in the second order spatial reconstruction and the non-negative water depth reconstruction—$\eta = h + z_b$.

Formulas (3) and (4) are solved herein by a finite volume Godunov-type numerical scheme proposed by Liang (2010), which has an emphasis of constructing a well-balanced scheme and evaluating the friction source term in order to ensure non-negativity of water depth [33,34]. The monotonic upstream-centered scheme for conservation laws (MUSCL) linear reconstruction method [35] with a MINMOD slope limiter [36] is used along with a Harten, Lax and van Leer approximate Riemann solver, with the contact wave restored (HLLC) for the calculation of interface fluxes [37]. The second-order temporal accuracy is achieved by using a Runge–Kutta time integration [38].

During the coupling process of the 1D and 2D models, two main aspects are involved: time synchronization and flow exchange. For time synchronization, this study used the 1D model time step as the benchmark and achieved time synchronization by adjusting the 2D model time step. For flow exchange, this study only considered vertical flow exchange, while lateral and frontal flow exchanges were neglected. Closed boundary conditions were adopted at the boundaries of the surface grid for the 2D model.

2.2. Time Synchronization in the Coupled 1D and 2D Models

Since the time step of the 1D model (Δt_{1D}) is often larger than that of the 2D model (Δt_{2D}), Δt_{1D} is taken as a benchmark. By adjusting Δt_{2D}, the synchronization of the time steps in the 1D and 2D models is achieved. During the simulation process, when the 1D model advances by one time step, the 2D model runs n steps. The first $n-1$ steps of the 2D model are calculated using the Courant–Friedrichs–Lewy (CFL) criterion, and the value of Δt_{2D} for the nth step is calculated as $\Delta t_{1D} - \sum_1^{n-1} \Delta t_{2D}$. The synchronization method for the 1D and 2D models is presented in Figure 1.

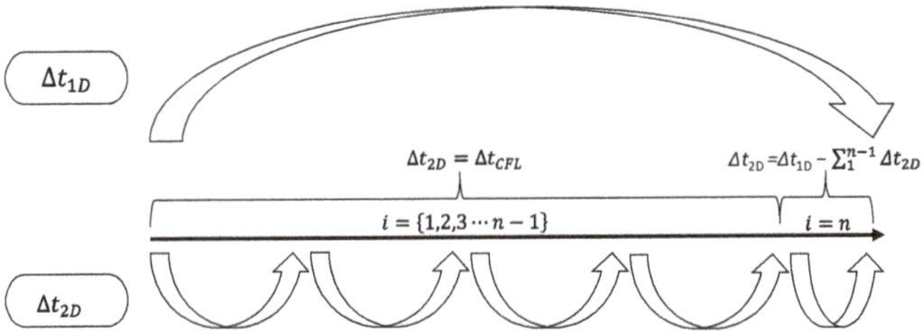

Figure 1. Time synchronization in the coupled 1D and 2D models.

3. Improved Vertical Flow Exchange Method

The three typical cases of vertical flow exchange at nodes are shown in Figure 2.

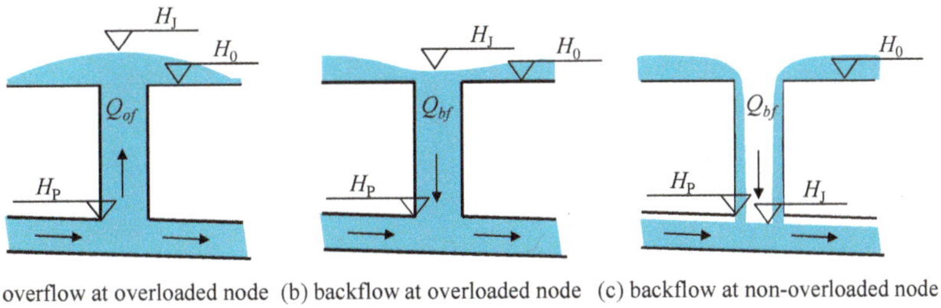

Figure 2. Three typical cases of vertical flow exchange at nodes.

In Figure 2, cases (a) and (b) represent situations of overloaded nodes, where the water head inside the node (same as the water level of the surface cell) is greater than the maximum at the upper edge of the connected conduit section, which is denoted as H_P. Since the 1D model generally ignores the volume inside node wells, this implies that the node water head H_J exceeds the ground elevation H_0. In such cases, the calculation of conduit flow in the 1D model should consider the surface water level of the 2D model at the node location. The inflow and outflow of the node are calculated according to the conduit flow values. The difference between the inflow and outflow is considered as the vertical flow exchange between the 1D and 2D models. If the value of the vertical flow exchange is positive, this indicates that the node is an overflow node. If the value is negative, this indicates that the node is a backflow node. If the value is equal to 0, this indicates that there is no vertical flow exchange in this node. Case (c) represents the situation of a non-overloaded node, where the node water head H_J is less than the maximum at the upper edge H_P of the section of connected conduits. In this case, the calculation of pipe flows in the 1D model was independent of the surface water level, and the backflow of the node was calculated using the orifice flow formula and was added as the inflow to the node.

3.1. Calculation of the Vertical Flow Exchange

The nodes of the 1D model are associated with the mesh cells of the 2D model in which they are located, with the mesh cell area serving as the ponded area for each node.

Figure 3 illustrates the relationship between the ponded area of the node and the mesh cell. The dark blue cell represents the mesh cell corresponding to the location of the node. Before the coupling simulation begins, the ponded area of the corresponding node is updated using the area of this cell. This cell not only participates in the calculations of the 2D model, but also contributes to the computation of the vertical flow exchange. The remaining cells are ordinary cells that only participate in the calculations of the 2D model.

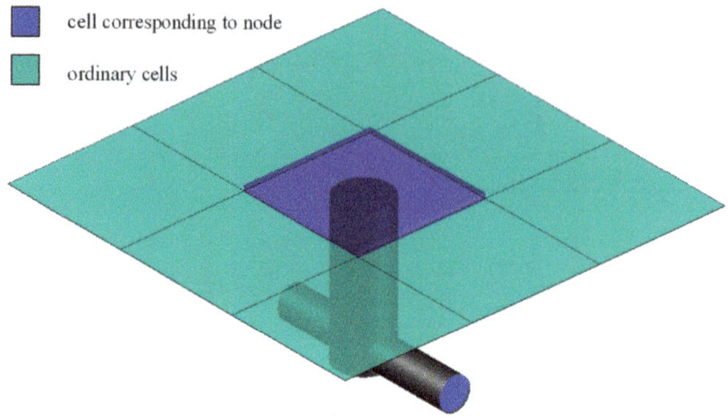

Figure 3. Relationship between the ponded area of a node and the cells in the 2D mesh grid.

3.1.1. Calculation of the Vertical Flow Exchange in Overloaded Nodes

When a node is in an overloaded state, the calculation of the flow in the connected conduits needs to consider the influence of the surface water level. In 1D hydraulic simulations, before the calculation of the dynamic wave in each time step begins, the water level of the mesh cell in which the node is located is used to update the node's water head in the 1D model. Then, the 1D hydraulic calculation of the dynamic wave is performed. The SWMM module uses an iterative method to calculate node water heads and conduit flow rates, with the convergence of node water heads as the stopping condition for iteration, and in each iteration, the vertical flow exchange is updated by applying the principle of water conservation. This ensures numerical harmony among the node water heads, conduit flows and vertical exchange flows. Thus, stability in the calculation of the vertical flow exchange is achieved. According to the principle of water conservation, the following relationship between the change in water volume within the ponded area and the flow in the connected conduits of the node exists:

$$V_{new} - V_{old} = \Delta t_{1D}\left(\sum_{i=1}^{k} Q_i + Q_{lat}\right) \tag{5}$$

where V_{old} represents the water volume within the ponded area of the node at the beginning of the time step (m^3); V_{new} represents the water volume within the ponded area of the node at the end of the time step (m^3); Q_i is the flow in the ith conduit connected to the node (m^3/s), with inflow as positive and outflow as negative; Q_{lat} is the sum of other inflows, such as surface runoff and dry weather flow (m^3/s); and k is the number of conduits connected to the node.

The change in the water volume within the ponded area of the node in the 1D model is the volume of vertical exchange between the 1D and 2D models, so the vertical exchange flow can be calculated as follows:

$$Q_{exf} = \frac{V_{new} - V_{old}}{\Delta t_{1D}} \tag{6}$$

where Q_{exf} represents the vertical exchange flow (m^3/s), with positive values indicating node overflow and negative values indicating node backflow.

3.1.2. Calculation of Vertical Flow Exchange in Non-Overloaded Nodes

When a node is not overloaded but there is ponding water on the surface cell, the flow calculation for the connected conduits does not need to consider the influence of the surface water level. Therefore, there is no need to update the node's water level, and the orifice flow formula is used here for the backflow calculation. During the backflow process, the surface water level may change. As a result, the following formula for the backflow calculation considering a variable head is used:

$$t_b = \frac{2S\left(\sqrt{H_1} - \sqrt{H_2}\right)}{c_o A_{mh}\sqrt{2g}} \quad (7)$$

where t_b represents the backflow time (s); S represents the ponded area (m^2); H_1 represents the surface water level at the node at the beginning of the backflow time t_b (m); and H_2 represents the surface water level at the node at the end of the backflow time t_b (m).

Assuming that $H_2 = 0$, Formula (7) is used to calculate the time t_b, which represents the time required for all of the ponding water in the cell to completely backflow. If $t_b > \Delta t_{1D}$, this indicates that within the current time step in the SWMM, not all of the ponding water in the cell backflows. Then, Formulas (8) and (9) are utilized in order to calculate the water level of the surface water and the backflow rate at the end of the current time step in the SWMM.

$$H_2 = \left(\sqrt{H_1} - \frac{c_o A \Delta t_{1D}\sqrt{2g}}{2S}\right)^2 \quad (8)$$

$$Q_{exf} = \frac{S(H_1 - H_2)}{\Delta t_{1D}} \quad (9)$$

If $t_b \le \Delta t_{1D}$, this indicates that within the current time step in the SWMM, all of the ponding water in the cell has completely returned into the underground network. Then, the backflow rate at the end of the current time step is calculated with Formula (10):

$$Q_{exf} = \frac{SH_1}{\Delta t_{1D}} \quad (10)$$

If the node is in an overloaded state at the end of Δt_{1D}, this indicates that it has transitioned from a backflow state to an overflow state within this time step. In this case, the vertical flow exchange of the node is calculated using the principle of conservation of inflow and outflow according to Formula (6).

3.1.3. Correction of Continuity Errors Caused by Node Water Head Updates

Figure 4 illustrates the meanings of various variables during the process of updating the node water head. In the coupling process, the 1D model's time step Δt^n_{1D} is calculated first. When a node is in an overflow state (shown in Figure 4a), the water head of the node changes from H_{old} to H_{new} within the time step. The water volume between these two levels represents the overflow volume of the 1D model. During the process of vertical flow exchange, this volume is transferred from the 1D model to the 2D model. Consequently, the volume of water in the 2D model increases accordingly. In order to ensure water conservation in the coupling process, this volume needs to be removed from the 1D model. Therefore, at the end of Δt^n_{1D}, the water head and ponding water volume corresponding to the node in the 1D model should be H_{old}.

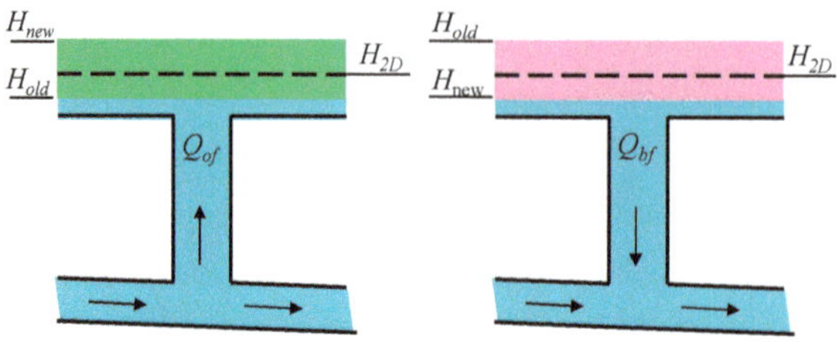

(a) Water level variables in overflow process

(b) Water level variables in backflow process

Figure 4. Diagram showing the continuity error caused by updating the node head.

After the overflow volume is added to the 2D model and the 2D model's corresponding time step is completed, the water level in the 2D cells becomes H_{2D}. Before the calculation of the next time step in the 1D model (Δt_{1D}^{n+1}) begins, it is necessary to update the node water head using H_{2D}. The water head at the node changes from H_{old} to H_{2D}, and the ponding water volume also increases accordingly. The increment in this ponding water volume is equivalent to adding additional virtual water to the 1D model without corresponding to any actual physical process, which may lead to continuity errors.

In order to eliminate this continuity error, it is necessary to eliminate the volume of water between H_{old} and H_{2D} during the calculation of the flow exchange in time step Δt_{1D}^{n+1}. The elimination method involves obtaining the initial result of the flow exchange Q_{exf0} using Formula (6) and then correcting it using Formula (11), in order to eliminate the continuity error.

$$Q_{exf} = Q_{exf0} - \frac{(H_{2D} - H_{old})S}{\Delta t_{1D}} \tag{11}$$

When the node is in an overloaded backflow state (shown in Figure 4b), the water head at the node changes from H_{old} to H_{new}. At the end of Δt_{1D}^{n} in the 1D model, the water head and ponding water volume of this node are calculated according to H_{new}. The water volume between H_{old} and H_{new} represents the vertical backflow volume, which will be removed from the corresponding 2D cells during the respective time step in the 2D model. Updating the node water head using H_{2D} before the calculation of the next time step in the 1D model Δt_{1D}^{n+1} would cause an increase in the ponding water volume in the 1D model's nodes. The water head changes from H_{new} to H_{2D}, and the ponding water volume changes accordingly. This increase in ponding water volume can be considered as the volume of surface water from the 2D model flowing back to the ponded area of the 1D model's nodes. This volume of water has actual physical significance and does not lead to continuity errors; therefore, no correction is needed.

3.2. Implementation of Model Coupling

In the coupled model utilizing the improved vertical flow exchange method, the SWMM requires the node ponding functionality to be enabled. The specific calculation steps during the coupling process corresponding to Δt_{1D} are as follows:

i. Determining which nodes are in an overloaded state;
ii. Updating the water heads for overloaded nodes;
iii. Calculating the backflow for non-overloaded nodes (with the method described in Section 3.1.2) and adding backflow to the corresponding node as inflow;
iv. One-step evolution of the 1D model (the vertical flow exchange for overloaded nodes is calculated with the method described in Section 3.1.1 by adding the corresponding functions);
v. Eliminating continuity errors;
vi. Transferring the vertical flow exchange values to the 2D model;
vii. Evolution of the 2D model until the end of Δt_{1D}. This is followed by step i.

Since the SWMM module has the function of calculating the ponding water volume at nodes, the vertical exchange flow under the condition of overloaded nodes can be directly obtained through the change in the ponding water volume at these nodes. Steps (i) to (v) can be implemented within the SWMM module by adding corresponding functions to it.

4. Case Study

4.1. Benchmark Software

In this study, InfoWorks ICM (from here on 'ICM') was used as a benchmark. ICM solves the complete 2D Saint Venant equations in a finite volume semi-implicit scheme [35] with a Riemann solver [39]. ICM combines a number of distinctive features such as the analysis and prediction of potential flood extent, depth and velocity, and the modeling of the interaction of surface and underground systems in a fully integrated environment [28].

Generally, in a 1D–2D coupled model created by ICM, the calculation of vertical flow exchange between 1D and 2D models is achieved through nodes of 2D and Gully 2D flood types. For 2D nodes, the exchange of water between the 1D and 2D models is calculated using weir equations, assuming a weir crest level at the node ground level and crest length equal to the node shaft circumference. For Gully 2D nodes, the exchange flow is calculated using a specific (customized) "head/discharge relationship". The Gully 2D flood-type nodes more accurately describe the water exchange between the 1D drainage system and the 2D surface at different rainwater inlets.

ICM can conduct both Fully Distributed simulations and Semi-Distributed simulations. In this study, only Semi-Distributed simulations were performed and discussed. For theoretical study cases, the nodes with vertical flow exchange do not represent any specific type of rainwater inlet. For practical study cases, the drainage network used was a reduced network, only conduits and computational nodes were included. Runoff volumes were estimated for the sub-catchment and then routed to the corresponding nodes as sub-catchments' outlets. In this case, only 2D flood type were used for water exchange calculation of both study cases, and the typical discharge value of 0.5 was used for all 2D flood type nodes.

4.2. Theoretical Study Case

The improved method was validated using a theoretical case study, as shown in Figure 5. The drainage system consisted of six nodes and six conduits, and the level surface area was a closed square with a side length of 200 m. Nodes 2 to 5 allowed vertical flow exchange. The elevation of the surface was 0 m, and the roughness coefficient (n value) was 0.025. Water entered the system through node 1, with the inflow gradually increasing from 0 to 1.0 m^3/s within the first 10 min of the simulation, remaining constant until the 20th hour and then dropping to 0. Water left the system through node 6, with the outfall set as a free outflow. The system was initially dry. The attributes of the conduits and nodes are shown in Table 1.

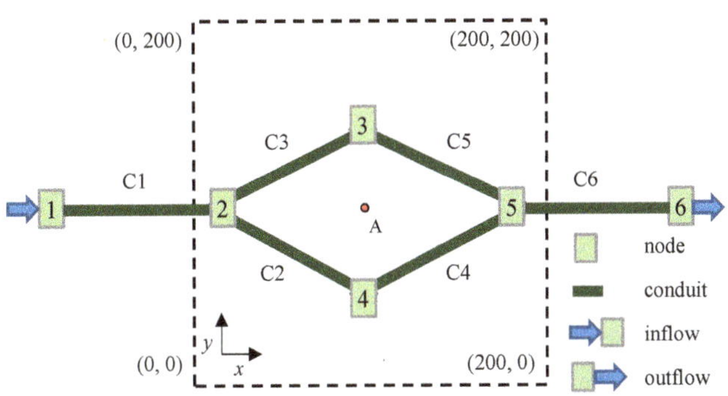

Figure 5. Diagram of the theoretical case.

Table 1. Properties of the elements of the drainage system.

ID	Conduit					Node				
	Upstream Node	Downstream Node	Dia /m	Len /m	Roughness	X Coordinate	Y Coordinate	Full Depth/m	Ground Elevation/m	Well Dia /m
1	1	2	1.0	100	0.013	−70.711	100.000	1.159	0	0.7
2	2	4	0.6	100	0.013	29.289	100.000	1.259	0	0.7
3	2	3	0.6	100	0.013	100.000	170.711	1.329	0	0.7
4	4	5	0.7	100	0.013	100.000	29.289	1.329	0	0.7
5	3	5	0.7	100	0.013	170.711	100.000	1.400	0	0.7
6	5	6	1.0	100	0.013	270.711	100.000	1.500	0	-

In the coupled model proposed in this study, the area of the mesh cell where the node is located was used as the ponded area of the node. Therefore, the impact of the grid mesh on the coupled process is important. Two types of grid meshing structures (2 m × 2 m (S4) and 4 m × 4 m (S16)) were used for the theoretical case.

Figure 6a shows the water balance in the improved model with grid meshing of cell areas of 16 m^2. During the process of vertical flow exchange, the accumulated overflow volume from the nodes should be equal to the sum of the surface water volume and the accumulated backflow volume from the nodes. Figure 6a shows that the improved model maintained water conservation during the vertical flow exchange process, as the accumulated overflow curve of the nodes perfectly overlapped with the curve of "accumulated backflow + surface water accumulation". In the process of the coupled 1D–2D simulation, the total water entering the coupled model (accumulated system inflow) needed to be equal to the sum of the water leaving the system (accumulated system outflow) and the water stored within the model (surface water). Figure 6b demonstrates that the overall water conservation of the model was also well satisfied.

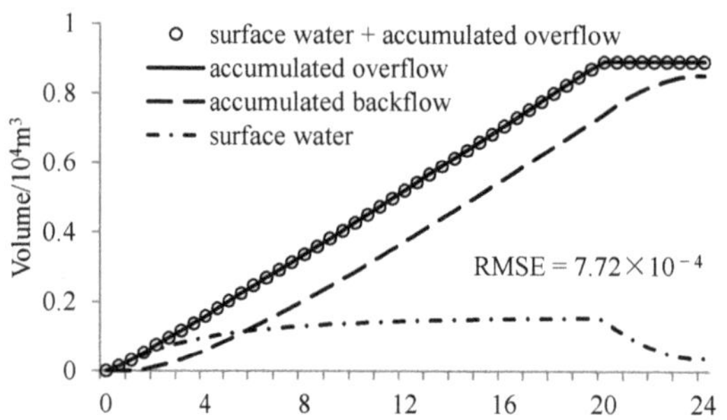

(a)

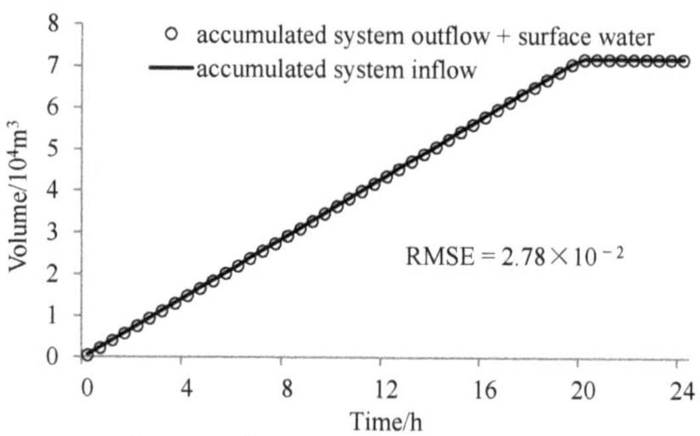

(b)

Figure 6. Analytical diagram of the water balance in the improved model. (**a**) Water conservation in the vertical flow exchange process. (**b**) Systematic water conservation in the coupled simulation process.

Root mean squared error (RMSE) was used as a statistical measure of deviation between "accumulated overflow" with "accumulated backflow + surface water accumulation" for Figure 6a, and "accumulated system inflow" with "accumulated system outflow + surface water" for Figure 6b.

Figure 7 takes node 2 as a typical overflow node, node 3 as a typical backflow node, and the locations of point A and node 3 as representatives for the surface water depth. It compares the calculation results of the improved model and the ICM model with two types of grid meshing. The Nash–Sutcliffe efficiency coefficient (E_{NS}) was used to quantitatively evaluate the degree of agreement between the simulation results of the two models. It can be seen that the calculation results of the improved vertical flow exchange method with different grid meshes had a high degree of agreement with the results of the ICM model. With the meter-level and ten-meter-level grid meshes, the coupled calculation results obtained with the improved vertical flow exchange method were basically consistent with the results obtained with the ICM model. It can be seen in Figure 7c that the difference in grid meshing had a certain impact on the simulation results for the surface water depth, but the improved method was able to obtain a water depth change curve that was almost consistent with that of the ICM model.

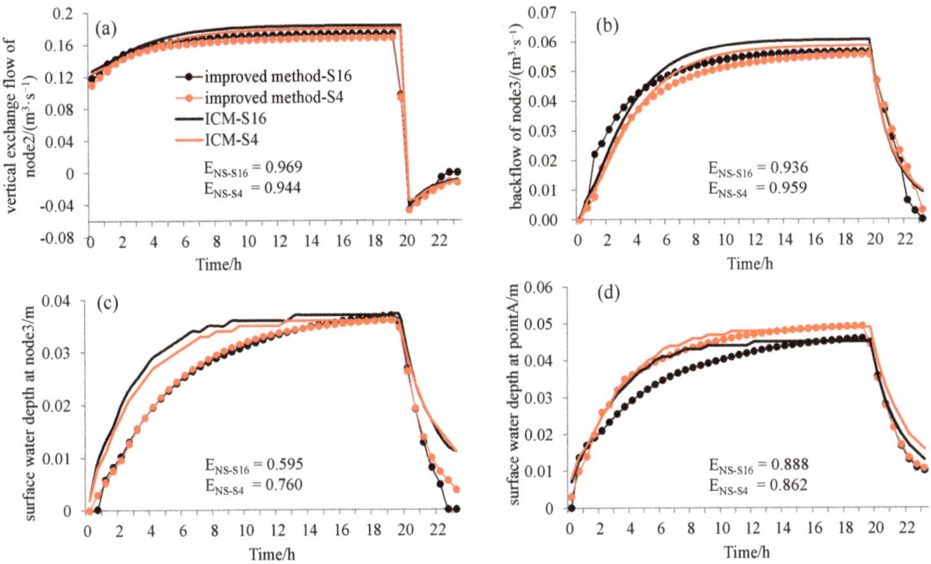

Figure 7. Comparison of the results of the improved model and the ICM model (**a**–**d**).

4.3. Practical Study Case

The Dongsha storm drainage system in Wuchang District, Wuhan City, was used as a practical case for the simulation of urban waterlogging; the study area covers a catchment range of 82.73 km^2. The terrain within the catchment area is relatively flat, with Shahu Lake and Donghu Lake located on the south side. These lakes are the main recipients of runoff. However, as the lake water levels remain relatively high throughout the year, it adversely affects the drainage of the stormwater drainage network, making local low-lying areas prone to waterlogging.

The terrain and drainage system of the study area are shown in the right subfigure of Figure 8. The ICM model and the improved model were both used to establish a coupled 1D–2D model for the study area. An unstructured grid with an average cell size of 450 m^2 was adopted as the ICM model's ground grid, and the total number of cells was 210,000. The improved model used a structured grid for the ground, with a cell size of 20 m $\times$ 20 m and a total of 240,000 cells. The rainfall data used were from two heavy rainstorms in the area, which occurred on 18 June and 6 July 2016 (hereinafter referred to as the 201606 and 201607 rainstorms). The cumulative rainfall amounts for the two storms were 143.5 mm and 125.6 mm, respectively, and the rainfall intensity curves for both storms is shown in the two left subfigures in Figure 8.

Both coupled models were used to simulate the waterlogging conditions for the two rainstorms. The inundated area of the improved model is shown in Figure 9. The 201606 rainstorm, with its heavy rainfall and short duration, led to severe urban waterlogging. The simulation results indicated many inundated areas and significant water depths on the ground. For the 201607 rainstorm, although it also had heavy rainfall, its longer duration and more uniform distribution of rainfall throughout the event, with a peak intensity approximately half of that of the 201606 rainstorm, resulted in less severe waterlogging and fewer recorded waterlogging points on that day. In both rainstorm simulations, the simulation results of the improved model were basically consistent with the actual recorded waterlogging points on the respective days. The simulation results demonstrated that the coupled model using the improved vertical flow exchange method was able to accurately simulate the inundation range and depth of urban waterlogging.

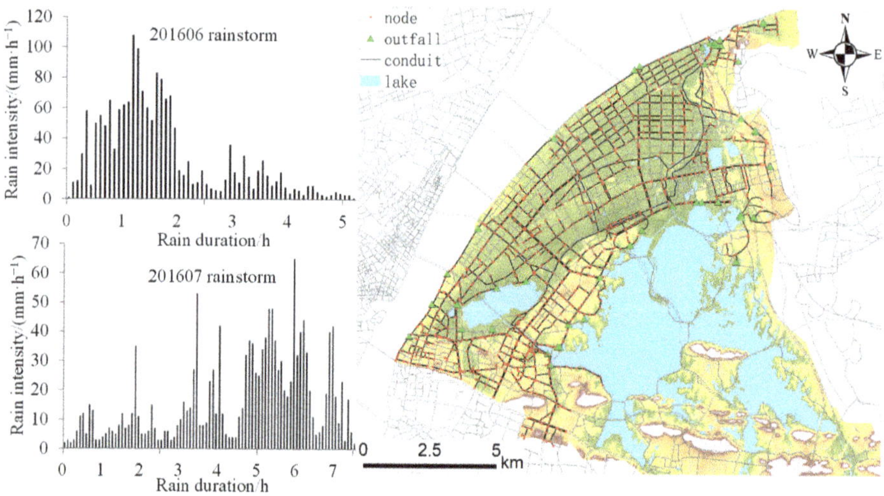

Figure 8. Storm drainage system and rainfall intensity curves of the selected storms.

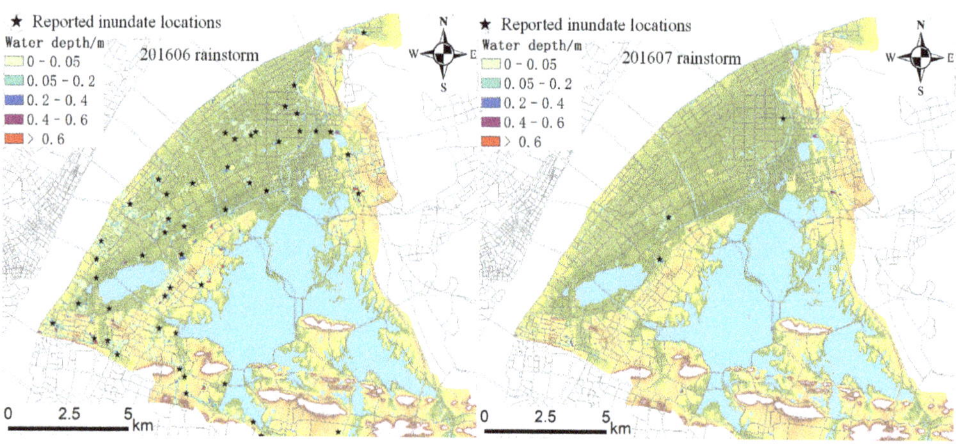

Figure 9. Inundated areas in the two rainstorms.

Figure 10 presents the temporal changes in the surface water volume for both the improved model and the ICM model during the two rainfall events. It can be observed that the vertical flow exchange calculations from the improved model closely aligned with those from the ICM model, as indicated by the high degree of overlap in the curves depicting the temporal variation in the volume of groundwater accumulation. This suggests that the vertical flow exchange calculations at the nodes using the improved method were highly consistent with those from the ICM model.

In the urban waterlogging simulations, the vertical flow exchange at overloaded nodes accounted for most of the overall vertical flow exchange. Figure 11 shows a comparison of the number of different types of vertical flow exchange nodes in the evolution of the 1D model during the coupling process. Based on the statistical analysis of the simulation results, the proportion of vertical flow exchange at overloaded nodes, with respect to the total vertical flow exchange at nodes during the two rainstorm simulations, was 81.43% and 88.66%, respectively. Using the improved method allowed for the direct calculation of the vertical flow exchange results at nodes using the SWMM module, based on the principle of water conservation in most cases. The calculation of the backflow volume using a variable-head orifice flow formula was only needed in a few instances of backflow at non-overloaded nodes. This significantly reduced the amount of vertical flow exchange calculations using the orifice or weir flow formulas, thereby improving the computational efficiency and stability of the coupled simulation.

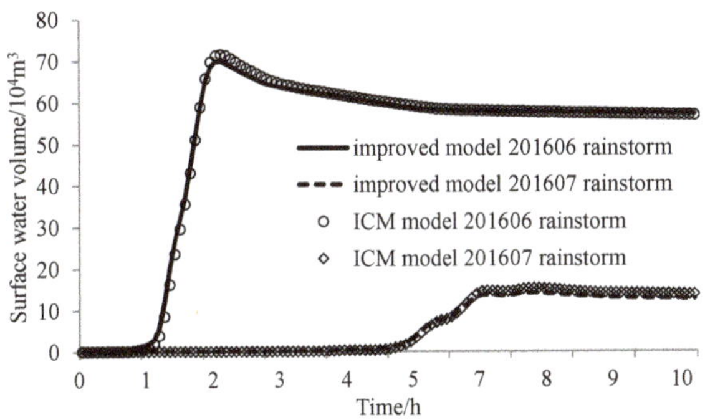

Figure 10. Comparison of temporal changes in the surface water volume of the different models.

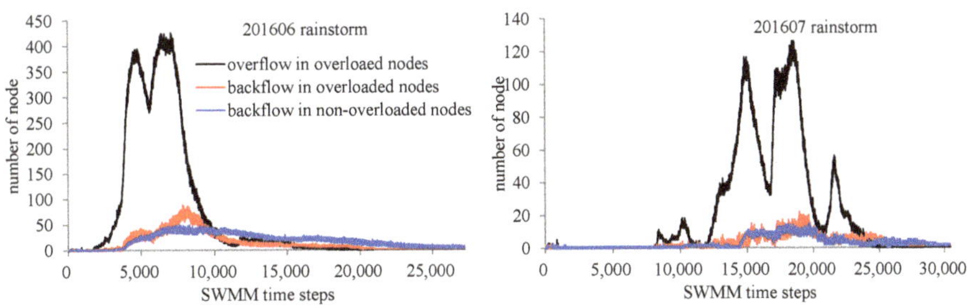

Figure 11. Numbers of different vertical flow exchange nodes in the 1D model during the coupling process.

5. Conclusions and Future Work

Based on the SWMM and an independently developed two-dimensional hydrodynamic model, a coupled hydrological and hydrodynamic model for urban waterlogging was constructed. Within this framework, the vertical flow exchange method was improved using the principle of node water balance. Both a theoretical drainage system and a real urban drainage system were used as case studies in order to validate the improved method in terms of simulation accuracy, reliability and water balance. The main conclusions are as follows.

Calculating the vertical flow exchange using the principle of water conservation can significantly reduce the calculations related to the orifice or weir formulas. Most calculations related to the vertical flow exchange can be implemented with the original functions in the SWMM. Only the backflow at non-overloaded nodes needs to be calculated outside the SWMM by using orifice or weir flow formulas.

The calculation of the vertical flow exchange is integrated into the SWMM calculation module. The SWMM module uses an iterative method to calculate the node water heads and conduit flow rates, with the convergence of node water heads as the stopping condition for each iteration. Thus, numerical harmony is achieved among the node water heads, conduit flows and vertical exchange flows. In addition, numerical stability of the simulation is achieved without any artificially set values in the calculation of the vertical flow exchange.

A comparison of the simulation results from the ICM model with those of the improved model demonstrates that the coupled model using the improved vertical flow exchange method exhibits good accuracy and reliability. Its results align closely with those of the ICM simulation across various grid sizes ranging from meters to hundreds of meters, making this method suitable for the numerical simulation of urban waterlogging.

However, this study still has certain limitations. It only discusses the calculation of vertical flow exchange in the case of Semi-Distributed Urban Stormwater Models. The nodes involved in vertical flow exchange are computational nodes, not specific rainwater inlets. Additionally, due to insufficient monitoring data, there has been no detailed calibration or discussion of the flow coefficients at the exchange nodes in both the ICM and proposed models. These shortcomings need to be addressed in future work.

Author Contributions: Methodology, X.J.; Software, X.J.; Validation, Y.M.; Writing—original draft, Y.M. All authors have read and agreed to the published version of the manuscript.

Funding: This research received no external funding.

Data Availability Statement: Data are contained within the article.

Conflicts of Interest: The authors declare no conflict of interest.

References

1. Xu, Z.X.; Chen, H.; Ren, M.F.; Cheng, T. Progress on disaster mechanism and risk assessment of urban flood/waterlogging disasters in China. *Adv. Water Sci.* **2020**, *31*, 713–724.
2. Huang, G.R.; Cheng, W.J.; Yu, H.J. Construction and evaluation of an integrated hydrological and hydrodynamics urban flood model. *Adv. Water Sci.* **2021**, *32*, 334–344.
3. Zhang, H.P.; Li, M.; He, R.M.; Cheng, G.S. Application scenarios and corresponding technical strategies of urban flood modeling. *Adv. Water Sci.* **2022**, *33*, 452–461.
4. Barreiro, J.; Santos, F.; Ferreira, F.; Neves, R.; Matos, J.S. Development of a 1D/2D Urban Flood Model Using the Open-Source Models SWMM and MOHID Land. *Sustainability* **2023**, *15*, 707. [CrossRef]
5. HELMIö, T. Unsteady 1D flow model of compound channel with vegetated floodplains. *J. Hydrol.* **2002**, *269*, 89–99. [CrossRef]
6. Yoshida, H.; Dittrich, A. 1D unsteady-state flow simulation of a section of the upper Rhine. *J. Hydrol.* **2002**, *269*, 79–88. [CrossRef]
7. Hsu, M.H.; Chen, S.H.; Chang, T.J. Inundation simulation for urban drainage basin with storm sewer system. *J. Hydrol.* **2000**, *234*, 21–37. [CrossRef]
8. Cao, X.; Lyu, H.; Ni, G.; Tian, F.; Ma, Y.; Grimmond, C.S.B. Spatial Scale Effect of Surface Routing and Its Parameter Upscaling for Urban Flood Simulation Using a Grid-Based Model. *Water Resour. Res.* **2020**, *56*, e2019WR025468. [CrossRef]
9. Cheng, T.; Xu, Z.; Yang, H.; Hong, S.; Leitao, J.P. Analysis of Effect of Rainfall Patterns on Urban Flood Process by Coupled Hydrological and Hydrodynamic Modeling. *J. Hydrol. Eng.* **2020**, *25*, 04019061. [CrossRef]
10. Huang, H.C.; Liao, W.; Lei, X.; Wang, C.; Cai, Z.; Wang, H. An urban DEM reconstruction method based on multisource data fusion for urban pluvial flooding simulation. *J. Hydrol.* **2023**, *617*, 128825. [CrossRef]
11. Shi, Y.Y.; Wan, D.H.; Chen, L.; Fu, G.X. Simulation of Rainstorm Waterlogging and Submergence in Urban Areas Based on GIS and SWMM. *Water Resour. Power* **2014**, *32*, 57–60+12.
12. Zhao, G.; Xu, Z.X.; Pang, B.; Wang, H.T. Estimation of urban flooding processes based on enhanced inundation model. *Adv. Water Sci.* **2018**, *29*, 20–30.
13. Zeng, Z.; Lai, C.; Wang, Z.; Wu, X.; Huang, G.; Hu, Q. Rapid simulation of urban rainstorm flood based on WCA2D and SWMM model. *Adv. Water Sci.* **2020**, *31*, 29–38. [CrossRef]
14. Bazin, P.H.; Nakagawa, H.; Kawaike, K.; Paquier, A.; Mignot, E. Modeling Flow Exchanges between a Street and an Underground Drainage Pipe during Urban Floods. *J. Hydraul. Eng.* **2014**, *140*, 04014051. [CrossRef]
15. Martins, R.; Leandro, J.; Chen, A.S.; Djordjević, S. A comparison of three dual drainage models: Shallow water vs. local inertial vs. diffusive wave. *J. Hydroinformatics* **2017**, *19*, 331–348. [CrossRef]

16. Kitsikoudis, V.; Erpicum, S.; Rubinato, M.; Shucksmith, J.D.; Archambeau, P.; Pirotton, M.; Dewals, B. Exchange between drainage systems and surface flows during urban flooding: Quasi-steady and dynamic modelling in unsteady flow conditions. *J. Hydrol.* **2021**, *602*, 126628. [CrossRef]
17. Wang, H.; Yue, S.; Zhang, Z.; Guo, F.; Wen, Y.; Chen, M.; Lü, G. Development of a component-based integrated modeling framework for urban flood simulation. *Environ. Model. Softw.* **2023**, *169*, 105839. [CrossRef]
18. Cosco, C.; Gómez, M.; Russo, B.; Tellez-Alvarez, J.; Macchione, F.; Costabile, P.; Costanzo, C. Discharge coefficients for specific grated inlets. Influence of the Froude number. *Urban Water J.* **2020**, *17*, 656–668. [CrossRef]
19. Rubinato, M.; Martins, R.; Kesserwani, G.; Leandro, J.; Djordjević, S.; Shucksmith, J. Experimental calibration and validation of sewer/surface flow exchange equations in steady and unsteady flow conditions. *J. Hydrol.* **2017**, *552*, 421–432. [CrossRef]
20. Chang, T.-J.; Wang, C.-H.; Chen, A.S. A novel approach to model dynamic flow interactions between storm sewer system and overland surface for different land covers in urban areas. *J. Hydrol.* **2015**, *524*, 662–679. [CrossRef]
21. Pina, R.D.; Ochoa-Rodriguez, S.; Simões, N.E.; Mijic, A.; Marques, A.S.; Maksimović, Č. Semi- vs. Fully-Distributed Urban Stormwater Models: Model Set Up and Comparison with Two Real Case Studies. *Water* **2016**, *8*, 58. [CrossRef]
22. Cardenas-Quintero, M.; Carvajal-SERNA, F. Hydraulic capacity of a grate inlet in supercritical flow: Weir equation. *Urban Water J.* **2023**, *20*, 763–768. [CrossRef]
23. Gomez, M.; Russo, B. Methodology to estimate hydraulic efficiency of drain inlets. *Proc. Inst. Civ. Eng.-Water Manag.* **2011**, *164*, 81–90. [CrossRef]
24. Russo, B.; Gomez, M.; Tellez, J. Methodology to Estimate the Hydraulic Efficiency of Nontested Continuous Transverse Grates. *J. Irrig. Drain. Eng.* **2013**, *139*, 864–871. [CrossRef]
25. Tellez-Alvarez, J.; Gómez, M.; Russo, B.; Amezaga-Kutija, M. Numerical and Experimental Approaches to Estimate Discharge Coefficients and Energy Loss Coefficients in Pressurized Grated Inlets. *Hydrology* **2021**, *8*, 162. [CrossRef]
26. Zaiter, A.; Sabtu, N.; Almaliki, D.F. Equations and methodologies of inlet drainage system discharge coefficients: A review. *Open Eng.* **2024**, *14*, 20220598. [CrossRef]
27. Russo, B.; Valentin, M.G.; Tellez-Alvarez, J. The Relevance of Grated Inlets within Surface Drainage Systems in the Field of Urban Flood Resilience. A Review of Several Experimental and Numerical Simulation Approaches. *Sustainability* **2021**, *13*, 7189. [CrossRef]
28. Russo, B.; Sunyer, D.; Velasco, M.; Djordjević, S. Analysis of extreme flooding events through a calibrated 1D/2D coupled model: The case of Barcelona (Spain). *J. Hydroinformatics* **2015**, *17*, 473–491. [CrossRef]
29. Fan, Y.; Ao, T.; Yu, H.; Huang, G.; Li, X. A Coupled 1D-2D Hydrodynamic Model for Urban Flood Inundation. *Adv. Meteorol.* **2017**, *2017*, 2819308. [CrossRef]
30. Peng, G.; Zhang, Z.; Zhang, T.; Song, Z.; Masrur, A. Bi-directional coupling of an open-source unstructured triangular meshes-based integrated hydrodynamic model for heterogeneous feature-based urban flood simulation. *Nat. Hazards* **2022**, *110*, 719–740. [CrossRef]
31. Gironás, J.; Roesner, L.A.; Rossman, L.A.; Davis, J. A new applications manual for the Storm Water Management Model (SWMM). *Environ. Model. Softw.* **2010**, *25*, 813–814. [CrossRef]
32. Jin, X.; Wang, F. Calculation Method of Urban Flooding One-dimensional/Two-dimensional Coupling Model Based on CUDA-based Parallel Implementation. *China Water Wastewater* **2020**, *017*, 36.
33. Liang, Q. Flood Simulation Using a Well-Balanced Shallow Flow Model. *J. Hydraul. Eng.* **2010**, *136*, 669–675. [CrossRef]
34. Liang, Q.; Marche, F. Numerical resolution of well-balanced shallow water equations with complex source terms. *Adv. Water Resour.* **2009**, *32*, 873–884. [CrossRef]
35. Van Leer, B. Towards the Ultimate Conservative Difference Scheme V. A Second Order Sequel to Godunov's Method. *J. Com. Phys.* **1979**, *32*, 101–136. [CrossRef]
36. Roe, P.L. Characteristic-Based Schemes for the Euler Equations. *Annu. Rev. Fluid Mech.* **1986**, *18*, 337–365. [CrossRef]
37. Toro, E.F. *Shock-Capturing Methods for Free-Surface Shallow Flows*; John Wiley: Hoboken, NJ, USA, 2001.
38. Yoon, T.H.; Kang, S.K. Finite volume model for two-dimensional shallow water flows on unstructured grids. *J. Hydraul. Eng.* **2004**, *130*, 678–688. [CrossRef]
39. Alcrudo, F.; Mulet, J. *Urban Inundation Models Based on the Shallow Water Equations*; Hermes Science Publishing: New Castle, PA, USA, 2005.

water

MDPI

Article

Optimization of Composite Cavitation Nozzle Parameters Based on the Response Surface Methodology

Gao Huang [1], Chengjun Qiu [1,2,3,*], Mengtian Song [1,2,*], Wei Qu [2,4], Yuan Zhuang [1], Kaixuan Chen [1], Kaijie Huang [1], Jiaqi Gao [1], Jianfeng Hao [1] and Huili Hao [1]

1 College of Mechanical, Naval Architecture & Ocean Engineering, Beibu Gulf University, Qinzhou 535011, China; huanggao2024@163.com (G.H.); zhuangyuan@bbgu.edu.cn (Y.Z.); chenkx@bbgu.edu.cn (K.C.); hkjhha@163.com (K.H.); znkjgl@sina.com (J.G.); 18435407763@163.com (J.H.); haohuili0022@163.com (H.H.)

2 Eastern Michigan Joint College of Engineering, Beibu Gulf University, Qinzhou 535011, China; quwei@bbgu.edu.cn

3 Guangxi Key Laboratory of Ocean Engineering Equipment and Technology, Qinzhou 535011, China

4 College of Electronics and Information Engineering, Beibu Gulf University, Qinzhou 535011, China

* Correspondence: qiuchengjun@bbgu.edu.cn (C.Q.); songmengtian@bbgu.edu.cn (M.S.); Tel.: +86-18177770651 (C.Q.); +86-18377181848 (M.S.)

Abstract: Cavitation is typically observed when high-pressure submerged water jets are used. A composite nozzle, based on an organ pipe, can increase shear stress on the incoming flow, significantly enhancing cavitation performance by stacking Helmholtz cavities in series. In the present work, the flow field of the composite nozzle was numerically simulated using Large Eddy Simulation and was paired with the response surface method for global optimizing the crucial parameters of the composite nozzle to examine their effect on cavitation behavior. Utilizing peak gas-phase volume percent as the dependent variable and the runner diameter, Helmholtz chamber diameter, and Helmholtz chamber length as independent variables, a mathematical model was constructed to determine the ideal parameters of the composite nozzle through response surface methodology. The optimized nozzle prediction had an error of only 2.04% compared to the simulation results, confirming the accuracy of the model. To learn more about the cavitation cloud properties, an experimental setup for high-pressure cavitation jets was also constructed. Impact force measurements and high-speed photography tests were among the experiments conducted. The simulated evolution period of cavitation cloud characteristics is highly consistent with the experimental period. In the impact force measurement experiment, the simulated impact force oscillates between 256 and 297 N, and the measured impact force oscillates between 260 N and 289 N, with an error between 1.5% and 2.7%. The simulation model was verified by experimental results. This study provides new insights for the development of cavitation jet nozzle design theory.

Keywords: composite submerged nozzle; large eddy simulation; response surface methodology; high-speed photography

1. Introduction

Cavitation refers to the process in a fluid flow where, due to changes in the pipeline structure or external conditions, the local fluid pressure drops below the current fluid's saturation vapor pressure. This drop causes tiny bubbles within the fluid to rapidly expand and collapse in a short period, releasing a large amount of energy. Johnson et al. [1] proposed to apply cavitation to water jet technology to improve the jet erosion effect. Relevant studies have shown that under the same conditions of pump pressure and flow rate conditions, the impact pressure of a cavitation jet is 8.6 to 124 times higher than that of a continuous jet [2–4]. Cavitation jet technology has wide applications in fields such as oil extraction, cleaning and rust removal, and metal surface strengthening [5–7].

The cavitation jet nozzle is the core component for generating cavitation and can have a significant impact on the cavitation effect [8–10]. The working conditions inside the cavitation jet nozzle are complex, involving phase transition, high pressure, and high speed. Therefore, the study of the internal phenomenon of the cavitation nozzle mainly relies on numerical simulation [11,12]. Chen et al. [13] confirmed through numerical simulations that cavitation jet corrosion of steel is primarily due to the generation of high-strength shock waves and instantaneous high temperatures when cavitation bubbles collapse. Dong et al. [14] modified the cavity wall contour of the Helmholtz nozzle into a circular arc shape, resulting in a 16.9% increase in turbulent flow energy. Yang et al. [15] captured the growth, shedding, and collapse of cavitation in three nozzle jets using high-speed photography, finding good agreement between experimental images and predicted model images. Dong [16] designed a diagonal nozzle with the k-ε model and researched the impact of intake constriction, parallel mid-section. Wang [17] used the proper orthogonal decomposition method (POD) to analyze cavitation cloud characteristics in organ pipe nozzles. Stanley et al. [18] used direct numerical simulation (DNS) to study the creation of a free jet flow and the non-constant flow properties of a free-jet-flow shear layer. The Reynolds-Averaged Navier–Stokes (RANS) method is commonly used for its lower computational demand, but it struggles to describe vortex dynamics and only simulates time-averaged flow characteristics, neglecting transient values in cavitated flow fields [19].

Large Eddy Simulation (LES) can decompose turbulent instantaneous motion into large- and small-scale vortices, which are calculated separately. Large-scale motion needs to be directly calculated through a numerical solution of the motion differential equation. The impact of small-scale motion on large-scale motion will be manifested in a stress term similar to Reynolds stress in the motion equation, which is called sub-grid Reynolds stress. They will be simulated by establishing a model that is a perfect combination of the direct simulation method and the Reynolds-averaged simulation method. With sufficient computational resources, more information on turbulence can be obtained. The method has been extensively used to study nozzle cavitation [20–23]. Kim et al. [24] used LES to investigate the influences of initial momentum thickness on free jet flow in a circular hole, discovering that these factors have a significant impact on the free jet's flow characteristics. Fang [25] compared LES results with experiments to better understand organ pipe nozzles. Li et al. [26] used LES to study the flow characteristics of highly under-expanded jets with several different nozzle geometries (circular, elliptical, square, and rectangular). The results showed significant differences in the formation and development of intercepted surges, with the elliptical jet having the slowest penetration rate. Yang et al. [27] used the RANS method, LES method, and the RANS-LES hybrid method to simulate the fluid field of submerged cavitation water jets and analyzed the accuracy of different turbulence models in predicting jet cavitation. The results showed significant deviations in RANS, while RANS-LES and LES models showed higher precision.

Recent research has begun integrating fluid dynamics with multi-objective parameter optimization [28]. Li et al. [29] used genetic algorithms to optimize curved nozzle structural parameters for maximum output power. Uebel et al. [30] utilized quenching conversion concepts based on CFD to optimize and improve system performance and obtained feasible design parameters for a quenching reactor. Qian et al. [31] conducted experiments to obtain the relationship between injection distance and nozzle diameter, establishing a mathematical model to identify optimal injection distances for each nozzle diameter. Edeling et al. [32] used a Bayesian algorithm, calibrated with the Launder–Sharma k-model, to predict new boundary layer flows and analyze errors.

Response surface methods are frequently employed to explore the nonlinear effects of numerous variables on the response variable in a particular range and alter them based on practical demands [33]. Han et al. [34] used the response surface methodology to optimize the internal structural parameters for the key part of a straight cone nozzle, with the flow coefficient used as the objective function. Zhang et al. [35] proposed a dual serpentine nozzle structure and analyzed the influence of key structural parameters on its aerodynamic performance using computational fluid dynamics and the response surface methodology. They established an axial thrust coefficient response model to understand interactions between exit aspect ratio, export area ratio, and nozzle outlet width ratio. Wang et al. [36] established a response surface method model with experimental impact force characteristics as dependent variables and nozzle structure parameters as separate variables. This model was used to calculate the optimal parameter solution. Using results from impact force and cleaning performance experiments, it was verified that the experimental errors were all less than 5%, indicating high reliability.

This paper establishes a response surface fitting model between the structural parameters of the cavitation nozzle and the peak gas-phase volume fraction inside the nozzle. The model, developed using Large Eddy Simulation and the response surface method, identifies optimal structural parameters. The optimized nozzle is then analyzed experimentally.

2. Numerical Simulation Model and Physical Model

2.1. Governing Equations of Large Eddy Simulation

Large Eddy Simulation decomposes turbulence into large-scale and small-scale vortices using a filter function. The large-scale vortices are directly solved using the N-S equation, while small-scale vortices are processed by the sublattice sub-scale model. In this study, the water jet cavitation phenomena, caused by the nozzle structure, results in a gas–liquid two-phase fluid. This fluid is considered incompressible and can be calculated by filtering the three-dimensional incompressible governing equations [37]:

$$\frac{\partial \overline{u}_i}{\partial x_i} = 0 \tag{1}$$

$$\frac{\partial \overline{u}_i}{\partial t} + \frac{\partial \overline{u}_i \overline{u}_j}{\partial x_j} = -\frac{1}{\rho}\frac{\partial \overline{p}}{\partial x_i} + v\frac{\partial^2 \overline{u}_i}{\partial x_i \partial x_j} - \frac{\partial \tau_{ij}}{\partial x_j} \tag{2}$$

$$\tau_{ij} - \frac{\delta_{ij}}{3}\tau_{kk} = -2v_\tau S_{ij} \tag{3}$$

where $\overline{u}_i$ and $\overline{u}_j$ are the instantaneous velocities of the fluid; x_i denotes the 3D coordinate direction; ρ and p represent the fluid density and fluid pressure, respectively; τ_{ij} is the sub-grid scale stress between the large- and small-scale vortices; δ_{ij} is Kronecker's delta; v_τ is the eddy viscosity; and S_{ij} is the resolved scale strain rate tensor.

The vortex viscosity coefficients are derived using the WALE model based on the velocity gradient's invariance. The governing equations of the WALE model are as follows [38]:

$$\mu_t = \rho L_s^2 \frac{\left(S_{ij}^d S_{ij}^d\right)^{\frac{3}{2}}}{\left(\overline{S}_{ij}\overline{S}_{ij}\right)^{\frac{5}{2}} + \left(S_{ij}^d S_{ij}^d\right)^{\frac{5}{4}}} \tag{4}$$

$$L_s = min\left(kd, C_w V^{\frac{1}{3}}\right) \tag{5}$$

$$S_{ij}^d = \frac{1}{2}\left(\overline{g}_{ij}^2 + \overline{g}_{ji}^2\right) - \frac{1}{3}\delta_{ij}\overline{g}_{kk}^2, \overline{g}_{ij} = \frac{\partial \overline{u}_i}{\partial x_j} \tag{6}$$

where L_s is the length of the sub-grid scale; C_w is the WALE model constant, taken as 0.35 in the current work; d represents the distance to the proximate wall; k represents the von Karman constant; and V is the computational cell volume.

2.2. Computational Grid and Boundary Conditions

In this study, an organ pipe was connected behind a Helmholtz nozzle to form a dual-chamber composite nozzle. When the fluid flows into the Helmholtz chamber at high speed from the inlet, due to the sudden increase in cross-section, the high-speed fluid will form a submerged free jet after entering the Helmholtz chamber. The original low-speed fluid in the chamber comes into contact with it and forms a shear layer at the outer boundary of the jet on the contact surface. The fluid within the boundary zone of a free jet undergoes entrainment with the nearby stationary fluid, causing the free jet to be in a reverse pressure state. At the same time, a shedding vortex is formed, which propagates downstream and interacts with the colliding wall to generate backflow, resulting in oscillation. When the jet enters the accordion tube and passes through the contracted section, it can generate pressure excitation in the fluid and feed back the pressure excitation to the resonant cavity, forming feedback pressure oscillation. According to transient flow theory, if the frequency of pressure excitation matches the natural frequency of the nozzle resonant cavity, the organ tube chamber will amplify the incoming flow from the Helmholtz chamber, which can transform the shear layer vortex of the jet into a large structure of separated annular vortex. The intermittent vortex circulation of this large structure can enhance cavitation, that is, generate cavitation jet at the outlet, thereby improving its erosion and cleaning ability. Figure 1 shows a geometric model of the Helmholtz–organ pipe composite nozzle, and the parameters are provided in Table 1.

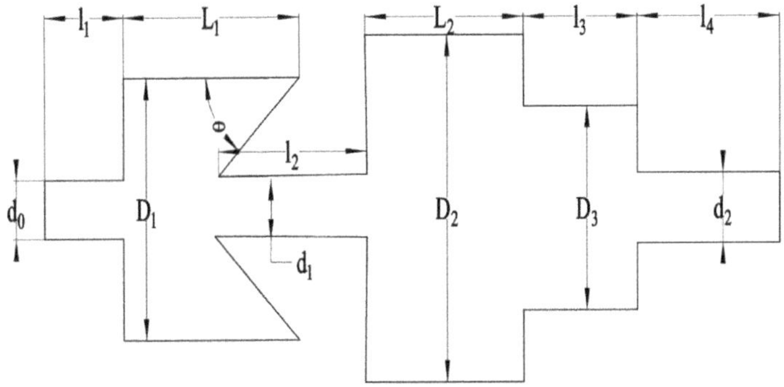

Figure 1. Composite nozzle model geometry.

Table 1. Composite nozzle structural parameters.

Position	Symbol/Unit	Dimensional Parameter
Inlet diameter	d_0/mm	6
Length of inlet straight section	l_1/mm	5
Helmholtz chamber diameter	D_1/mm	12–16
Helmholtz chamber length	L_1/mm	7–11
Helmholtz chamber angle	$\theta/^\circ$	60
Connection runner diameter	d_1/mm	2–8
Connection runner length	l_2/mm	6
Organ pipe resonant cavity diameter	D_2/mm	10
Diameter of the middle chamber of the organ pipe	D_3/mm	6
Organ pipe resonant cavity length	L_2/mm	6
Length of the cylindrical section of organ pipe	l_3/mm	4
Length of diffusion section	l_4/mm	5
Outlet diameter	d_2/mm	2

The computational domain for the submerged flow field is a cylinder having the diameter of 30 mm with the length of 100 mm, where it must be ensured that the fluid field develops properly in the composite nozzle. In this study, CFD grid generation was carried out using Fluent Meshing in ANSYS 2021R1 software. Polyhedral grids were used for meshing the geometry, and a denser grid was created in the composite nozzle jet's core region, as well as shear layers near the exit. To verify grid independence, the model was simulated for four sets of grids with different grid sizes. The results of the peak gas-phase volume fraction, as shown in Figure 2, indicate that as the amount of grid cells grew from 1,861,356 to 2,792,034 altogether, the change of the peak gas-phase volume fraction was less than 2.9%. Hence, the grid with 1,861,356 cells was used in the later models. The overall mesh quality of the selected grid was about 0.7, which meets the computational requirement of Large Eddy Simulations.

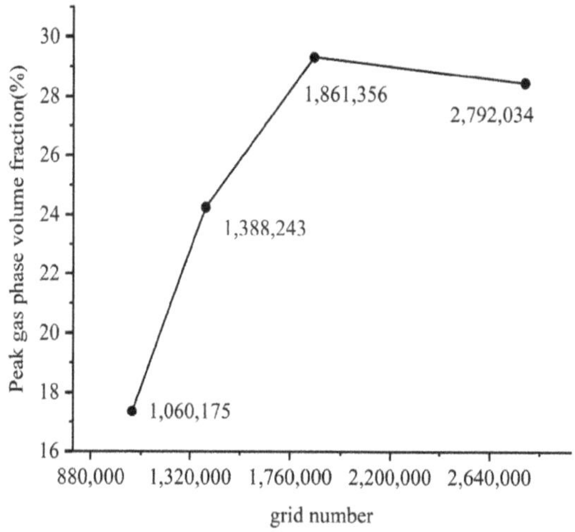

Figure 2. Effect of grid quantity on peak gas-phase volume fraction.

The final mesh of computational domain and boundary settings are shown in Figure 3. The pressure inlet was set at 10 MPa, while the output was set to 0 MPa. The ambient pressure was set at 101,325 Pa, with a turbulence intensity of 5% and a turbulent viscosity ratio of 10; additionally, a non-sliding wall surface was used. The fluid medium was pure water with a saturation vapor pressure of 3540 Pa. The WALE model was selected for LES as a sub-grid model, and the pressure–velocity coupled momentum equation was calculated using the PISO algorithm. Because cavitation is a complex unstable process, transient simulation was used in this research. The grid number of 1,861,356 was taken as the model, and the time step was 1×10^{-4} s, 1×10^{-5} s, 1×10^{-6} s, and the time independence test was conducted for the selected grid number. The calculated results are shown in Table 2. The simulation results changed greatly when the time step size changed from 1×10^{-4} s to 1×10^{-5} s, but when the time step is further reduced, the change in the results is relatively small. It can be concluded that when the transient time step is less than 1×10^{-5} s, the results of CFD simulation are independent of the time step, resulting in the time step of 1×10^{-5} s and the iteration step of 1000. The computational residuals for each time step of each conservation equation are less than 1×10^{-6}.

Table 2. Results of the time step size independence tests.

Time step size (s)	1×10^{-4}	1×10^{-5}	1×10^{-6}
Peak gas-phase volume fraction (%)	29.06	29.32	29.33

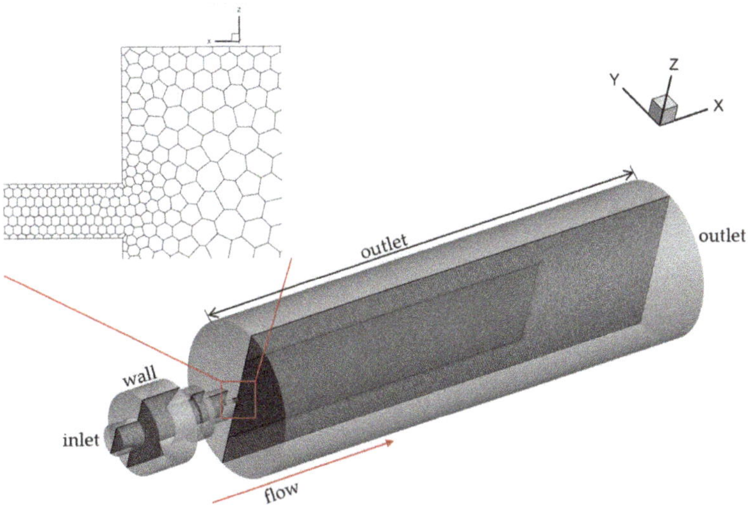

Figure 3. Mesh of computational domain and boundary settings.

3. Response Surface Method Design

The response surface technique aims to find the optimal solution by specifying a range of values for each variable. It involves conducting a finite number of tests and building a polynomial fit to approximate the relationship function between each factor [39,40]. The formula for the response surface method is as follows [41]:

$$y = \beta_0 + \sum_{i=1}^{n} \beta_i x_i + \sum_{i=1}^{n} \beta_{ii} x^2 + \sum_{i<j}^{n} \beta_{ij} x_i x_j + \varepsilon \tag{7}$$

The above equation includes the constant, linear, squared, and interaction terms, along with an error term. In Equation (7), n is the number of independent variables; ε is the total error; β_0, β_i, β_{ii}, and β_{ij} are the regression coefficients; and x is the investigation factor.

This study adopted the Box–Behnken Design (BBD) suitable for 2–5 factors in response to surface methodology. The experiments considered the peak gas-phase volume fraction as the dependent variable because the volume fraction of the gas phase can effectively reflect the effectiveness of cavitation; it can also reflect the cleaning ability of the composite nozzle, while the independent variables were taken as the diameter (D_1), length (L_1), and connecting channel diameter (d_1) of the Helmholtz chamber in the composite nozzle. The design included three factors, A, B, and C, at three levels: −1, 0, and 1. The coding and actual values for each design factor in the composite nozzle are displayed in Table 3.

Table 3. Levels of independent variables.

Variable	Coded Variables	Variable Levels		
		−1	0	1
L_1/(mm)	A	7	9	11
D_1/(mm)	B	12	14	16
d_1/(mm)	C	2	5	8

According to Table 3, the experimental plan was designed with 17 sets of test points. The initial 12 sets were used for factor analysis, and the remaining 5 sets were used to estimate the experimental error. The final experimental plan and response values by simulation are shown in Table 4.

Table 4. Experimental design and results.

Run	Variables			Peak Gas-Phase Volume Fraction V (%)
	A: L_1/(mm)	B: D_1/(mm)	C: d_1/(mm)	
1	7	12	5	16.1516
2	11	12	5	26.2346
3	7	16	5	26.4647
4	11	16	5	20.3565
5	7	14	8	18.75
6	11	14	8	13.2654
7	7	14	2	24.364
8	11	14	2	30.5335
9	9	12	8	12.415
10	9	16	8	17.369
11	9	12	2	24.2589
12	9	16	2	25.1698
13	9	14	5	29.123
14	9	14	5	29.328
15	9	14	5	32.1356
16	9	14	5	28.7596
17	9	14	5	30.2115

4. Results and Discussion

4.1. Response Surface Analysis

Regression fitting was performed using the data from Table 4, involving multiple interactions to establish a predictive regression equation. This equation relates the peak gas-phase volume fraction of the composite nozzle to several factors:

$$V = 29.91 + 0.5825A + 1.29B - 5.32C - 4.05AB + 2.91AC + 1.01BC - 2.84A^2 - 4.77B^2 - 5.34C^2 \tag{8}$$

ANOVA was used to analyze the regression equation, and the results are presented in Table 5. Analysis reveals that the primary term C, and the secondary terms B^2 and C^2 have significant effects on the peak gas-phase volume fraction ($p \leq 0.01$). The primary term B, interaction terms AB and AC, and secondary term A^2 also significantly affect it ($p \leq 0.05$). Other terms were found to have no significant effect. The F values in Table 5 indicate the degree of influence of three factors on the peak gas-phase volume fraction, as follows: connection channel diameter (C) > Helmholtz chamber diameter (B) > Helmholtz chamber length (A). The p-value of this overall model ($p < 0.001$) reached a high significance level, and the error of the out-of-fit term ($p = 0.6848 > 0.05$) was found to be not significant. This indicates that the model is quite accurate and stable. The regression coefficient R^2 of the model is 0.9838. This indicates that 98.38% of the variability in the data are accounted for by the model, suggesting a strong fit for substituting the experimental results. The adjusted R^2 was 0.9630, which suggests that the model can explain 96.30% of the variance in the peak gas-phase volume fraction after adjusting for the number of predictors in the model. The difference between adjusted R^2 and the predicted R^2 is lower than 0.2. This indicates that the predicted values of the model are very close to experimental values, further indicating its high reliability.

Figure 4 shows the normal probability distribution of residuals and the distribution of residuals and predicted values, respectively. From Figure 4a, it can be seen that the normal probability distribution of residuals approximates a straight line, indicating reliable fitting results for the model's independent variables and response values. Figure 4b displays the externally studentized residuals of the model predictions against the predicted values of the model. The figure shows that the residuals are distributed randomly and irregularly. Based on the above judgment conditions, it can be concluded that the peak gas-phase volume fraction model of the composite nozzle, obtained via the response surface method, exhibits good adaptability and conforms to the inspection principles.

Table 5. ANOVA results of regression model.

Source	Sum of Squares	df	Mean Square	F Value	p Value	
Model	622.42	9	69.16	47.26	<0.0001	Significant
A	2.71	1	2.71	1.85	0.2155	
B	13.26	1	13.26	9.06	0.0197	
C	226.07	1	226.07	154.47	<0.0001	
AB	65.54	1	65.54	44.78	0.0003	
AC	33.95	1	33.95	23.20	0.0019	
BC	4.09	1	4.09	2.79	0.1386	
A^2	34.02	1	34.02	23.24	0.0019	
B^2	95.70	1	95.70	65.39	<0.0001	
C^2	120.11	1	120.11	82.07	<0.0001	
Residual	10.24	7	1.46			
Lack of fit	2.92	3	0.9729	0.5312	0.6848	Not significant
Pure Error	7.33	4	1.83			
Cor Toal	632.66	16				
R^2					0.9838	
Adjusted R^2					0.9630	
Predicted R^2					0.9081	

Figures 5–7 show the surface response plots for the peak gas-phase volume fraction of the composite nozzle at different values of A, B, and C, with one variable set at a level of 0. A larger slope indicates greater factor influence. It can be seen from Figure 5 that with a fixed diameter of the connecting runner d_1, the peak gas-phase volume fraction has a tendency to first increase and then decrease with the increase in the diameter of the Helmholtz chamber, peaking at a chamber diameter of 13.6 mm. When the diameter of the Helmholtz chamber is less than 13 mm, due to limited chamber space, cavitation bubbles become blocked in the chamber, energy is disrupted, and the vortex ring cannot effectively feedback upstream, resulting in a small gas-phase volume. Until the diameter of the Helmholtz chamber reaches 13.6 mm, the degree of cavitation significantly increases, and the gas-phase volume fraction reaches its peak. However, when the diameter exceeds 13.6 mm, the chamber space and feedback distance of the bubbles increase, and more bubbles stay near the collision wall, causing energy dissipation and reducing cavitation performance. Figure 6 shows a similar increase and decrease effect of the connecting runner diameter d_1 on the peak gas-phase volume fraction. As the length of the Helmholtz chamber increases, the peak gas-phase volume fraction also increases. Due to the increase in length, there is enough space inside the chamber to develop and grow, forming large-scale vortex rings. However, when the length is large, it can also lead to a reduction in cavitation. The slope of the response surface in Figure 7 is greater than that in Figures 5 and 6, showing that modifications in B and C have a considerable effect on the peak gas-phase volume percent. Because the connection runner diameter is less than 2.9 mm, the energy inside the chamber is disrupted, which is not conducive to cavitation. When the connection runner diameter increases from 2 mm to 2.9 mm, the gas-phase volume fraction also increases.

After conducting response surface analysis and solving the fitted predictive regression equation, the value range of the three factors was set. The objective function was set to the greatest value of the peak gas-phase volume fraction, leading to the determination of the optimal size of the composite nozzle, as shown in Table 6. The predicted peak gas-phase volume fraction under these conditions was 31.747%. In order to validate the accuracy of this fitting model, numerical simulations of the optimized structure were carried out, resulting in an actual peak gas-phase volume fraction of 32.41%. The peak gas-phase volume fraction of the optimized nozzle was significantly improved; compared to the same type of Helmholtz single-chamber nozzle [14], the gas-phase volume fraction increased by 189% and the relative error between the predicted and simulated values obtained by the response surface method under the same operating conditions was 2.04%. This small error confirms the reliability and accuracy of the model obtained by the response surface method.

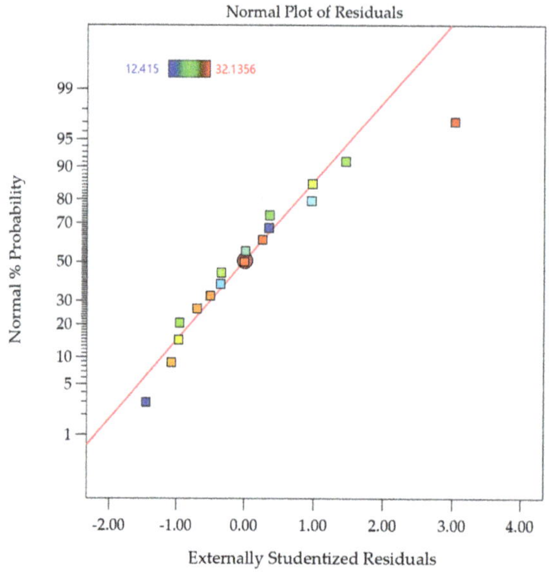

(a)

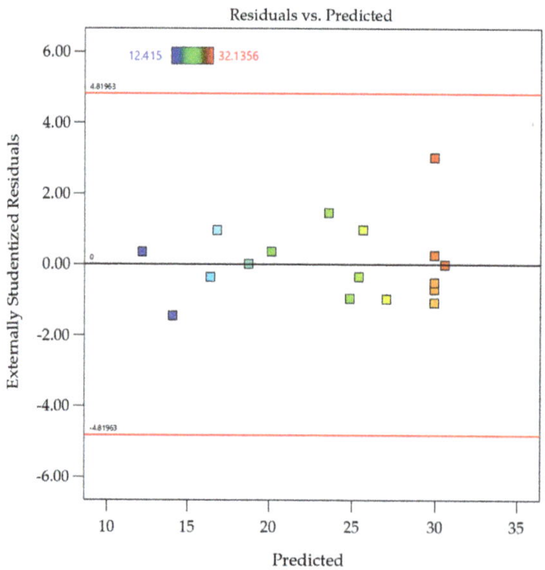

(b)

Figure 4. Residual plots. (**a**) Normal probability distribution of residual; (**b**) residuals vs. predicted values.

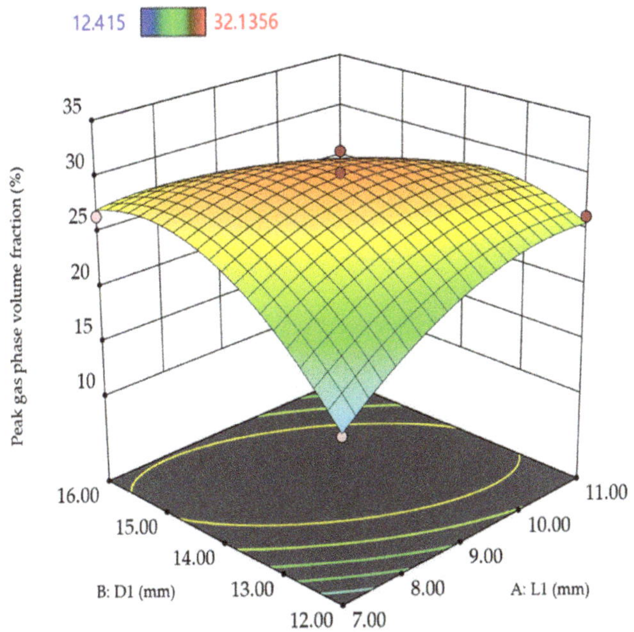

Figure 5. Peak gas-phase volume fraction of D_1–L_1 response surface.

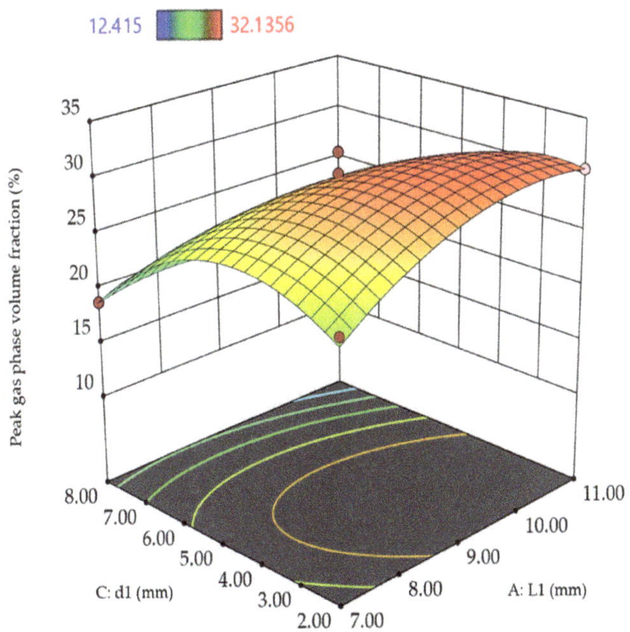

Figure 6. Peak gas-phase volume fraction of d_1–L_1 response surface.

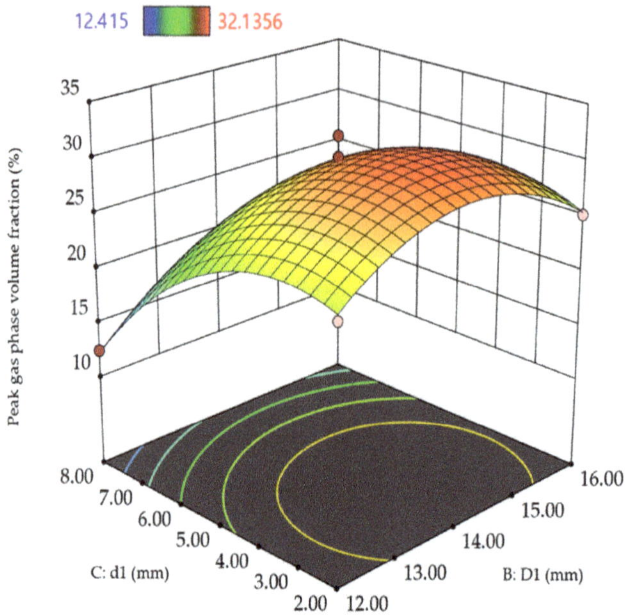

Figure 7. Peak gas-phase volume fraction of d_1–D_1 response surface.

Table 6. Parameters after nozzle optimization.

Structure	d_1 (mm)	D_1 (mm)	L_1 (mm)
Size	2.9	13.6	10.1

4.2. Experimental Analysis of Cavitation Cloud Characteristics

Figure 8a presents a schematic diagram of the cavitation jet experimental system. This system was mainly composed of a plunger pump, water tank, composite nozzle, motor, and high-speed camera. The experiment required high pressure, which was supplied by an AR high-pressure piston pump. It has a maximum operating pressure of 20 MPa. The experiments used clean water at 25 °C as the medium, with an inlet pressure of 10 MPa, and were conducted under submerged conditions. The composite nozzle adopted the optimized parameters—listed in Table 6—obtained using the response surface method. The nozzle along with the test setup are shown in Figure 8b. A high-speed camera was used to record the development and evolution of cavitation clouds at the submerged nozzle, while a pressure sensor was used to record the impact force. The experimental photos are shown in Figure 9.

Visualization experiments were conducted on the composite nozzle using the aforementioned experimental equipment. The cavitation clouds, as captured in Figure 9, are organized according to their development over time, as shown in Figure 10a. This process can be divided into three stages: the nascent stage (0–100 μs), the developmental stage (100–1500 μs), and the abscission and collapse stage (1500–1900 μs). In the nascent stage of cavitation, the fluid is ejected from the nozzle into a stationary flow field. The interaction between the jet and the stationary fluid generates a pressure difference and shear flow, promoting the generation and development of bubbles. The minimum cavitation range is concentrated at the nozzle outlet. In the developmental stage, as fluid continues to eject from the jet, the length of the cavitation cloud increases gradually. Under the influence of jet shear, the high-speed flow diffuses to the surrounding area, leading to symmetric bubble distribution. The cavitation volume ratio in the red area increases to 0.7. With the continued impact of subsequent jets on the tail of the cavitation bubbles, the tails begin to collapse inward. In the abscission and collapse stage, the cavitation cloud, at its maximum of around 100 mm, begins to detach from the jet, as shown by the red dashed line in Figure 10a. The maximum cavitation volume fraction decreases, and the cavitation begins to collapse. Figure 10b shows the simulation results using the Large Eddy Simulation with the WALE model.

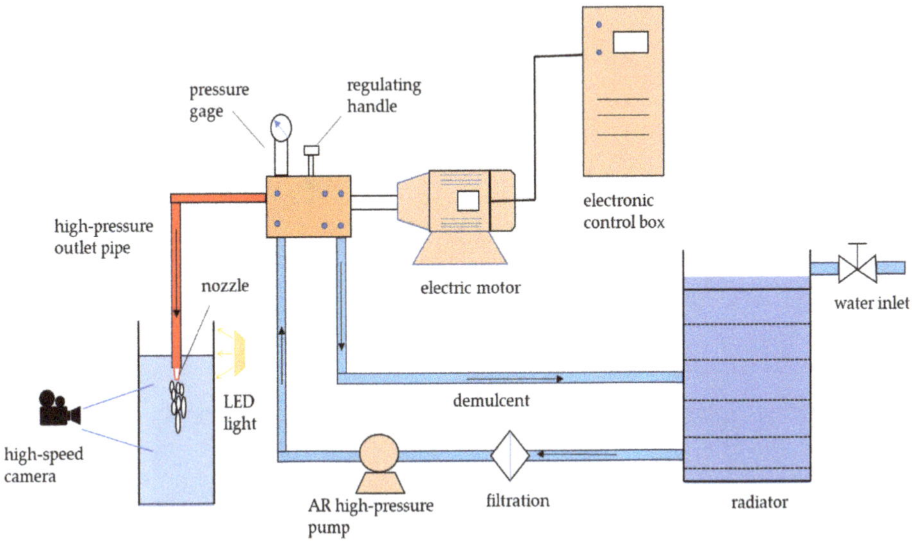

(a)

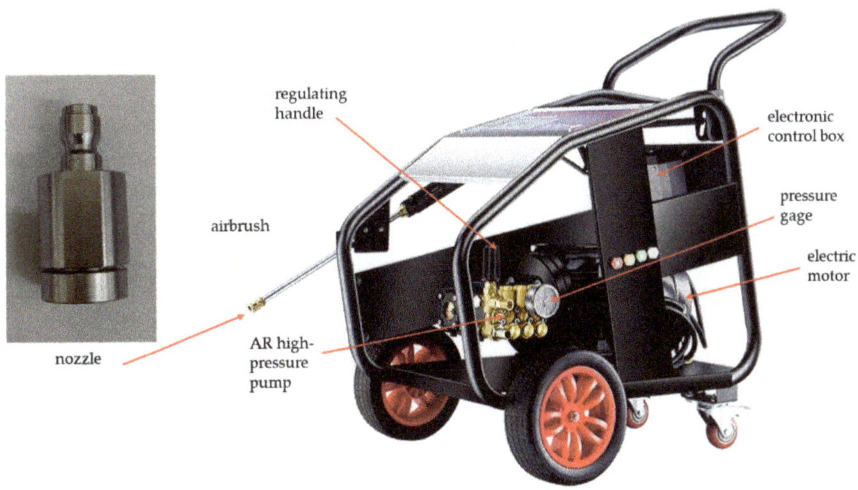

(b)

Figure 8. Test platform for the high-pressure submerged jet. (**a**) Schematic diagram. (**b**) Photo of the experimental platform.

Figure 9. The experimental photos.

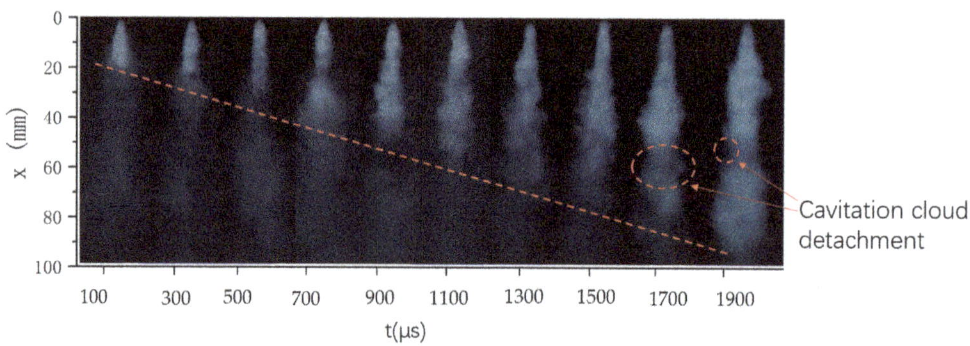

(**a**) Test results

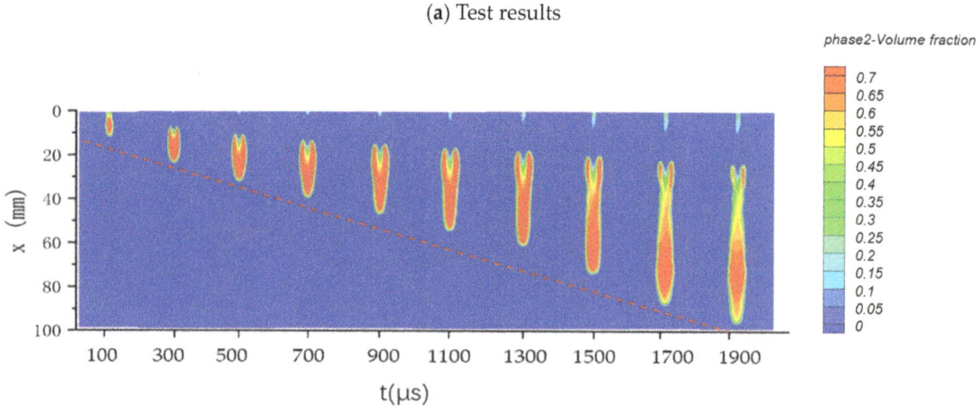

(**b**) Simulation results

Figure 10. The evolution process of cavitation clouds of composite nozzles. (**a**) Test results. (**b**) Simulation results.

By comparing the numerical simulation with the experimental images, as can be seen from the red sloping dotted line, it is found that the evolution process of cavitation clouds is roughly the same. However, differences in grid numbers and numerical simulation models prevent the experimental images from aligning perfectly with the simulations.

The normalized Q-criterion, where the value of Q denotes the level of vortex strength, was used to form simulations in order to identify and display the vortex's features. The vortex structure's temporal evolution is depicted in Figure 11. At Q = 30,000, the emphasis is on the creation of coherent structures inside the jet, which contains several scale vortices. At t = 100 μs, in the nascent stage, high flow velocity at the nozzle outlet results in dense, small-scale vortices. At t = 1300 μs, during the period of cavitation development, small-scale vortices near the nozzle increase, and the conical large-scale vortex at the rear end elongates and fractures. At t = 1900 μs, in the abscission and collapse stage of cavitation, the collapse of cavitation bubbles generates energy that causes disturbance to the fluid, breaking down the vortices into smaller scales and leading to an increase in the overall length of vortices. As illustrated in Figure 11, the flow velocity decreases outward from the central axis throughout the entire process, with small-scale vortices distributed along the axis and large-scale vortices present on the outer side. The Helmholtz structure causes the uniform inlet flow to oscillate, forming a "Ring Vortices" structure at the nozzle outlet.

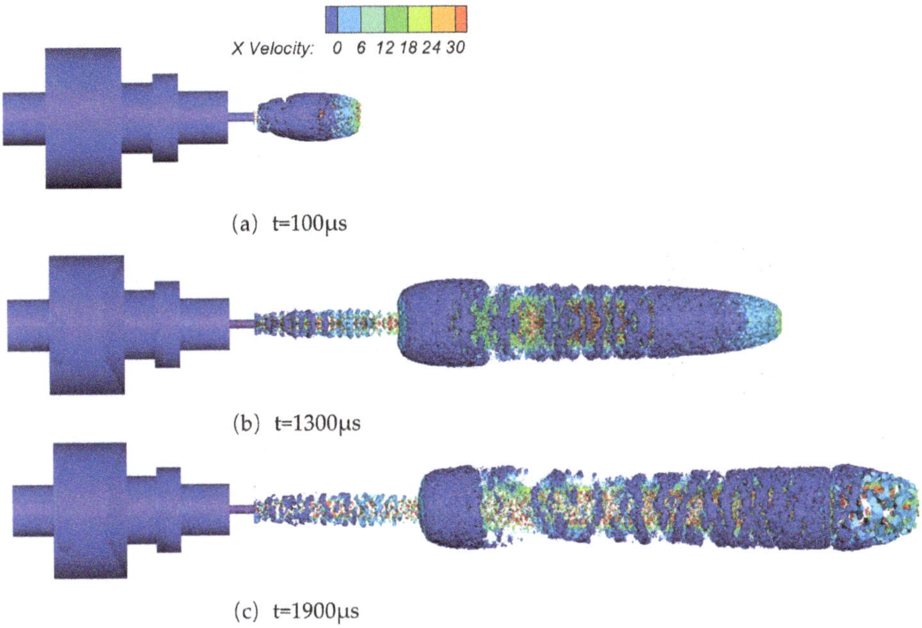

Figure 11. Vortex structure evolution over time.

Figure 12 shows the curve of impact force variation over time in tests and simulations with an inlet pressure of 10 MPa and a target distance of 20 mm. Here, F is the impact force on the target surface, and t is the test time. Between 0.05 and 0.25 S, the simulated impact force oscillates between 256 and 297 N, with a pulsation amplitude of 41 N. The impact force measured in the test oscillates between 260 and 289 N, with a pulsation amplitude of 29 N. The error between the test and simulation is in the range of 1.5% to 2.7%, which verifies the reliability of the optimized nozzle. The higher simulated values compared to test results are attributed to the measurement losses of transducer surface forces during testing.

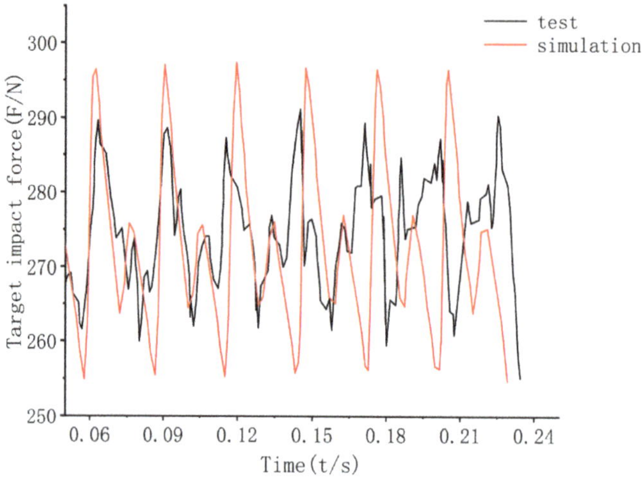

Figure 12. Comparison of target impact force between test and simulation results.

5. Conclusions

This paper optimized the composite nozzle, which is composed of a Helmholtz–organ pipe. It combined CFD Large Eddy Simulation with the response surface method. A high-pressure water jet cavitation visualization experiment system has been established, and a comparative study of the cavitation cloud evolution has been carried out using a high-speed camera. Through design optimization and experimentation, the following conclusions were reached:

(1) The BBD response surface methodology was used to construct a predictive regression equation between the composite nozzle parameters and the peak gas-phase volume fraction. A total of 98.38% of the data in the regression equation fitted well with the actual simulation results, reaching a highly significant level. This can replace numerical calculations to predict the results.

(2) Among the single factors, the diameter of the connecting channel and the diameter of the Helmholtz chamber significantly influenced the peak gas-phase volume fraction. Taking the peak gas-phase volume fraction as the response value, the influence of three factors, namely the diameter of the Helmholtz chamber, the length of the Helmholtz chamber, and the diameter of the connecting flow channel, on the response value can be sorted as follows: connection channel diameter > Helmholtz chamber diameter > Helmholtz chamber length.

(3) The optimized structural parameters of the composite nozzle at an inlet pressure of 10 MPa were determined using the response surface method: the diameter of the connecting runner is 2.98 mm, the diameter of the Helmholtz chamber is 13.6 mm, and the length of the Helmholtz chamber is 10.1 mm. The peak gas-phase volume fraction achieved using the optimized parameters was 32.41%, with a relative error of 2.04% from the model prediction. The optimized composite nozzle showed a significant improvement in cavitation performance.

(4) Visual experiments on the flow field of composite nozzles indicate that the simulation of the cavitation period matches the experimental photos, consistent with existing research. The nozzle-generated vortex is mainly distributed along the axis, with small-scale vortices at the nozzle outlet and a conical large-scale vortex at the far end of the nozzle outlet. Over time, large-scale vortices collapse into small-scale vortices. In the impact force testing experiment, the calculation results of the optimized composite nozzle scheme aligned well with the experimental model test results, verifying the optimization results and simulation accuracy.

Further experimental and numerical study is needed, such as observing the properties of cavitation clouds using particle image velocimetry (PIV) and using artificial neural network methods to solve the nonlinear multiparameter coupling problem of nozzles, establishing the relationship between the structure of cavitation jet nozzles and cavitation effects.

Author Contributions: Conceptualization, G.H. and C.Q.; Data curation, G.H., K.H. and J.G.; Investigation, W.Q., Y.Z., J.H. and H.H.; Methodology, G.H., M.S. and K.C.; Project administration, G.H. and C.Q.; Validation, W.Q. and M.S.; Writing—review and editing, G.H. and M.S. All authors have read and agreed to the published version of the manuscript.

Funding: This research was funded by the CSSC Guangxi Shipbuilding and Offshore Engineering Technology Collaboration Project: ZCGXJSB20226300222-06; Project of Qinzhou Science and Technology Source: 202116622; High-end Foreign Expert Introduction Plan: G2022033007L.

Data Availability Statement: The data presented in this study are available on request from the corresponding author.

Acknowledgments: We appreciate the anonymous reviewers' ideas to improve the manuscript. We appreciate the editors' work on the manuscript.

Conflicts of Interest: The authors declare no conflict of interest.

Nomenclature

$\overline{u}_i, \overline{u}_j$	Instantaneous velocities of the fluid
x_i	3D coordinate direction
ρ	Fluid density
p	Fluid pressure
τ_{ij}	Sub-grid scale stress
δ_{ij}	Kronecker's delta
v_τ	Eddy viscosity
S_{ij}	Resolved scale strain rate tensor
L_s	Length of the sub-grid scale
C_w	WALE model constant
d	Distance to the proximate wall
k	von Karman constant
V	Computational cell volume
n	The number of independent variables
ε	Total error
β_0, βi, βii, and βij	Regression coefficients
x	Investigation factor
Df	Freedom
Q	Magnitude of vortex intensity

References

1. Johnson, V.E.; Kohl, R.E.; Thiruvengadam, A.; Conn, A.F. Tunnelling, fracturing, drilling, and mining with high speed water jets utilizing cavitation damage. In Proceedings of the 1st International Symposium on Jet Cutting Technology, Coventry, UK, 5–7 April 1972.
2. Wen, J.W.; Chen, C.; Campos, U. Experimental research on the performances of water jet devices and proposing the parameters of borehole hydraulic mining for oil shale. *PLoS ONE* **2018**, *13*, 32. [CrossRef] [PubMed]
3. Soyama, H. Cavitating Jet: A Review. *Appl. Sci.* **2020**, *10*, 7280. [CrossRef]
4. Soyama, H. Cavitation Peening: A Review. *Metals* **2020**, *10*, 270. [CrossRef]
5. Szolcek, M.; Cassineri, S.; Cioncolini, A.; Scenini, F.; Curioni, M. CRUD removal via hydrodynamic cavitation in micro-orifices. *Nucl. Eng. Des.* **2019**, *343*, 210–217. [CrossRef]
6. Shi Khan, N.S.; Hussanan, A.; Kumam, W.; Kumam, P.; Suttiarporn, P. Accessing the thermodynamics of Walter-B fluid with magnetic dipole effect past a curved stretching surface. *Zamm J. Appl. Math.* **2023**, *103*, e202100112. [CrossRef]
7. Luo, Y.; Zang, J.; Zheng, H. Flow Field and Gas Field Distribution of Non-Submerged Cavitation Water Jet Based on Dual-Nozzle with Concentric Configuration. *Water* **2023**, *15*, 2904. [CrossRef]
8. Jian-Jie, W.; Li-Yu, C.; Xia-Ming, Y.; Zhen-Ying, X.U.; Yun, W. Research Progress of the Cavitation Water-jet Application. *J. Netshape Form. Eng.* **2016**, *8*, 156–162.

9. Zhao, F.; Wang, X.; Xu, W.; Zhu, H. Study on Different Parameters of the Self-Excited Oscillation Nozzle for Cavitation Effect under Multiphase Mixed Transport Conditions. *J. Mar. Sci. Eng.* **2021**, *9*, 1159. [CrossRef]
10. Wan, C.H.; Wang, R.H.; Zhou, W.D.; Li, L.P. Experimental study on viscosity reduction of heavy oil by hydrogen donors using a cavitating jet. *RSC Adv.* **2019**, *9*, 2509–2515. [CrossRef]
11. Wang, Z.L.; Lei, Y.; Wu, Z.H.; Wu, J.; Zhang, M.L.; Liao, R.Q. Structure Size Optimization and Internal Flow Field Analysis of a New Jet Pump Based on the Taguchi Method and Numerical Simulation. *Processes* **2023**, *11*, 24. [CrossRef]
12. Cui, L.; Ma, F.; Cai, T. Investigation of Pressure Oscillation and Cavitation Characteristics for Submerged Self-Resonating Waterjet. *Appl. Sci.* **2021**, *11*, 6972. [CrossRef]
13. Haosheng, C.; Jiang, L.; Darong, C.; Jiadao, W. Damages on steel surface at the incubation stage of the vibration cavitation erosion in water. *Wear* **2008**, *266*, 69–75. [CrossRef]
14. Dong, J.; Meng, R.; Chen, J.; Liu, M.; Zhong, X.; Pan, X. Numerical Investigation on Cavitation Jet in Circular Arc Curve Chamber Self-Excited Oscillation Nozzle. *J. Mar. Sci. Eng.* **2023**, *11*, 1391. [CrossRef]
15. Yang, Y.; Li, W.; Shi, W.; Zhang, W.; El-Emam, M.A. Numerical Investigation of a High-Pressure Submerged Jet Using a Cavitation Model Considering Effects of Shear Stress. *Processes* **2019**, *7*, 541. [CrossRef]
16. Dong, W.; Yao, L.; Luo, W. Numerical Simulation of Flow Field of Submerged Angular Cavitation Nozzle. *Appl. Sci.* **2023**, *13*, 613. [CrossRef]
17. Wang, G.W.; Yang, Y.F.; Wang, C.; Shi, W.D.; Li, W.; Pan, B. Effect of Nozzle Outlet Shape on Cavitation Behavior of Submerged High-Pressure Jet. *Machines* **2022**, *10*, 17. [CrossRef]
18. Stanley, S.A.; Sarkar, S.; Mellado, J.P. A study of the flow-field evolution and mixing in a planar turbulent jet using direct numerical simulation. *J. Fluid Mech.* **2002**, *450*, 377–407. [CrossRef]
19. Im, K.S.; Cheong, S.K.; Powell, C.F.; Lai, M.C.D.; Wang, J. Unraveling the Geometry Dependence of In-Nozzle Cavitation in High-Pressure Injectors. *Sci. Rep.* **2013**, *3*, 2067. [CrossRef]
20. Li, L.D.; Xu, Y.; Ge, M.M.; Wang, Z.C.; Li, S.; Zhang, J.L. Numerical Investigation of Cavitating Jet Flow Field with Different Turbulence Models. *Mathematics* **2023**, *11*, 18. [CrossRef]
21. Liu, Z.M.; Li, Z.M.; Liu, J.B.; Wu, J.C.; Yu, Y.S.; Ding, J.W. Numerical Study on Primary Breakup of Disturbed Liquid Jet Sprays Using a VOF Model and LES Method. *Processes* **2022**, *10*, 20. [CrossRef]
22. Zhang, L.; Wang, Q.; Guan, W.; Cao, T.Y.; Duan, L.; Pachiannan, T.; He, Z.X. Large eddy simulations on asymmetrical atomization of the elliptical jet with cavitation. *Atom. Sprays* **2019**, *29*, 177–198. [CrossRef]
23. Khan, N.S.; Sriyab, S.; Kaewkhao, A.; Thawinan, E. Hall current effect in bioconvection Oldroyd-B nanofluid flow through a porous medium with Cattaneo-Christov heat and mass flux theory. *Sci. Rep.* **2022**, *12*, 19821. [CrossRef] [PubMed]
24. Kim, J.; Choi, H. Large eddy simulation of a circular jet: Effect of inflow conditions on the near field. *J. Fluid Mech.* **2009**, *620*, 383–411. [CrossRef]
25. Fang, Z.; Hou, W.; Xu, Z.; Guo, X.; Zhang, Z.; Shi, R.; Yao, Y.; Chen, Y. Large Eddy Simulation of Cavitation Jets from an Organ-Pipe Nozzle: The Influence of Cavitation on the Vortex Coherent Structure. *Processes* **2023**, *11*, 2460. [CrossRef]
26. Li, X.; Zhou, R.; Yao, W.; Fan, X. Flow characteristic of highly underexpanded jets from various nozzle geometries. *Appl. Therm. Eng.* **2017**, *125*, 240–253. [CrossRef]
27. Yang, Y.; Shi, W.; Tan, L.; Li, W.; Chen, S.; Pan, B. Numerical Research of the Submerged High-Pressure Cavitation Water Jet Based on the RANS-LES Hybrid Model. *Shock Vib.* **2021**, *2021*, 6616718. [CrossRef]
28. Wang, Z.; Sun, X.; Chen, S. Multi-objective parameters optimization design of self-excited oscillation pulsed atomizing nozzle. *J. Braz. Soc. Mech. Sci. Eng.* **2019**, *41*, 510. [CrossRef]
29. Li, W.H.; Liu, K.; Guo, Q.H.; Zhang, Z.M.; Ji, Q.L.; Wu, Z.J. Genetic Algorithm-Based Optimization of Curved-Tube Nozzle Parameters for Rotating Spinning. Front. *Bioeng. Biotechnol.* **2021**, *9*, 10. [CrossRef]
30. Uebel, K.; Rössger, P.; Prüfert, U.; Richter, A.; Meyer, B. CFD-based multi-objective optimization of a quench reactor design. *Fuel Process. Technol.* **2016**, *149*, 290–304. [CrossRef]
31. Qian, Y.L.; Bi, Y.X.; Zhang, Q.; Chen, H.Y. The optimized relationship between jet distance and nozzle diameter of a pulse-jet cartridge filter. *Powder Technol.* **2014**, *266*, 191–195. [CrossRef]
32. Edeling, W.N.; Cinnella, P.; Dwight, R.P.; Bijl, H. Bayesian estimates of parameter variability in the k-ε turbulence model. *J. Comput. Phys.* **2014**, *258*, 73–94. [CrossRef]
33. Wu, D.Z.; Wang, J.Y.; Miao, T.; Chen, K.Y.; Zhang, Z.L. Performance Optimization of FA-GGBS Geopolymer Based on Response Surface Methodology. *Polymers* **2023**, *15*, 1881. [CrossRef] [PubMed]
34. Yang, H.; Zhang, X.R.; Yuan, G.; Li, Z.L.; Zhang, F.B. Structure optimization of straight cone nozzle and its effect on jet impingement heat transfer. *Numer. Heat Tranf. A Appl.* **2023**, 13. [CrossRef]
35. Zhang, X.Y.; Shan, Y.; Zhang, J.Z. Optimization and sensitivity analysis of double serpentine nozzles within confined layouts by response surface methodology. *Therm. Sci. Eng. Prog.* **2023**, *43*, 22. [CrossRef]
36. Wang, Y.; Wang, X.L.; Zhang, Z.L.; Li, Y.; Liu, H.L.; Zhang, X.; Hocevar, M. Optimization of a Self-Excited Pulsed Air-Water Jet Nozzle Based on the Response Surface Methodology. *Stroj. Vestn. J. Mech. Eng.* **2021**, *67*, 75–87. [CrossRef]
37. Piomelli, U. Large eddy simulations in 2030 and beyond. *Philos. Trans. R Soc. A* **2014**, *372*, 20130320. [CrossRef]
38. Ahmed, B.H.T.; Kanfoudi, H.; Zgolli, R.; Ennouri, M. Investigation of Subgrid-Scale Models in Large Eddy Simulation on the Unsteady Flow Around a Hydrofoil Using OpenFOAM. *Iran. J. Sci. Technol. Trans. Mech. Eng.* **2019**, *44*, 465–480.

39. Wang, J.; Wei, M.; He, J.; Wang, Y.; Ren, C. Optimization of Repair Process Parameters for Open-Arc Surfacing Welding of Grinding Rolls Based on the Response Surface Method. *Processes* **2022**, *10*, 321. [CrossRef]
40. Le, H.S.; Galal, A.M.; Alhamrouni, I.; Aly, A.A.; Abbas, M.; Saidi, A.S.; Truong, T.H.; Dahari, M.; Wae-hayee, M. Heat transfer efficiency optimization of a multi-nozzle micro-channel heat sink utilizing response surface methodology. *Case Stud. Therm. Eng.* **2022**, *37*, 18. [CrossRef]
41. Jiang, S.H.; Li, D.Q.; Zhou, C.B.; Zhang, L.M. Capabilities of stochastic response surface method and response surface method in reliability analysis. *Struct. Eng. Mech.* **2014**, *49*, 111–128. [CrossRef]

water

MDPI

Article

Improving Flash Flood Hydrodynamic Simulations by Integrating Leaf Litter and Interception Processes in Steep-Sloped Natural Watersheds

Gergely Ámon [1,*], Katalin Bene [1,*], Richard Ray [2], Zoltán Gribovszki [3] and Péter Kalicz [3]

[1] National Laboratory for Water Science and Water Security, Department of Transport Infrastructure and Water Resources Engineering, Széchenyi István University, Egyetem ter 1., H-9026 Győr, Hungary

[2] Department of Structural and Geotechnical Engineering, Széchenyi István University, Egyetem ter 1., H-9026 Győr, Hungary; ray@sze.hu

[3] Institute of Geomatics and Civil Engineering, Faculty of Forestry, University of Sopron, Bajcsy-Zsiliszky u. 4, H-9400 Sopron, Hungary; gribovszki.zoltan@uni-sopron.hu (Z.G.); kalicz.peter@uni-sopron.hu (P.K.)

* Correspondence: amon.gergely@sze.hu (G.Á.); benekati@sze.hu (K.B.)

Abstract: More frequent high-intensity, short-duration rainfall events increase the risk of flash floods on steeply sloped watersheds. Where measured data are unavailable, numerical models emerge as valuable tools for predicting flash floods. Recent applications of various hydrological and hydrodynamic models to predict overland flow have highlighted the need for improved representations of the complex flow processes that are inherent in flash floods. This study aimed to identify an optimal modeling approach for characterizing leaf litter losses during flash floods. At a gauged watershed in the Hidegvíz Valley in Hungary, a physical-based model was calibrated using two distinct rainfall–runoff events. Two modeling methodologies were implemented, integrating canopy interception and leaf litter storage, to understand their contributions during flash flood events. The results from the model's calibration demonstrated this approach's effectiveness in determining the impact of leaf litter on steep-sloped watersheds. Soil parameters can estimate the behavior of leaf litter during flash flood events. In this study, hydraulic conductivity and initial water content emerged as critical factors for effective parametrization. The findings underscore the potential of a hydrodynamic model to explore the relationship between leaf litter and flash flood events, providing a framework for future studies in watershed management and risk-mitigation strategies.

Keywords: hydrodynamics; numerical modeling; leaf litter; flash flood; watershed modeling; overland flow; interception; depth-averaged modeling

1. Introduction

Climate change has produced an upsurge in high-intensity, short-duration rainfall events, increasing the threat of flash flooding through steeply sloped watersheds. These floods pose a hazard to natural watersheds and urban areas. Higher rainfall intensities increase the overland flow and may cause substantial damage to an area. Smaller (area = 5–50 km^2) steep-sloped watersheds or the upper regions of larger watersheds present a more significant risk due to the lack of hydrological data and a scarcity of management resources for preventative action [1,2]. When measured data are unavailable, numerical models can help predict flash floods. Several hydrological and hydrodynamic models were recently applied to predict overland flow [3–8]. Many of these models do not represent the complex flow processes of flash floods and have reduced complexity, increased parameter sensitivity, and reduced prediction accuracy [9].

Natural, steep-sloped watersheds contain stream beds with woodland and meadow areas. Rainfall–runoff processes for these types of watersheds include rainfall, rainfall losses, and the remaining excess rainfall that turns into overland flow [10,11]. The excess rainfall comes from subtracting losses from the rainfall in the watershed. These losses are the interception (storage capacity of the canopy), surface storage, and infiltration into the unsaturated soil layer. In natural watersheds, rainfall intensity, land cover, and the unsaturated upper soil are the most important parameters to define the losses from a rainfall event [1,12,13]. Surface storage in natural watersheds includes leaf litter interception and storage as well. During short, intense rainfalls, canopy interception, leaf litter, and near-surface infiltration may also reduce the short-term runoff and affect the overland flow ratio [14,15]. Accurately determining the runoff hydrograph at the outlet of these watersheds requires a better understanding of the physical processes of the losses from interception and leaf litter storage.

Forested watersheds have leaf litter layers with parameters that change over time (layer width, humidity, and storage capacity). The mechanism of leaf litter interception and how it stores and releases water is poorly understood [16]. Most studies on leaf litter storage are empirical lab experiments based on a given watershed's specific fauna [17–21]. Further studies have combined lab experiments with different numerical approaches [22–24]. Although studies have experimented with different leaf litter types worldwide, local vegetation can be compared to these measurements. The water retention in leaf litter depends on its current composition and thickness [25]. The decomposition process will reduce a leaf litter's storage capacity [26]. The impact of leaf litter on runoff remains difficult to quantify; therefore, back calculations based on measured runoff are needed. This paper focuses mainly on leaf litter's storage capacity as a function of initial conditions (dry or saturated) and rainfall intensity.

Various model structures, such as physics-based, index-based, and conceptual models, are available to describe rainfall–runoff processes in ungauged watersheds [1]. Physics-based models can effectively assess and predict overland flow [27]. Two-dimensional models applying shallow water equations may accurately predict runoff and flood occurrence timing and duration [28]. They are computationally and data-intensive, and there is a need to evaluate the input parameters and rainfall–runoff processes to improve modeling success and avoid unnecessary uncertainty.

A proper model requires several components: (1) a precise topographic description because the slope geometry dictates the flow velocity; (2) detailed land use that defines the dominant vegetation, as well as seasonal land surface changes (e.g., cultivation, plowing, and harvesting); and (3) the temporal condition of land cover that may impede surface runoff, such as leaf litter or shallow channeling, or sediment deposition from previous runoff events. These components affect the quantity and rate of runoff on steep slopes [29]. In addition to the fluid movement, the model must calculate the natural losses [30,31]. Including quick infiltration in the unsaturated upper soil layer can improve the shallow surface flow [6,32–34]. For calculating losses, different infiltration models have been included (Green and Ampt, CN) [35]. Flash floods are not affected by slower processes such as transpiration or long-term infiltration.

Costabile et al., 2020 [28], studied rainfall-induced floods in watersheds using the Hydrologic Engineering Centre River Analysis System (HEC-RAS). The current HEC-RAS version provides a valuable tool for complex modeling. The fully 2D computational model includes full hydrodynamical equations with an additional eddy-viscosity model. This model requires many input parameters to reproduce flow paths, times to concentration, and behavior of overland flow [35–38].

Findings from prior research were synthesized into a comprehensive model. While numerous studies propose applicable methodologies, they focus on watersheds that are larger than those examined. For larger watersheds, existing methods for evaluating overland flow are likely appropriate [4–9]. However, customizing the infiltration and surface storage volume is necessary for smaller, steeply inclined woodland watersheds. While hydrodynamic calculations for larger watersheds can be simplified [4], the complexities of numerical methods and eddy viscosity play critical roles in ensuring accuracy for smaller, steep catchments [6,12,32]. As previously discussed, the storage capacities of leaf litters show variability across different studies and experiments. These findings are instrumental in devising an approximation that can be seamlessly incorporated into a hydrodynamic model.

In this study on a gauged watershed, a complex depth-integrated model was evaluated using two rainfall–runoff events. In addition to the complex overland flow simulation, the model includes an infiltration add-in. Applying two modeling methodologies to incorporate interception and leaf litter storage, this study aimed to determine a better model for describing leaf litter losses during flash floods. The following section summarizes the forest rainfall–runoff processes and briefly introduces the study region. Section 4 describes the rainfall–runoff methodology applied during modeling; Section 5 presents prediction results and challenges.

2. Forest Rainfall–Runoff Processes

A natural, mainly forested watershed will store much water during and after rainfall. The canopy will intercept some rain while the rest continues as throughfall. The leaf litter then detains and stores some throughfall, while another portion passes through the leaves and reaches the ground's surface. Finally, the remaining water may infiltrate the soil, collect in surface depressions (storage), or continue downhill as overland flow.

The quantity of rain involved in each process will depend on the rainfall intensity, the size and type (deciduous or coniferous) of the trees, the thickness and condition of the leaf litter, the irregularities in the ground surface, and the antecedent moisture in the soil. As the storage capacities in the canopy, leaf litter, and soil reach their limits, rainfall will convert directly into overland flow, initiating flooding conditions. Evaporation and transpiration processes are slow relative to short, violent bursts of rain, so they do not significantly reduce the quantity of overland flow during such events. Figure 1 presents the conceptual model for forest rainfall hydrology.

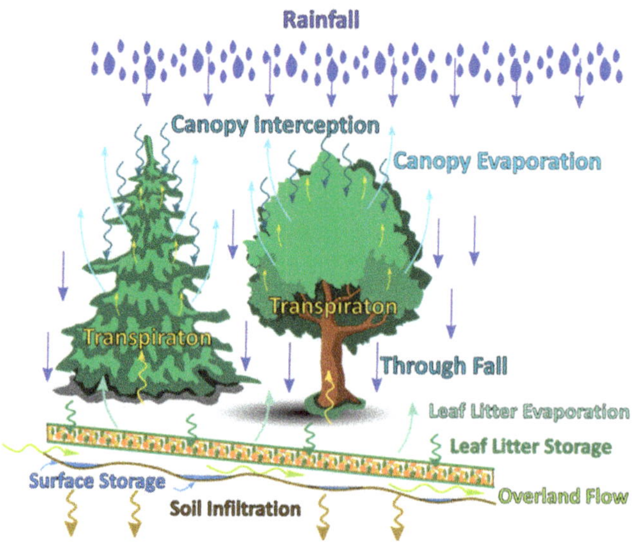

Figure 1. Forest rainfall–runoff process.

This study focused on short-term, high-intensity rainfall events. Since watersheds are forested and have significant slopes, the main processes occur over a short period, in which soil infiltration has only a limited impact. Rainfall intensity and duration, canopy which soil infiltration has only a limited impact. Rainfall intensity and duration, canopy and leaf litter interception, surface storage, and slope steepness have the most significant effects on the occurrence and intensity of flash flooding. In small watersheds, short-duration, high-intensity rainfalls can cause floods.

A leaf litter layer forms in watersheds covered with forested areas. It has storage and infiltration capacities and becomes saturated during rainfall as it stores water. This water may then infiltrate the soil, remain on the surface, evaporate, or slowly release into overland flow. The processes occurring in leaf litters usually require a more extended period than a rainfall event. A leaf litter's capacity changes seasonally and depends on its thickness and area distribution. In these cases, infiltration dominates the water's movement.

In steep-sloped watersheds, the runoff volume and peak flow increase while the time to peak decreases. These types of watersheds are more susceptible to flash floods. The interception, depression storage, and soil infiltration capacity also decrease. In this paper, the authors considered watersheds with a 5–10% slope or higher and observed that, with an increasing slope, the infiltration decreased [29].

Typical Sequence of Rainfall–Runoff Events

The rainfall–runoff process in a natural watershed can be visualized in three steps, as shown in Figure 2. During a high-intensity rainfall (T1), rainfall breaks through the canopy and fills the leaf litter layer. Without a leaf litter layer (meadows, agricultural land), rainfall infiltrates the soil layer. If no leaf litter covers the ground, soil infiltration and surface ponding may begin. The depressions fill up on an uneven surface, and the water surface in these depressions starts to behave as a temporary impervious surface. At the same time, overland flow initiates, driven by the rainfall not captured by the other processes. A significant quantity of rainfall may directly convert to overland flow if the rainfall intensity is high.

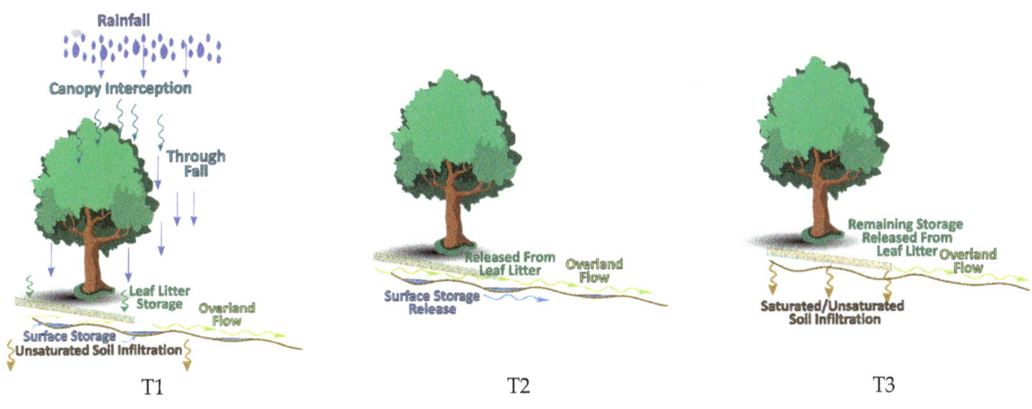

Figure 2. Rainfall–runoff processes.

At time T2, when the rainfall stops, the leaf litter layer slowly releases the stored water on the surface, producing overland flow. The water in the depressions filled during the rainfall starts infiltrating the soil layer.

Days after the rainfall (T3), the stored water in the leaf litter layer continues to release on the surface, and infiltration continues into the unsaturated soil layer. The water in the depressions also continues to infiltrate into the soil layer.

3. Study Area

3.1. Watershed Description

The Hidegvíz Valley extends over ~6 km^2 in the northwestern corner of Hungary. It is a small experimental watershed with dedicated hydrometeorological, overland flow, and infiltration measurement stations. The watershed consists of three subbasins formed by the confluence of three streams (Figure 3).

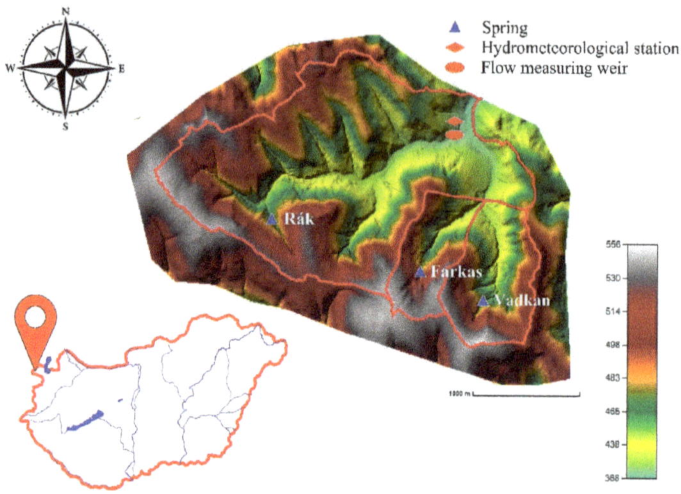

Figure 3. Location of the Hidegvíz Valley watershed.

From west to east, the Rák creek is the primary water body; the tributaries are the Farkas and the Vadkan are streams. The main channel length is ~3000 m, and the average bed slope is 4%. The Vadkan and Farkas Streams' lengths are ~1400 and ~990 m, with an average 4–6% slope. More than 2/3 of the area of the three subwatershed belongs to the Rák Creek, 4.4 km^2; the Vadkan Stream comprises 0.93 km^2 and the Farkas Stream comprises 0.6 km^2. The entire watershed area is 6 km^2 [39].

The Hidegvíz Valley lies within the Sopron Mountains. It has an average annual precipitation of 700–900 mm, with peak rainfall in July. The annual temperature averages 8–8.5 °C. The climate exhibits a strong subatlantic effect [40]. The Köppen system classifies this region as Cfb (warm temperate, without dry season, and warm summer).

Although the watersheds are small, the streams are perennial, fed by springs. While researchers have not measured the springs' flow rates, they have observed similar flows from all three.

The geology of the catchment is crystalline bedrock and Tertiary (Miocene) fluvial sediment deposited in five general layers. Podzolic brown forest soils, non-podzolic highly acidic soils, and lessivated brown soils (soil structure broken down by clay minerals leaching into the layer) have evolved in this area. These soil types are typical for cool, humid forest environments.

The natural forested areas in the watershed appear in Figure 4. The primary woodland type was beech, with smaller oak (dashed line) and negligible pine areas (red) [41]. In the 19th century, due to intensive afforestation, 50% of the woodland areas became needle-leaf wood areas [42].

The meadow areas are generally open, containing smaller vegetation. The stream beds are natural; the only artificial structures are the culverts at the crossings of the forestry routes. Because of the erosion, the beds are winding, and the bed material is mainly rocky.

The watershed has two major soil types [43]. "Magasbérci" formation soils cover the stream beds, while "Brennbergi" formation soils cover the rest of the watershed. Figure 5 shows the distribution of the main soil types.

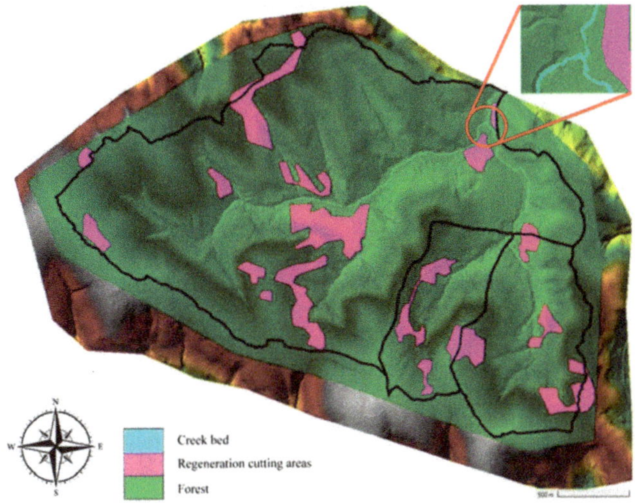

Figure 4. Land use distribution on the watershed.

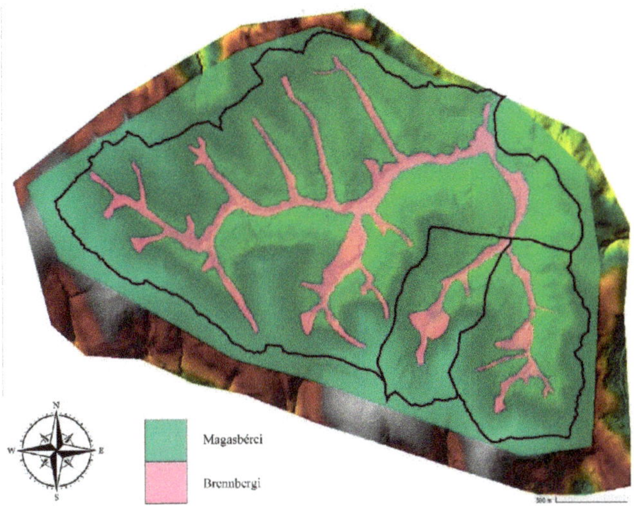

Figure 5. Distribution of near surface soil types.

The "Magasbérci" (magenta) layer is primarily an alluvial loamy sand layer with fluvial gravel pockets. This soil covers only areas near stream beds [44].

The "Brennbergi" (green) layer represents the higher parts of the watershed over the "Magasbérci" layer. The formation contains mostly blocks, with an average size of over 1 m; the blocks are sharp and cracked. Alluvial soil types have washed into the cracks, mainly loam, sand, and gravel debris. The soil layer's proper composition is heavily unclassified [44].

The root zone is 2–3 m deep. Since porosity is greater in woodland areas than in meadows, they have a higher water storage capacity [44].

3.2. Measurements

Researchers have conducted meteorological, hydrological, and forest-related measurements and experiments on the watershed since 1983, with reliable data for modeling since

2010. Located near the Rák–Farkas Creek junction, the hydrometeorological station measures precipitation, temperatures, and wind speed. A broad-crested weir with a trapezoidal section measures the discharge flow nearby in the bed of Rák Creek, which is the combined flow from the Rák, Farkas, and Vadkan streams. Water levels in the stream are measured using pressure transducers with an absolute precision of ±2 mm, a sensitivity of ±0.1 mm, and a maximum depth of 1.5 m. Level readings occur at 1-minute intervals. The pressure transducers are manufactured by Dataqua Ltd. [45]. The in situ and laboratory experiments for leaf litter interception also occurred at the station [41].

4. Case Studies

Rainfall data at the Hidegvíz station was collected daily and at 10-minute intervals. The events between 2010and 2013 were evaluated based on the following criteria:

- The cumulative one-day 10 min and daily measurements should match.
- The intensity should be high enough to cause overland flow.
- Measurement errors must be avoided; the full data set for rainfall and runoff and the time measurement of the events had to be correct.

Two events in July, Event 1 in 2010 and Event 2 in 2012, were selected for model prediction. Their rainfall intensities, durations, and volumes appear in Table 1. Runoff volumes and rainfall/runoff volume ratios also appear in the table.

Table 1. Measured rainfall and outflow values for Events 1 and 2.

	i_{max} (mm/h)	i_{avg} (mm/h)	Rainfall Duration (min)	t_c (min)	Rainfall Volume (m^3)	Flow Volume (m^3)	Ratio
Event 1 2010	190.2	35.6	70	170	276,438	5257	1.90%
Event 2 2012	85.5	45.7	100	210	318,052	8133	2.56%

4.1. Event 1

The rainfall started on 15 July 2010 at 20:20 and ended on 15 July 2010 at 21:30 (Figure 6). The cumulative rainfall was 45.7 mm, totaling 70 min. Overland flow at the measuring station started at 21:47, and peak flow was reached between 22:38 and 22:59. During this time, the measurement weir did not record any flow. For this duration, we assigned a constant value of 0.44 m^3/s. The recession limb has two distinct inflection points. The first inflection point occurred on 15 July at 23:49; after this time, rapid overland flow drainage turned more gradual as interflow reached the outflow of the creek. There is a less-distinct change in the slope on 16 July at 3:39, when interflow starts to play a more dominant role. On 16 July at 16:00, the hydrograph reaches its minimum; the flow is close to the previous baseflow value.

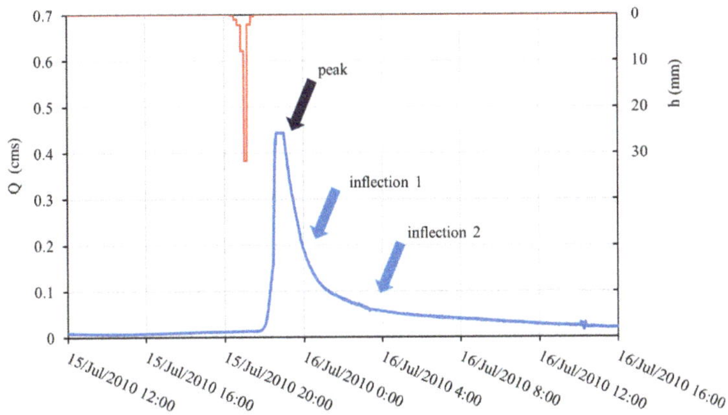

Figure 6. Event 1, between 15 and 16 July 2010.

Estimated time of concentration: 170 min (measured from the center of rainfall to the inflection point on the recession limb).

4.2. Event 2

The rainfall started on 7 June 2012 at 15:20 and ended on 7 June 2012 at 17:00 (Figure 7). The cumulative rainfall was 53.4 mm, lasting 1 hour and 40 min. Overland flow at the measuring station started at 16:40, and peak flow was reached between 17:53 and 18:08. Between the two peaks, flow decreased for 8 min and increased again to reach the same peak 7 min later. The inflection point in the recession limb of the hydrograph is 19:09. The flow suddenly jumps lower at 22:00, then rises back up on 7 July at 1:53 a.m. The jump indicates a possible disturbance in flow measurement around the measuring weir. On the following day (7 July 2012) at 8:00, the hydrograph levelled off to the baseflow level. The estimated time of concentration is 210 min (center of rainfall, inflection point).

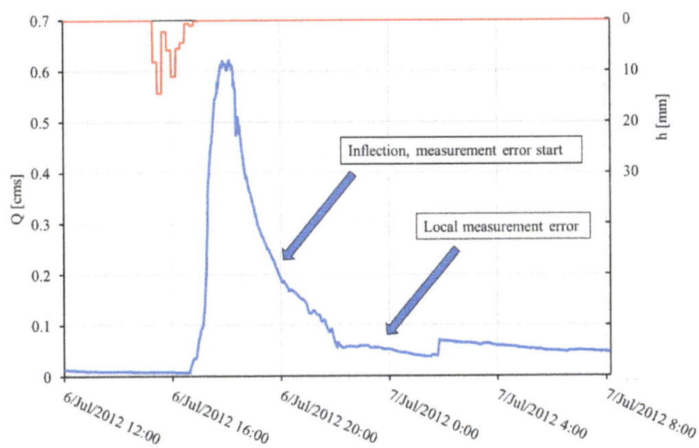

Figure 7. Event 2 between 6 and 7 July 2012.

Even though both events occurred in July, there are distinct differences between the two hyetographs and hydrographs. For Event 2, precipitation had a longer duration but lower intensity. For Event 1, the maximum rainfall intensity was twice that of Event 2.

The runoff hydrograph shapes have similarities and differences; the rising limb increases sharply for both hydrographs, but the peaks and recession limbs differ. Since measurement error during Event 1 forced an assumed peak at 0.44 m^3/s, comparing peak flows becomes problematic. The recession limb descends more gradually for Event 2 than for Event 1. There can be several reasons for the difference between the two events, such as (1) antecedent conditions in the catchment, (2) dissimilar rainfall intensities, and (3) different rainfall durations.

Several meteorological conditions influence antecedent conditions, but rainfall is the most significant. Two days before Event 2, two smaller rainfall events occurred, while no rainfall event occurred twenty days before Event 1. Therefore, the watershed was drier before Event 1 than Event 2. The rainfall before Event 2 produced measurable runoff, as shown in Figure 8.

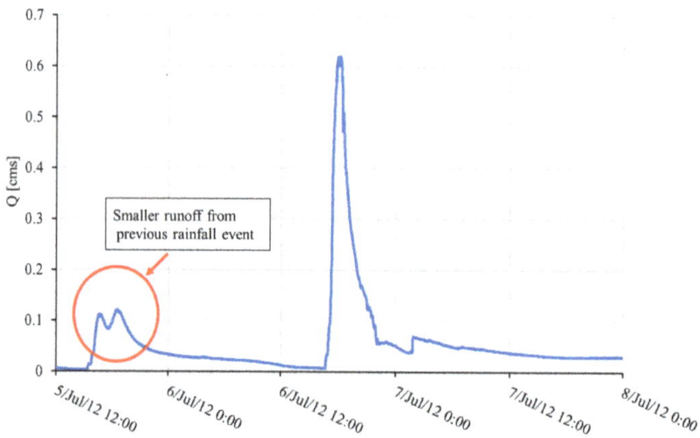

Figure 8. Flow event occurrence before Event 2.

A smaller runoff event started on 5 July 2012 at 15:50. The decreasing flow becomes baseflow around 6 July 2012 at 12:00. According to Kiss [41], the leaf litter releases water after rainfall. After the first 24 hours, there is a continuous outflow from the leaf litter. The flow decreases every day, but it lasts for several days. As a result, one can assume that for Event 2, the leaf litter storage was not empty, and the soil was not dry.

5. Materials and Methods

5.1. Canopy Interception and Leaf Litter Storage

For canopy interception, the following equation was used [39]:

$$I_c = S \cdot \left(1 - e^{-\frac{p}{S}}\right) + 0.07 \cdot h \quad (1)$$

where I_c [mm] = interception, S [mm] = storage capacity of the canopy, and h [mm] = rainfall height. The formula was designed for daily rainfalls and required adjustment.

Experimental data show that canopy interception decreases linearly with time during rainfall [22]. The linear decline is important for long-duration rainfalls; however, this study investigated short rainfall events, and the canopy interception capacity remained constant.

Various experiments determined leaf litter interception parameters using needle-leaf and broadleaf litter materials. Both types exist in the Hidegvíz Valley watershed. According to experiments presented in [41], leaf litter behaves like a sponge, drawing up and releasing water during and after rainfall. Csáfordi et al. [39] estimated litter interception to be a linear process with time.

Kiss [41] determined a formula for leaf litter storage from experiments in the Hidegvíz watershed. The following equation was derived based on leaf litter in the Hidegvíz Valley:

$$E_s = S_a \cdot \left(1 - \frac{1}{\left(1 + \frac{T_h}{S_a \cdot n}\right)^n}\right) \quad (2)$$

where E_s [mm] = leaf litter interception, S_a [mm] = maximum storage capacity of the leaf cover, T_h [mm] = rainfall height, and n = is the dimensionless factor. Equation (6) does not consider the increased leaf litter storage with time. The loss of the stored water from leaf litter is around 0.5 mm in a day. Similarly to Equation (1), this formula considers daily rainfalls [41].

Zhao et al.'s [22] experiments have shown a connection between rainfall duration and litter storage capacity in different sloped areas. They later investigated forests with two wood types: Magnolia grandiflora and Pinus massonlana [22]. They found that leaf litter storage depended on the watershed slope. The experiment used different litter masses for slopes 5° and 20°. At the Hidegvíz Valley, the average slope of the watershed area is around 10°.

Li et al. [23] showed experimentally that the leaf litter storage changes with time, and storage capacity is greater for high rainfall intensity. In their work, they investigated the following woodland types:

- Platycladus orientalis
- Pinus tabulaeformis
- Quercus variabilis
- Acer truncatum

The Quercus variabilis and Pinus tabuleaformis tree varieties are similar to the Hungarian broadleaf and needle types. They performed experiments to measure leaf litter interception during a 60 min rainfall event using 10, 50, and 100 mm/h intensities. Their results for Q. variabilis and P. tabulaeformis appear in Figure 9.

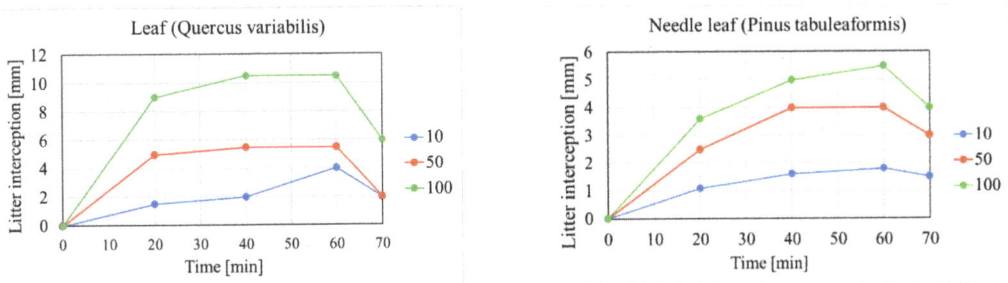

Figure 9. Leaf litter storage capacity for different leaf types on a 60-minute rainfall event with 10, 50, and 100 mm/h intensities.

For 50 mm/h rainfall intensity, litter interception is constant after 40 min of rainfall for both leaf and needle leaf: 5.8 mm for leaf and 4 mm for needle leaf.

Kiss [41] determined the release of stored water from the leaf litter. With Equation (3) below, 0.2–1.7 mm of interception can be determined on average for oak leaves; meanwhile, with beech and needle leaf types, around 3–4 mm of interception can be determined on average through a rainfall event. Equation (3) used an exponential draining process for three different leaf litter types:

$$w = w_{min} + (S_a - w_{min}) \cdot e^{(\alpha' \cdot t')} \quad (3)$$

where w [mm] = the current water content of leaf litter, w_{min} [mm] = the minimal water content, S_a [mm] = the maximum storage capacity, α' [1/day] = the coefficient of the amount of the release of water content, t' [day] = elapsed time from the end of the rainfall.

In the experiment, the collected samples were 0.75–1.25 kg/m^2, and the litter layer was around 1 cm thick. According to Kucsara and Gribovszki [46], a leaf litter layer's storage capacity grows with its thickness. The leaf litter releases its water rapidly in the first days, and then the release becomes slower (Figure 10). According to the measurements [41], the leaf litter has a sponge-like behavior. The water intake is fast, and the release is significantly slower. Therefore, laboratory experiments provided a more comprehensive picture of the leaf litter draining process. The experiment focused on the seasonal drainage in winter and summer. Based on experiments, the leaf litter drainage lasted 15–25 days.

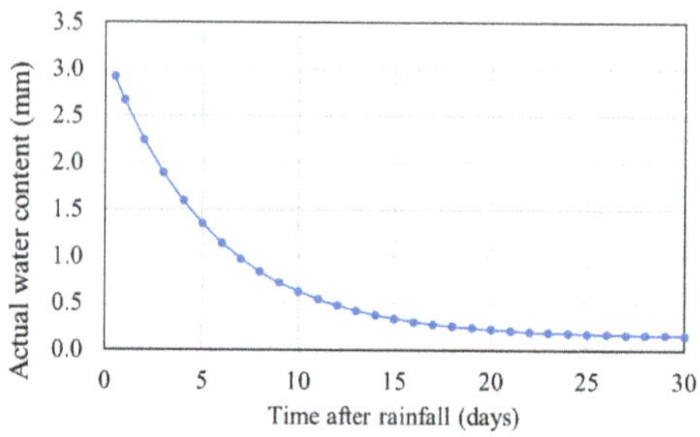

Figure 10. Leaf litter storage (mm) vs. time (day) [46].

5.2. Soil Infiltration

In this study, the Green and Ampt equations modeled the soil infiltration process [47]. The 3D hydrosoil [48] soil map provided estimates of the soil texture and hydraulic conductivity distribution in the Hidegvíz valley in the top 10 cm soil layer. The other soil parameters for the Green-Ampt equation were estimated based on the hydraulic conductivity, using values reported by Rawls et al. [49].

5.3. Overland Flow Ratio

Watersheds with natural land cover have a significant variation in overland flow ratio (excess rainfall/total rainfall). Typical overland flow ratios used in Hungary for forested watersheds are 5–15%; meadows: 15–30% [50].

For step-sloped watersheds, the ratio of overland flow can increase with increased intensity of rainfall [51].

5.4. Modeling Approach

Two modeling approaches, Methods A and B, were developed and evaluated (Figure 11). For Method A, canopy and leaf litter interceptions are subtracted from the rainfall time series. The remaining rainfall then infiltrates into the unsaturated soil layer. Excess rainfall generates overland flow:

$$h_{eff}(t) = h(t) - I_c(t) - I_l(t) - S_d(t) - F_s(t) \quad (4)$$

where h_{eff} [mm] = effective rainfall, h [mm] = rainfall, I_c [mm] = canopy interception, I_l [mm] = leaf litter interception, S_d = depression storage, and F_s [mm] = soil infiltration.

For Method B, only canopy interception is subtracted from rainfall. A pseudo-soil layer with high porosity and conductivity replaces the leaf litter, provides storage, and allows infiltration. The remaining excess rainfall generates overland flow:

$$h_{eff}(t) = h(t) - I_c(t) - F_l(t) - S_d(t) - F_s(t) \quad (5)$$

where h_{eff} [mm] = effective rainfall, h [mm] = rainfall, I_c [mm] = canopy interception, F_l [mm] =litter infiltration, S_d = depression storage, and F_s [mm] = soil infiltration. The numerical model incorporated the two methods using the following approach.

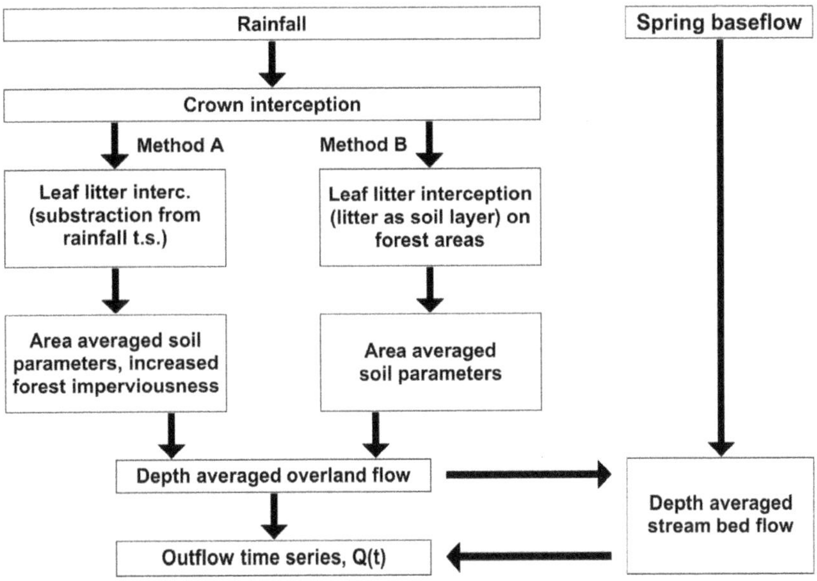

Figure 11. Modeling approaches.

The crown interception was calculated similarly in both models using Equation (1). Infiltration losses were calculated differently. Method A subtracted the leaf litter interception from the rainfall time series. The woodland areas were covered with leaf litter and assumed to be saturated and impervious throughout the simulation. Infiltration was calculated only in the remaining meadows (Brennbergi), lowlands, and stream beds (Magasbérci).

Method B modeled the leaf litter as a soil layer throughout the woodland area. The remaining meadow and lowlands consisted of Brennbergi and Magasbérci soils similar to Method A.

All the simulated events were about one day long, during which the leaf litter released perhaps 10% of its retained water [41]. Therefore, the simulation did not consider water released from the litter. Excess rainfall resulted from subtracting canopy, leaf litter, and infiltration losses from rainfall. The hydrodynamic numerical model calculated overland flow based on excess rainfall. The selected two events allowed for model prediction and comparison between Methods A and B.

5.5. Governing Equations

The HEC-RAS model simulated surface runoff. The Finite Volume Method (FVM) used a 2D mesh to calculate overland flow. The most accurate solution is the complete hydrodynamical equation (shallow-water equations, SWE).

$$\frac{\partial V}{\partial t} + (V\cdot\nabla)V + f_c k \times V = -g\nabla z_s + \frac{1}{h}\nabla\cdot(v_t h \nabla V) - \frac{\tau_b}{\rho R} + \frac{\tau_s}{\rho h} \tag{6}$$

where V [m/s] = velocity in x–y directions, $\nabla = \left(\frac{\partial}{\partial x}, \frac{\partial}{\partial y}\right)$ = gradient operator, f_c = Coriolis parameter, k = unit vector in the vertical direction, g [m/s^2] = gravity, z_s [m, mBf] = elevation of water surface, h [m] = water depth, v_t [m^2/s] = horizontal eddy coefficient, τ_b [N/m^2] =

bed shear stress, ρ [kg/m^3] = water density, R [m] = hydraulic radius, τ_{sy} [N/m^2] = wind shear stress on the water surface.

This study did not apply the wind shear stress factor. Due to its chaotic motion, turbulent mixing transfers the momentum between computational cells. In time-averaged velocity fields, dispersed mixing transfers momentum. Huang [6] suggested that an additional turbulence model could simulate overland flow more accurately. With an additional eddy viscosity component, the size of the computation cells and the time step define the degree of numerical diffusion. The added eddies represent the flow more appropriately, producing more accurate results.

The Smagorinsky-Lilly turbulence model calculated the eddy viscosity by:

$$\vartheta_S = Du_*h + (C_s\Delta)^2|\underline{S}| \tag{7}$$

where ν_S [m^2/s] = eddy viscosity, D = matrix of dispersion coefficients, u^* [m/s] = averaged velocity around the stream bed, h [m] = water depth in a cell, C_s = Smagorinsky-coefficient (~0.005–0.2), Δ [m] = filter length (equal to the cell side length), S [1/s] = temporal deformation vector

The modified SWE equation's final form with the additional eddy viscosity model becomes:

$$\frac{\partial V}{\partial t} + (V\cdot\nabla)V + f_c k \times V = -g\nabla z_s + \frac{1}{h}\nabla\cdot(v_t h \nabla V) - \frac{\tau_b}{\rho R} + \nabla\cdot(\vartheta_S\cdot\nabla V) \tag{8}$$

The complete SWE method exhibited unstable behavior, even with the shortest possible time steps. Therefore, a simplified local inertia method (SWE-LIA) was adopted. The model's stability increased by eliminating the nonlinear advection terms [35]:

$$\frac{\partial V}{\partial t} + f_c k \times V = -g\nabla z_s - \frac{\tau_b}{\rho R} + \nabla\cdot(\vartheta_S\cdot\nabla V) \tag{9}$$

Almeida and Bates [32] stated that the LIA method produces similar results to the SWE method when the Froude number is less than 0.5. Both methods (SWE and LIA) produced stable outputs during simulations of Event 2; therefore, the LIA method was adopted.

6. Numerical Grid Structure

The calculation algorithm generated an adaptive mesh to discretize the watershed (Figure 12). The mesh's resolution depended on local geometrical parameters: where the effective overland flow ratio is high (creek beds, steep slopes, roads), the resolution is finer, while other areas did not need high resolution due to their high storage and slower flow velocity. The grid size ranged between 4×4 and 20×20.

The mesh's resolution depends on the geometry of the terrain. The terrain types were divided into creek beds, steep slopes, and plain areas.

- Creek beds: The steeper gradient necessitated the use of smaller cells. Accurate calculations in smaller creeks require the midpoints of the cells to be positioned within the bed. Otherwise, the calculations may introduce numerical errors from inappropriate mesh geometry in other areas.
- Steep slopes: The model could quickly become unstable due to potential high velocities and corresponding Froude numbers. The average slope of the watershed areas is around 10%. Smaller cells improve the calculation's stability.
- Plain areas: The upland areas and areas around the bed creeks decline gently. The flow velocity decreases in these areas and does not require small cells.

The average time step of 0.4 s produced stable results.

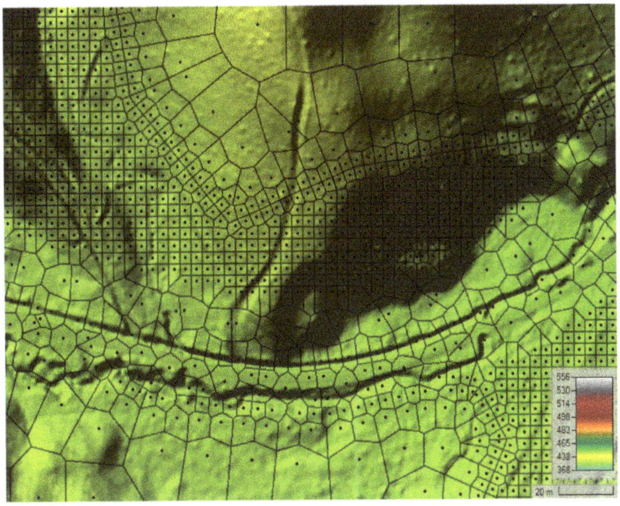

Figure 12. The adaptive grid mesh.

7. Interception and Litter Storage Calculation

The crown interception was calculated for both methods using Equation (5). Crown storage capacity changes with different woodland types. The capacity varies from 1.5 to 4 mm and averages around 2–3 mm. During high-intensity rainfall, needle-type leaves can hold up to 6–8 mm [46].

Method A calculated leaf litter storage for every time step of the measured rainfall data. Table 2 lists input parameters for storage capacity. Calculating leaf litter storage on a watershed scale presented uncertainties due to variations in litter depth, area distribution through seasons, and its initial storage capacity before an event. The initial leaf litter storage was calculated based on the time after the preceding rainfall event. For Event 1, it was 20 days; for Event 2, it was 0.9 days.

Table 2. Antecedent water content in the leaf litter for both events (Equation (3)).

	t' [day]	Sa [mm]	w_{min} [mm]	a'	w [mm]
Event 1	20	3.184	0.147	−0.185	0.22
Event 2	0.9	3.184	0.147	−0.185	2.72

7.1. Event 1

Time series interval: 15 July 2010 20:20–21:20, duration 70 min.

Figure 13a compares rainfall to effective rainfall after crown and leaf litter interception. It shows that only the highly intensive parts (20 min) of the rainfall reach the surface, creating overland flow and soil infiltration. Only the crown interception was subtracted in Figure 13b. The effective rainfall time is 40 min, with slight rainfall loss.

The total rainfall volume was 271299 m^3, the excess volume for Method A was 210,606 m^3, and for Method B was 232,676 m^3.

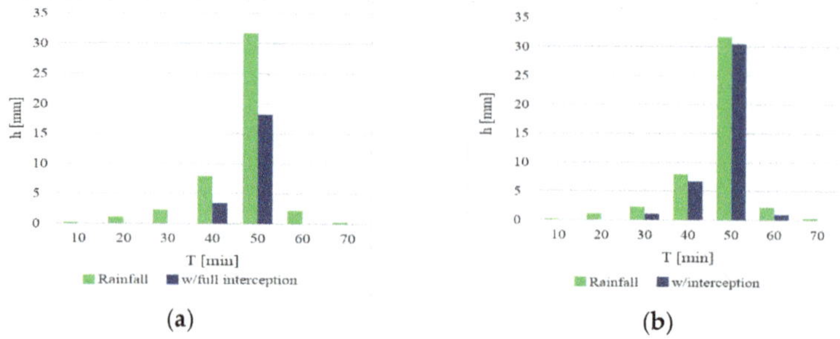

Figure 13. Event 1 effect of interceptions, Method A (**a**) and Method B (**b**).

7.2. Event 2

Time series interval: 7 July 2012 14:40–16:50, duration 100 min.

Method A (Figure 14a): The rainfall time series is 100 min long. By adding both crown and leaf litter interception, the length of the rainfall that reaches the surface was reduced to 70 min. In Method B (Figure 14b), when only the crown interception is subtracted from the rainfall time series, the rainfall that effectively reaches the surface is 90 min.

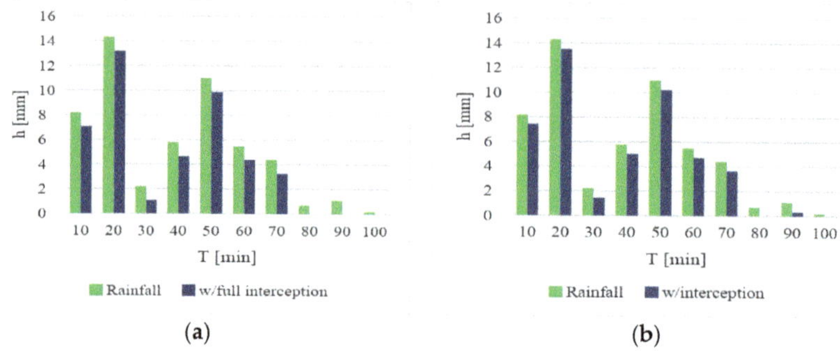

Figure 14. Event 2 effect of interceptions, Method A (**a**) and Method B (**b**).

The total rainfall volume was 317,011 m^3, the excess volume for Method A was 258,324 m^3, and for Method B was 275,188 m^3.

8. Results

A comparison of results from Methods A and B follows in the next section. The discussion will first focus on the simulated discharge hydrographs at the watershed outlet for the two events. Then, the following sections discuss the difficulties and nuances of calibrating some parameters for the modeling process.

8.1. Soil Calibration Process

The calibration focused on determining soil parameters for the Hidegvíz Valley watershed. Analysis Method A required parameters for Bennberg and Magasberc soils that underlie the meadow and lowland (streambed) areas. Method B required a third pseudo-soil, the leaf litter, located in the woodland area. Figure 15 shows the two meadowland/lowland soils with Magasberc in magenta and Breenberg in grey. Magasberc is prevalent along the streambeds and lowlands, while Breenberg is upland. The wooded area covers the majority of the valley in green. In order to distinguish between woodland and meadow areas, the land cover map was laid over the soil map. Where woodland areas were found (green), leaf litter was assumed and considered a soil layer or interceptor, as described earlier.

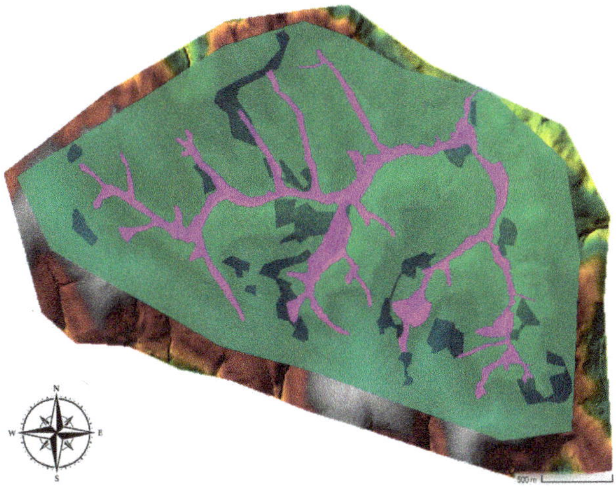

Figure 15. Modeled soils (green—woodland (leaf litter as soil); dark blue—Magasbérc soil (cutting areas); magenta—Brennberg soil (meadows, creeks)).

The woodland areas are the areas where leaf litter accumulates. Therefore, the focus during calibration was on this area. The calculation for Method A was as follows: In the woodlands, litter deficit is subtracted from the rainfall, and no further infiltration occurs (impervious). The rainfall in non-woodland areas interacts with the Magasberc or Brennberg surficial soils. The Green and Ampt equation determined the degree of infiltration. Green and Ampt parameters for the two soils were adjusted to produce reasonable behavior in the watershed.

The calculation for Method B was as follows: In the woodlands, a soil layer was approximated through leaf litter interception and storage using the Green and Ampt equations. In non-wooded areas, the same soils from Method A provided infiltration and runoff behavior.

The models required three steps for calibration. For Method A, the two events were evaluated, and parameters were adjusted until they reached optimal results. The interception and the leaf litter storage were subtracted from the rainfall time series and did not change during calibration. The soil parameters for the Green and Ampt equation were adjusted and remained the same for both events, except for initial water content. The final values appear in Table 3.

Table 3. Calibrated soil parameters.

Soil Type	S_f [mm]	K [mm/h]	Initial Soil Water Content	Saturated Soil Water Content	Residual Soil Water Content	Pore Size Distribution Index
Magasberc	261.76	15.67	0.20	0.33	0.04	0.35
Brennberg, cutting areas	219.04	26.67	0.18	0.35	0.03	0.40

In method B, where the leaf litter was approximated by soil, the two events were evaluated separately since the initial conditions of the litter were different. The soil parameters in the other regions (Brennberg and Magasberc) were identical to Method A. Table 4 shows the modeled soil types with the selected parameters for the Green and Ampt equations.

Table 4. Calibrated soil parameters with the added pseudo-soil of the leaf litter layer.

Area Type	Wetting Front Suction S_f [mm]	K_{sat} [mm/h]	Initial Soil Water Content (-)	Saturated Soil Water Content (-)	Residual Soil Water Content (-)	Pore Size Distribution Index (-)
			Magasberc soil (loam–sandy loam)—orange			
Meadows areas, streambed	261.76	15.67	0.20	0.33	0.04	0.35
			Brennberg soil (loamy sand–sandy loam)—grey			
Meadow areas-	219.04	26.67	0.18	0.35	0.03	0.40
			Leaf litter (loamy sand)—green, event 1			
Woodland areas	134.24	115	0.13	0.41	0.02	0.56
			Leaf litter (loamy sand)—green, event 2			
Woodland areas	20	63	0.38	0.41	0.02	0.56

The most recent rainfall before Event 1 occurred 20 days before, while for Event 2, rain occurred two days before. Therefore, the initial soil water contents for Event 2 were set at values higher than those set for Event 1. When a soil layer approximated the leaf litter, varying only the initial moisture content could not approximate the measured flow. The composition of the leaf litter was unknown, and its thickness may differ for the two events. The Green and Ampt equations could approximate leaf litter behavior, but it would require experimental determination since it does not behave like typical soil. During the evaluation, the leaf litter in the model was not sensitive to the initial water content; instead, the saturated hydraulic conductivity was the key. Earlier measurements showed similar scales in hydraulic conductivity [52]. The saturated hydraulic conductivity was higher in Event 1 than in Event 2. The different conditions of the leaf litter, such as thickness, composition, and age, may explain this. Since several parameters interact and cannot be easily measured, the saturated hydraulic conductivity value might approximate the cumulative behavior.

The other parameter considered for calibration was the Manning n number. The n value depends on the structure of the surface. The overland flow's depth is small; therefore, the shear stress has a higher impact on the flow; a rougher surface must be defined to approximate the surface roughness properly. For overland flow, Manning's n-values were assumed uniform during each event. The surface roughness was determined based on the land use distribution (Figure 4) and applied based on literature values. During calibration, the n values remained constant since the current HEC-RAS version only allows a linear change from an initial roughness to a final one. Possibly, if the roughness can vary nonlinearly throughout the simulation, the calibration would be more accurate. Manning's n-values are listed in Table 5 for the four terrain types.

Table 5. Calibrated Manning's n-values.

Terrain Type	Applied Manning's n-Value
Creek bed	0.03
Cutting area	0.1
Forest	0.4
Forest covered with leaf litter	0.2

Method A assumed the forested areas were impervious, and the recommended n-value was applied to the forested surface. When a soil approximated the leaf litter (Method B), its effect on the overland flow differed. During the calibration, the overland flow on the leaf litter surface was smoother than on the natural forested surface; therefore, the forest area with litter received a smaller n-value.

8.2. Model Predictions

8.2.1. Event 1

Event 1 was modeled using Methods A and B. The results appear in Figure 16. Using Method A, the peak flow and the recession limb followed the measured data well. Method B slightly overestimated the flow.

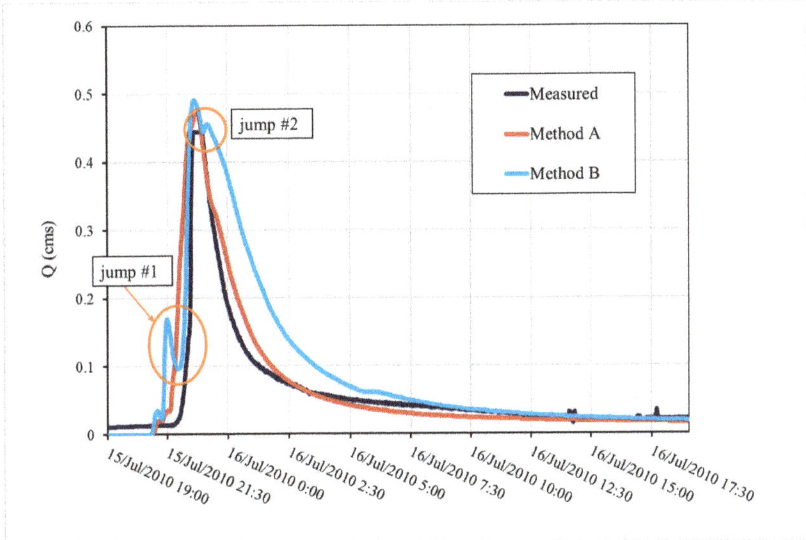

Figure 16. Prediction results for Event 1, for Methods A and B.

The time to peak of the modeled hydrographs is correct with both methods. There are jumps in the rising limb of the modeled hydrograph (Method B). Another jump occurs soon after the peak as well. Method B predicted a much more gradual recession than the measured flow.

Overprediction of overland flow velocities and volumes produced small jumps in the model hydrograph (Figure 16). In Method B, when only the canopy interception is subtracted from the rainfall time series, the time of the concentration is shorter, and the runoff volume is greater than that in Method A. Due to the faster surface flow, runoff from the two small watersheds (Vadkan and Farkas Creek) arrived very quickly at the outlet, producing jump #1 in the numerical model. The modeled flow reduced after this peak for a short time until the flow from the larger, more dominant watershed (Rak Creek) arrived, and the flow increased again. The model accurately predicted the rising limb of the measured hydrograph after this time.

The second jump predicted by Method B comes from the overflows of the temporary ponding areas. This effect has also occurred during smaller flood events, as evidenced by flow measurements.

Method A estimated the measured hydrograph well. Also, in method A, the overland flow is slower due to the higher roughness of the forest area.

To illustrate the flow dynamics during Event 1, mainly when jump #1 occurs, Figure 17 displays the spatial distribution of overland and channel flow depths. Method B exhibits a larger surface ponding area near the junction of Rák Creek (location 2, white circle) due to creek overflow. Channel flow (white rectangle) is smaller because the local peak flow has already moved downstream. In contrast, the creek channel is deeper in Method A because the surface runoff moves slower, allowing the creek to contain and convey the flow without breaching its banks. The difference originates from the choice of Manning's n-values for Methods A and B, reflecting the differing surface roughness of soil and leaf litter, respectively. Method B allows faster overland flow due to the smoother leaf litter surface.

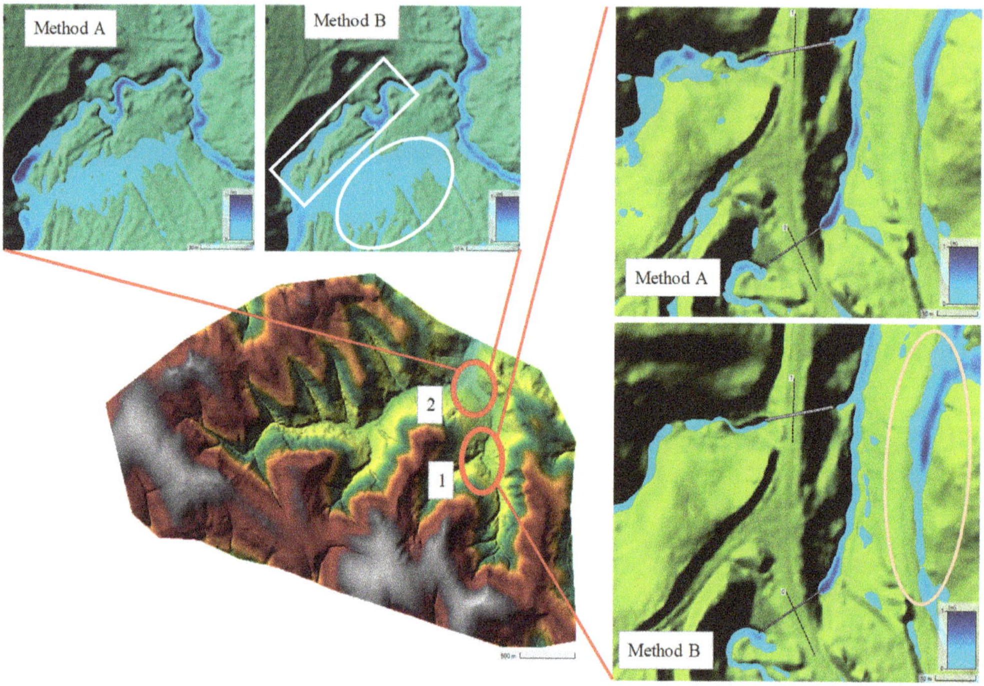

Figure 17. State of modeled overland flow at jump #1 in both methods around the culverts (location 1) and the lower Rák Creek junction area (location 2) during Event 1.

Near the Farkas and Vadkan culverts (location 1), both overland and creek flow depths are depicted for the upper region. Method B shows less water depth since the overland flow velocity is higher, indicating that the peak flow from these two smaller watersheds has passed this location. In local surface depressions, such as shown on the right side of the drainage system (orange circle), Method B accumulates more water in local storage since its more significant surface runoff has not yet infiltrated. The culverts across the road caused backwater effects along the stream since, in most cases, sediments fill the culverts and reduce flow. There was no information about the conditions of the culverts during the investigated events, so it was difficult to represent them accurately in the numerical model. In Method A, the average depth in the creek bed is lower, thus reducing the backwater effect at the culverts compared to Method B.

For Method B, recalibrating the culvert geometry may reduce jump #1 (Figure 16) by simulating the effect of accumulated sediment. Since the actual conditions of the culverts were unknown, no further adjustments were attempted.

8.2.2. Event 2

Methods A and B showed promising results for Event 2, 7 June 2012 (Figure 18). The predicted time of concentration, peak flow, and the recession limb agreed with the measured hydrograph. Method A slightly underestimated and Method B slightly overestimated peak flow.

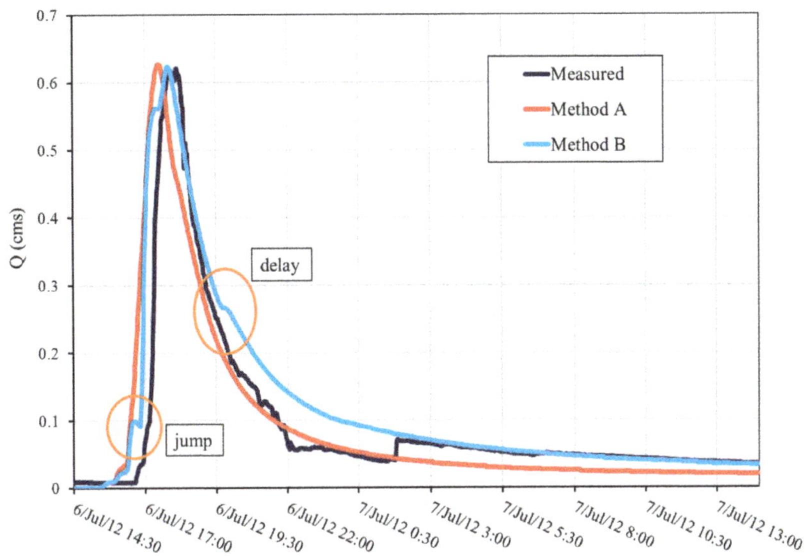

Figure 18. Prediction results for Event 2, for Methods A and B.

Both methods modeled the shape of the rising limb, but they rose earlier than the measured hydrograph and reached their peak earlier than the measured value. Method B showed the same jump response in the rising limb, similar to Event 1, but to a much lesser degree. A small jump appears in the measured flow but is much less significant. As with Event 1, it comes from the faster time of concentration of the Farkas and Vadkan Creek watersheds.

Both of the methods overpredicted the flow velocity for the two smaller watersheds, but the modeled flow from Rák Creek appears to follow the measured hydrograph well. Both of the methods' recession limbs are more gradual until their inflection points. After the inflection point, Method A reaches baseflow earlier and slightly underestimates it. With Method B, the flow slows slightly after the inflection point and then reduces to the measured flow, slightly overestimating it. The slight delay in the recession limb may result from the saturating leaf litter layer or numerical error. The numerical error originated from the constraints of the model itself. Throughout the modeling process, numerical sensitivity to the mesh and time steps increased in regions where steeper slopes transitioned to milder ones, particularly after reaching peak flow. Enhancing the average resolution of the cells mitigated these errors. However, reducing the time step beyond a specific threshold (0.1 s) was not feasible without introducing instability into the entire model. During modeling, larger excess rainfall contributed to stabilizing the model.

The adjusted model with Method B nicely fits into the baseflow curve after the jump in the measured hydrograph.

Table 6 shows the summary statistics of the calibrated models using Methods A and B. They both approximated peak flows well. Method B predicted the time to peak better than Method A for Event 2, where a soil layer defines the leaf litter. The volume error was unreliable since there was an error in the measurement on the recession limb of the hydrograph. Blockages in the weir could have produced the observed decrease in the measured data, perhaps driftwood or sediment accumulation. After 07.07, at 01:30, Method B fits the measured data better. If the measurement error is not considered, the decreasing limb of Method B fits better for the possible real flow time series. Therefore, the NSE value should be higher.

Table 6. Statistics of the model results for both methods.

		Peak Flow [cms]	Time to Peak [hr]	NSE (-)	Volume Error (-)
Event 1 15 July 2010	Measured	0.44	2.47		
	Method A	0.48	2.42	0.72	0.11
	Method B	0.49	2.33	0.43	0.51
Event 2 7 June 2012	Measured	0.62	2.95		
	Method A	0.62	2.37	0.74	0.05
	Method B	0.62	2.67	0.81	0.25

9. Conclusions

The present study focused on applying canopy interception and leaf litter retention in the hydrodynamic 2D modeling of a flash flood event in a small, forested watershed. Earlier models simulated flash flood events but did not include these components. Two modeling approaches were applied to study the influence of canopy and leaf litter on estimating surface runoff for two rainfall events.

Both models predicted runoff better for Event 2, where the runoff volume was larger, creating a greater depth in overland flow. The shallower depths from Event 1 tended to be less stable and less predictable. Unfortunately, these two events were the largest in the database. A higher-intensity event with substantial duration would have improved the prediction process.

Another significant challenge in the modeling process was the variability in the condition of leaf litter (such as its thickness and moisture content), which was greatly influenced by preceding rainfall events and prevailing meteorological conditions. This variability added a layer of complexity to the accurate prediction of the watershed's response to rainfall events, particularly in establishing the initial conditions for each simulation. The leaf litter's storage capacity was very sensitive to its condition, requiring a more meticulous reassessment before each event than ordinary soil conditions.

This study averaged soil and leaf litter parameters over the computational grid areas. However, increasing the refinement of the soil grid created additional instabilities in calculating surface runoff. The model's sensitivity to the mesh and time steps, combined with the inherent limitations of the modeling software regarding time step adjustments, meant that any further increase in the resolution of the soil and leaf litter distribution reduced numerical stability. Employing area averaging for loss approximations in smaller watersheds could prove more effective in achieving accurate results.

This behavior also occurred for culvert structures within the model. The sensitivity of culverts' parameters varied with the slope and length of the structure and had minimal impact on the outcomes. In contrast, modifications in the diameter, representing sediment accumulation inside the culvert, had a pronounced effect by creating backwater conditions.

Method A subtracted the canopy and leaf litter interceptions from the rainfall and assumed that the areas covered with leaf litter were impervious for the event duration. In Method B, leaf litter interception and storage processes were modeled as soil using the Green and Ampt equations. From the modeling, the following conclusions can be stated:

- The linear formula based on woodland types estimates the crown interception to a reasonable degree of accuracy.
- If the Green and Ampt soil parameters model is applied to the leaf litter calculations, then it must account for antecedent moisture conditions (saturation). Events 1 and 2 began with different initial moisture conditions in the leaf litter and underlying soil. Before either event, there was no information about the leaf litter's thickness, current composition, or in situ moisture content.

- The results also showed that predicting the decreasing limb had less error for the smaller rainfall event when the leaf litter was modeled as a soil layer (Method B).
- For both events, Method A performed well. It was less sensitive to the initial conditions of the leaf litter, rainfall duration, and intensity.

The models showed that a 2D hydrodynamical model can be used to estimate the effect of leaf litter on a steep-sloped watershed. The calculation with both methods is more stable if the excess rainfall is higher. These results suggest that the accuracy improves with flash-flood events. Therefore, both methods may correctly model flash floods.

When leaf litter interception was subtracted from the rainfall time series, the leaf litter storage capacity was connected to the rainfall intensity. These subtractions were the basis for calculating the losses in Method A. Even a rudimentary estimate improves the modeling results. This method is ideal for ungauged watersheds, planning processes, or water damage prevention projects. During a rainfall event on an ungauged forested watershed, where the land use is known, leaf litter storage volumes can be predicted as a function of rainfall intensity. However, this method cannot calculate the saturation of the leaf litter. Therefore, it should be applied if measured data are available, or if the leaf litter layer is thick and mostly dry.

Method B is more precise than Method A for predicting the flow's physical processes. Soil parameters can estimate the behavior of the leaf litter during a flash flood event since litter interception and storage in a short period work similarly to unsaturated soil. Of course, the behavior of leaf litter is not the same as that of the soil, but field or laboratory measurements can provide estimates. Since proper leaf litter data will not always be available for a given watershed, the modeling of the litter should depend on as few parameters as possible. This study found that hydraulic conductivity and the initial water content were the two most significant parameters.

The results showed that, during the modeling of the overland flow, the land use controls the time to peak and the slope of the rising and the recession limb. Compared with Method A, only crown interception modifies the rainfall time series. The process of surface storage calculation in this method is more complex, and the surface roughness is more accurate since the numerical model does not neglect leaf litter. Therefore, Method B's complex physical model structure gives a more accurate simulation for predicting flood events in a forested watershed. Method B can be more accurate for ungauged watersheds since the land use and the vegetation type are the model's key elements (in addition to a proper DEM).

This study applied a new approach to overland flow modeling compared to previous studies. Both crown and leaf litter interception significantly affect overland flow, although the conditions change with time. Therefore, combining complex overland flow simulation and substituting leaf litter with a soil layer can effectively predict flash floods in forested watersheds.

The main accomplishment of this study was in demonstrating the feasibility of constructing a model for steeply sloped, forested watersheds. Such watersheds, often situated in the upper reaches of larger catchments, are typically prone to high risks of flash flooding. Detailed and complex models are essential to understand the behavior of these types of watersheds. This study focused on integrating numerical solutions with empirical methods, illustrating how various available data can be utilized within a single model framework. The dual approach of numerical and empirical methods also lays the groundwork for comparative studies. Although the methodology for calculating overland flow remains consistent, the computation of losses varies, which is particularly valuable since many of these smaller watershed areas lack gauging stations. In ungauged watersheds, comparative modeling is a vital strategy for reducing uncertainties. Consequently, one of the aims of this study was to establish a foundation for modeling approaches that can be applied in planning processes, especially in ungauged watersheds with a significant risk of flash flooding.

Author Contributions: Conceptualization, G.Á., K.B. and R.R.; methodology, G.Á. and K.B.; software, G.Á.; validation, R.R. and Z.G.; formal analysis, G.Á.; investigation, G.Á., K.B., R.R. and Z.G.; resources, G.Á., R.R. and Z.G.; data curation, Z.G. and P.K.; writing—original draft preparation, G.Á.; writing—review and editing, K.B. and R.R.; visualization, G.Á. and R.R.; supervision, K.B.; project administration K.B. All authors have read and agreed to the published version of the manuscript.

Funding: This study is part of ongoing research entitled "Microscale influence on runoff", supported by the National Research, Development, and Innovation Office (OTKA project grant number SNN143972) and the Slovenian Research and Innovation Agency (N2-0313). As a joint project, the TKP2021-NKTA-43 project also supported the preparation of this paper: TKP2021-NKTA-43 has been implemented with the support provided by the Ministry of Innovation and Technology of Hungary (successor: Ministry of Culture and Innovation of Hungary) from the National Research, Development, and Innovation Fund, financed under the TKP2021-NKTA funding scheme. Additional research presented in the article was carried out within the framework of the Széchenyi Plan Plus program with the support of the RRF 2.3.1 21 2022 00008 project.

Data Availability Statement: The data presented in this study are available on request from the corresponding author. The data are not publicly available due to current research and running data processes.

Conflicts of Interest: The authors declare no conflict of interest.

References

1. Blöschl, G. Runoff Prediction in Ungauged Basins, Synthesis across Processes, Places and Scales. In *Runoff Prediction in Ungauged Basins*; Chapter: A Data Acquisition Framework for Predictions of Runoff in Ungauged Basins; Cambridge University Press: Cambridge, UK, 2013.
2. Rana, V.K.; Suryanarayana, T.M.V. Estimation of flood influencing characteristics of watershed and their impact on flooding in data-scarce region. *Ann. GIS* **2021**, *27*, 397–418. [CrossRef]
3. Garambois, P.A.; Larnier, K.; Roux, H.; Labat, D.; Dartus, D. Analysis of flash flood-triggering rainfall for a process-oriented hydrological model. *Atmos. Res.* **2013**, *137*, 14–24. [CrossRef]
4. Hromadka, T.V.; Rao, P. Application of Diffusion Hydrodynamic Model for Overland Flows. *Open J. Fluid Dyn.* **2019**, *9*, 334–345. [CrossRef]
5. Weinghuo, H.; He, H. Flood Forecasting Model Using the Combination Approach. *Comput. Sci. IT Res. J.* **2020**, *1*, 59–64. [CrossRef]
6. Huang, W.; Cao, Z.-X.; Qi, W.-J.; Pender, G.; Zhao, K. Full 2D Hydrodynamic Modelling of Rainfall-induced Flash Floods. *J. Mt. Sci.* **2015**, *12*, 1203–1218. [CrossRef]
7. Bizhanimanzar, M.; Leconte, R.; Nuth, M. Catchment-Scale Integrated Surface Water-Groundwater Hydrologic Modelling Using Conceptual and Physically Based Models: A Model Comparison Study. *Water* **2020**, *12*, 363. [CrossRef]
8. Lago, C.A.F.; Giacomoni, M.H.; Bentivoglio, R.; Taormina, R.; Gomes, M.N.; Mendiondo, E.M. Generalizing rapid flood predictions to unseen urban catchments with conditional generative adversarial networks. *J. Hydrol.* **2023**, *618*, 129276. [CrossRef]
9. Xia, X.; Liang, Q.; Ming, X.; Hou, J. An efficient and stable hydrodynamic model with novel source term discretization schemes for overland flow and flood simulations. *Water Resour. Res.* **2017**, *53*, 3730–3759. [CrossRef]
10. Liao, Y.; Zhao, H.; Jiang, Z.; Li, J.; Li, X. Identifying the risk of urban nonpoint source pollution using an index model based on impervious-pervious spatial pattern. *J. Clean. Prod.* **2020**, *288*, 125619. [CrossRef]
11. Schneidewind, U.; Schneidewind, U.; Vandersteen, G.; Joris, I.; Seuntjens, P.; Batelaan, O. Delineating groundwater-surface water interaction. *Hydrol. Process.* **2015**, *30*, 203–216. [CrossRef]
12. Ámon, G.; Bene, K. Impact of different rainfall events on overland flow using a 2D hydrodynamical model on a steep-sloped watershed. In Proceedings of the EGU23-9179, Vienna, Austria, 24–28 April 2023. [CrossRef]
13. Acharya1, S.; McLaughlin, D.; Kaplan, D.; Cohen, M.J. A proposed method for estimating interception from near-surface soil moisture response. *Hydrol. Earth Syst. Sci.* **2020**, *24*, 1859–1870. [CrossRef]
14. Gribovszki, Z. Comparison of specific-yield estimates for calculating evapotranspiration from diurnal groundwater-level fluctuations. *Hydrogeol. J.* **2018**, *26*, 869–880. [CrossRef]
15. Zagyvai-Kiss, K. Az Avarintercepció Vizsgálata a Soproni-Hegységben (Hungarian, an Examination of the Leaf Litter Interception in the Mountains of Sopron). Ph.D. Thesis, Pál Kitaibel Doctoral School of Environmental Sciences, University of West Hungary, Sopron, Hungary, 2014.
16. Deng, W.; Zheng, X.; Xiao, S.; Chen, Q.; Gao, Y.; Zhang, L.; Huang, J.; Bai, T.; Xie, S.; Liu, Y. Effects of leaf type, litter mass and rainfall characteristics on theinterception storage capacity of leaf litter based on process simulation. *J. Hydrol.* **2023**, *624*, 129943. [CrossRef]
17. Sato, Y.; Kumagai, T.; Kume, A.; Otsuki, K.; Ogawa, S. Experimental analysis of moisture dynamics of litter layers—The effects of rainfall conditions and leaf shapes. *Hydrol. Process.* **2004**, *18*, 3007–3018. [CrossRef]
18. Putuhena, W.M.; Cordery, I. Estimation of interception capacity of forest floor. *J. Hydrol.* **1995**, *180*, 283–299. [CrossRef]

19. Bulcuck, H.H.; Jewitt, G.P.W. Field data collection and analysis of canopy and litter interception in commercial forest plantations in the KwaZulu-Natal Midlands, South Africa. *Hydrol. Earth Syst. Sci.* **2012**, *16*, 3717–3728. [CrossRef]
20. Zagyvai-Kiss, K.A.; Kalicz, P.; Szilágyi, J.; Gribovszki, Z. On the specific water holding capacity of litter for three forest ecosystems in the eastern foothills of the Alps. *Agric. For. Meteorol.* **2019**, *278*, 107656. [CrossRef]
21. Urbanik, M.; Olejnik, J.; Miler, A.T.; Krysztofiak-Kaniewska, A.; Ziemblinska, K. Rainfall Interception for Sixty-Year-Old Pine Stand at the Tuczno Forest District. *Infrastrukt. Ekol. Teren. Wiej.* **2015**, *2*, 377–384. [CrossRef]
22. Zhao, L.; Hou, R.; Fang, Q. Differences in interception storage capacities of undecomposed broadleaf and needle-leaf litter under simulated rainfall conditions. *For. Ecol. Manag.* **2019**, *446*, 135–142. [CrossRef]
23. Li, Q.; Lee, Y.E.; Im, S. Characterizing the Interception Capacity of Floor Litter with Rainfall Simulation Experiments. *Water* **2020**, *12*, 3145. [CrossRef]
24. Du, J.; Xie, S.; Xu, Y.; Xu, C.; Singh, V.P. Development and testing of a simple physically-based distributed rainfall-runoff model for storm runoff simulation in humid forested basins. *J. Hydrol.* **2006**, *336*, 334–346. [CrossRef]
25. Kim, T.; Kim, J.; Lee, J.; Kim, H.S.; Park, J.; Im, S. Water Retention Capacity of Leaf Litter According to Field Lysimetry. *Forests* **2023**, *14*, 478. [CrossRef]
26. Cheng, W.; Tie, L.; Zhou, S.; Hu, J.; Ouyang, S.; Huang, C. Effects of Soil Arthropods on Non-Leaf Litter Decomposition: A Meta-Analysis. *Forests* **2023**, *14*, 1557. [CrossRef]
27. Kees, C.E.; Band, L.E.; Farthing, L.W. *Effects of Dynamics on Hillslope Water Balance Models;* North Carolina State University, Center for Research in Scientific Computation: Raleigh, NC, USA; pp. 27695–28205.
28. Costabile, P.; Costanzo, C.; Ferraro, D.; Macchione, F.; Petaccia, G. Performance of the New HEC-RAS Version 5 for 2-D Hydrodynamic-Based Rainfall-Runoff Simulations at Basin Scale: Comparison with a State-of-the Art Model. *Water* **2020**, *12*, 2326. [CrossRef]
29. Chen, L.; Young, M.H. Green-Ampt infiltration model for sloping surfaces. *Water Resour. Res.* **2006**, *42*, 1–9. [CrossRef]
30. Tokunaga, T.K. Simplified Green-Ampt Model, Imbibition-Based Estimates of Permeability, and Implications for Leak-off in Hydraulic Fracturing. *Water Resour. Res.* **2020**, *56*, e2019WR026919. [CrossRef]
31. Tzimopoulos, D. New Explicit Form of Green and Ampt Model for Cumulative Infiltration Estimation. *Res. J. Env. Sci.* **2020**, *14*, 30–41.
32. Almeida, G.; Bates, P. Applicability of the local inertial approximation of the shallow water equations to flood modeling. *Water Resour. Res.* **2013**, *49*, 4833–4844. [CrossRef]
33. Ballesteros, C.J.A.; Eguibar, M.; Bodoque, J.M.; Díez-Herrero, A.; Stoffel, M.; Gutiérrez-Pérez, I. Estimating flash flood discharge in an ungauged mountain catchment with 2D hydraulic models and dendrogeomorphic palaeostage indicators. *Hydrol. Process.* **2011**, *25*, 970–979. [CrossRef]
34. Cea, L.; Garrido, M.; Puertas, J.; Jácome, A.; Del Río, H.; Suárez, J. Overland flow computations in urban and industrial catchments from direct precipitation data using a two-dimensional shallow water model. *Water Sci. Technol.* **2010**, *62*, 1998–2008. [CrossRef]
35. Brunner, G.W. *HEC-RAS, River Analysis System Hydraulic Reference Manual;* US Army Corps of Engineers, Hydrologic Engineering Center: Davis, CA, USA, 2020.
36. Zhao, J.; Liang, Q. Novel variable reconstruction and friction term discretisation schemes for hydrodynamic modelling of overland flow and surface water flooding. *Adv. Water Resour.* **2022**, *163*, 104187. [CrossRef]
37. Ferro, V.; Guida, G. A theoretically-based overland flow resistance law for upland grassland habitats. *Catena* **2022**, *210*, 105863. [CrossRef]
38. Szatmári, G.; Pásztor, L. Comparison of various uncertainty modelling approaches based on geostatistics and machine learning algorithms. *Geoderma* **2018**, *337*, 1329–1340. [CrossRef]
39. Csafordi, P.; Eredics, A.; Gribovszki, Z.; Kalics, P.; Koppan, A.; Kucsara, M. *Hidegvíz Valley Experimental Watershed;* Department of Hydrology, Institute of Geomatics and Civil Engineering, Faculty of Forestry, University of West Hungary: Sopron, Hungary, 2012.
40. Tímár, G.; Szmorad, F. Új adatok a Soproni-hegység flórájához, (Hungarain, New Data on the Flora of the Soproni Mountains), KITAIBELIA I. *J. Pannonian Bot.* **1996**, 17–24, ISSN 2064-4507 (Online) ISSN 1219-9672 (Print).
41. Kiss, K.A. Az Avarintercepció Vizsgálata a Soproni-Hegységben. Ph.D. Thesis, University of Sopron, Sopron, Hungary, 2012.
42. Király, G.; Csapody, I.; Szmorad, F.; Tímár, G. A Soproni-hegység edényes flórája (Vascular Flora of the Sopron Hills). *Flora Pannonica J. Phytogeogr. Taxon.* **2004**, *2*, 1589–7788. (In Hungarian)
43. Kárpáti, Z. Die Florengrenzen in der Umgebung von Sopron und Florendistrikt Laitaicum. *Acta Bot. Hung.* **1956**, *2*, 281–307.
44. Vendel, M. Sopron Környékének Geológiája II. Rész: A Neogén és a Negyedkor üledékei Erdészeti Kísérletek. 1930, pp. 1–16. Available online: https://epa.oszk.hu/01600/01635/00221/pdf/EPA01635_foldtani_kozlony_1977_107_34_256-265.pdf (accessed on 19 February 2024).
45. Kucsara, M.; Gribowszki, Z. *Az Intercepció Vizsgálata és Számszerűsítése, Csapadékeseményhez Kötött Modellel,(Hungarian, Examination and Quantification of Rainfall Interception, with a Model Tied to a Rainfall Event) Research Report;* University of Soprom: Sopron, Hungary, 2014.
46. Feldman, A.D. *Hydrologic Modeling System HEC-HMS;* Technical References Manual; US Army Corps of Engineering, Hydrologic Engineering Center: Davis, CA, USA, 2000.
47. Laborczi, A.; Szatmári, G.; Kaposi, A.D.; Pásztor, L. Comparison of soil texture maps synthetized from standard depth layers with directly compiled products. *Geoderma* **2018**, *352*, 360–372. [CrossRef]

48. Rawls, W.J.; Brakensiek, D.L.; Saxton, K.E. Estimation of Soil Water Properties. *Trans. ASAE* **1982**, *25*, 1316–1320. [CrossRef]
49. Jancsó, B.; Kaveczki, G.; Kóczán, G.; Laborczi TMarcell, K. Csapadékvíz-Gazdálkodás Tervezési Követelményei Laszlo Raum. Requirements of Rainwater-Resources Planning. 2019. Available online: https://mernokvagyok.hu/vizgazdalkodas/wp-content/uploads/sites/23/2022/05/Csapade%CC%81kvi%CC%81zgazda%CC%81lkoda%CC%81s-terveze%CC%81si-ko%CC%88vetelme%CC%81nyei_201-VVT.pdf (accessed on 19 February 2024). (In Hungarian)
50. Ámon, G.; Bene, K. Rainfall Duration and Parameter Sensitivity on Flash-Flood at A Steep Watershed. *Pollack Period* **2023**, *18*, 54–59. [CrossRef]
51. Orfánus, T.; Zvala, A.; Čierniková, M.; Stojkovová, D.; Nagy, V.; Dlapa, P. Peculiarities of Infiltration Measurements in Water-Repellent Forest Soil. *Forests* **2021**, *12*, 472. [CrossRef]
52. Engman, E.T. Roughness Coefficients for Routing Surface Runoff. *J. Irrig. Drain. Eng.* **1986**, *112*, 39–53. [CrossRef]

water

MDPI

Article

The Impact of Decreased Atmospheric Pressure on Forced Aeration of Discharged Flow

Yijiao Guo, Luchen Zhang *, Lei Yu, Shaoze Luo, Chuang Liu and Yuan Liu

State Key Laboratory of Hydrology-Water Resources and Hydraulic Engineering, Nanjing Hydraulic Research Institute, Nanjing 210029, China; real_gyj@163.com (Y.G.); lyu@nhri.cn (L.Y.); szluo@nhri.cn (S.L.); kobelc1999@163.com (C.L.); 18115103063@163.com (Y.L.)

* Correspondence: lczhang@nhri.cn

Abstract: To account for changes in the performance of spillway aerator structures of high-altitude dams, depressurization generalized model experiments and theoretical analyses were conducted in this study. Measurements were taken for ventilation hole air velocity, cavity subpressure, cavity length, and air concentration in crucial regions. The study proposed correction formulas for the aeration coefficient and water air concentration in aerator devices operating under low atmospheric pressure. The pressure range of the experiments was between 26.3 kPa and 101.3 kPa. The results indicated that with decreasing atmospheric pressure, ventilation hole air velocity, ventilation volume, cavity subpressure, and water air concentration all showed a decreasing trend. For every 15 kPa decrease in pressure, ventilation hole air velocity decreased by approximately 24%. When the atmospheric pressure dropped from 101.3 kPa to 26.3 kPa, the cavity subpressure decreased and eventually approached zero. The maximum reduction in air concentration was 14.9% in the cavity backwater area, 38.5% at the cavity end, and 38.3% in the downstream bubble escape segment. The proposed correction formulas could be used for a rapid estimation of ventilation volume and air concentration in low-pressure environments. This research provides a scientific basis for the design of aeration devices in water projects located in high-altitude regions.

Keywords: low atmospheric pressure; depressurization model experiment; ventilation volume; air concentration

1. Introduction

Cavitation damage caused by high-speed water flow can pose a serious threat to the safe operation of spillway structures. Spillway outlets of projects such as the Boulder Dam in the United States, the Liujiaxia Hydropower Station and the Ertan Hydropower Station in China, and the Aldeadavila Dam in Spain have all experienced severe cavitation damage during operation [1–4]. Currently, three main damage reduction measures are employed in engineering practices: reducing the incipient cavitation number through the optimization of the overflow boundary body shape and the control of the unevenness of the overflow surface; using materials with strong cavitation resistance; and using air entrainment to alleviate cavitation, aerating areas where cavitation is likely to occur in the water flow [5]. Extensive engineering practice has shown that aeration is one of the effective methods for preventing cavitation damage on overflow surfaces [6,7]. Due to its simple technical structure and noticeable effects, it is widely used in various high-head spillway structures. With the continuous development of water resources in China, the western high-altitude regions rich in hydropower resources are gradually becoming hotspots for hydropower development. An increasing number of high dams will be built in areas at altitudes of 3000 m and above. Besides characteristics such as high head and large flow, the environmental conditions of high altitude and low atmospheric pressure are gradually becoming a focus of attention for their impact on hydraulic safety, especially regarding cavitation in spillway structures. Due to the decrease in atmospheric pressure and reduction in cavitation number with increasing altitude, spillway structures are prone to cavitation and subsequent cavitation damage [8,9].

Under normal atmospheric pressure, researchers have conducted extensive research on using air entrainment to alleviate cavitation. Research directions mainly encompass the mechanisms of using air entrainment to alleviate cavitation [10,11], the characteristics of forced aeration in water flow [12–15], the design of aeration devices [16], the downstream bubble characteristics of aeration devices [17], and the scale effects of aeration ratios [18,19]. The ventilation volume is an important parameter for measuring the effectiveness of air entrainment devices. A significant amount of research has been conducted by previous researchers on the ventilation volume of aeration devices. There are numerous influencing factors, including incoming flow hydraulic characteristics, channel and aeration device geometry, and affected aeration phenomena [20–22]. However, relatively limited research has been conducted on aeration under low atmospheric pressure. Falvey [3] developed computer programs to support the design of aerator structures at different altitudes based on local atmospheric pressure. This approach has been applied to an altitude range of approximately 375 m to 2280 m. However, there was no generalized study of the extent and mechanisms of the impact of air pressure on aeration and cavity hydraulic characteristics. Currently, commonly used formulas for ventilation volume are mostly empirical formulas based on prototype or model data [12,23,24], without considering the influence of atmospheric pressure on ventilation volume.

Existing research indicates that as atmospheric pressure decreases, the distribution of gas nuclei in water undergoes changes, the surface tension coefficient of water increases, the cavitation number of the water flow decreases, and it becomes more difficult to generate backwater in the cavity [8,9,25,26]. Pang et al. [26], through the collection of gas nucleus images under different pressures, found that as the environmental air pressure decreased, the number of gas nuclei decreased, and the proportion of small-sized gas nuclei increased. Wang et al. [8], using the capillary method at different air pressures and water temperatures, measured surface tension. The results showed that at the same water temperature, lower air pressure corresponded to a larger surface tension coefficient. At a maximum vacuum pressure of 0.1 MPa, the surface tension of 20 °C water increased by approximately 18% compared to atmospheric pressure. Additionally, the influence of air pressure and temperature on the water surface tension was essentially independent as the air pressure decreased. Pang [9] utilized theoretical equations to calculate the variations in the critical pressure of a single gas nucleus under different atmospheric pressure conditions. The results demonstrated that under low atmospheric pressure, the incipient cavitation number of gas nuclei decreased, whereas the reduction in the flow cavitation number was significantly enhanced compared to that of the incipient cavitation number. For every increase in altitude of 500 m, the cavitation number of the water flow decreased by 2.5% to 3.5%. In high-altitude areas, the risk of cavitation and cavitation erosion increased in fast-flowing water. Wu [27] analyzed the changing patterns of air content in water under different atmospheric pressure and temperature conditions. As the atmospheric pressure decreases, the air content will also decrease. If there are differences in air content between the prototype and the model, the incipient cavitation numbers will also differ, leading to a scale effect. During experiments on a 1:35 scale model of a spillway at a hydropower station, Zhao and Hu [28] conducted observational comparisons and found that when the depressurization chamber reached a vacuum level of 93%, the cavity length decreased by approximately 10% compared to normal atmospheric pressure conditions. Using a computational fluid dynamics (CFD) numerical simulation, Pang et al. [29] calculated the cavity length in an aeration device under different atmospheric pressure conditions. The results indicated that as the atmospheric pressure decreases, the cavity length in high-velocity water flows becomes smaller. For each 10 kPa decrease in atmospheric pressure, the average cavity length decreased by approximately 1.7% to 2.9%. Xue et al. [30] conducted steady-state numerical simulations of a specific hydraulic turbine in high-altitude regions to analyze the effects of high-altitude environments on hydraulic machinery performance. They found that as the atmospheric pressure decreased, the low atmospheric pressure area of the elbow section of the draft became larger. Simultaneously, the cavitation coefficient of the hydraulic turbine increased with decreasing atmospheric pressure, making cavitation more likely to occur. In summary, research on the aeration characteristics in low atmospheric pressure environments is still in its initial stages, while the current calculation formulas for using air entrainment to alleviate cavitation have not yet provided regulations for the relevant hydraulic calculations and designs under low atmospheric pressure conditions.

Therefore, this paper presents depressurization generalization model experiments and theoretical analysis conducted with a primary focus on hydraulic characteristics such as ventilation hole air velocity, ventilation volume, and air concentration. Considering four discharges and six atmospheric pressures, a total of twenty-four sets of experimental conditions were established. This study investigated the ventilation characteristics of aeration devices under different low atmospheric pressure conditions and the downstream water air concentration characteristics. Furthermore, we proposed a correction method for using air entrainment to alleviate cavitation characteristics under low atmospheric pressure, with the aim of providing a scientific basis for the design of aeration devices in high-altitude hydraulic engineering projects.

2. Research Methods

To simulate a series of low atmospheric pressures, experiments were conducted inside a depressurization chamber. The generalized model was based on a typical design of a tunnel spillway, which included a control section, a gentle slope section, a steep slope section, and a jet bucket section. The total length was 4.14 m, the net width was 0.188 m, and the height was 1.66 m. The arrangement of the model within the depressurization chamber is shown in Figure 1. A WES weir was used for the control section. The gentle slope section had a gradient of 3.0% and a length of 0.33 m. The steep chute section followed the gentle slope section and was composed of a parabolic curve, a straight line, and a counter-arc curve. To study the forced aeration characteristics under different low atmospheric pressure conditions, an aeration device was installed at the transition between the parabolic curve and the straight line. For the aeration device, a configuration combining a ramp and a slot was adopted, as shown in Figure 2. The slope of the ramp relative to the downstream straight section was 1:10, and the longitudinal length of the ramp was l = 20 cm. The total depth of the aeration slot was d = 4.5 cm, and the total length was 4.23 cm. The air vent hole was located within the chute and had a diameter of 3.0 cm.

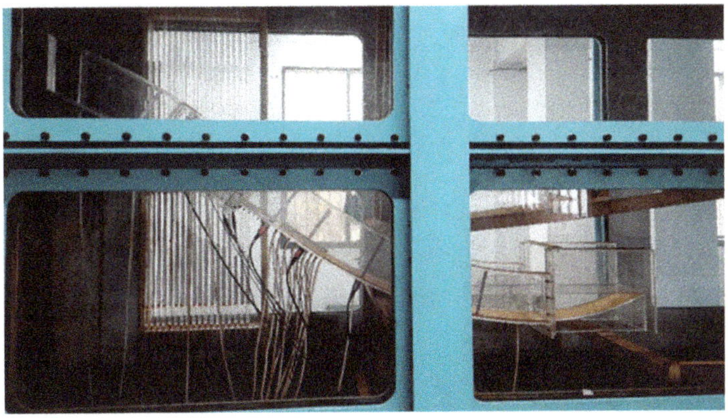

Figure 1. Generalized model deployment.

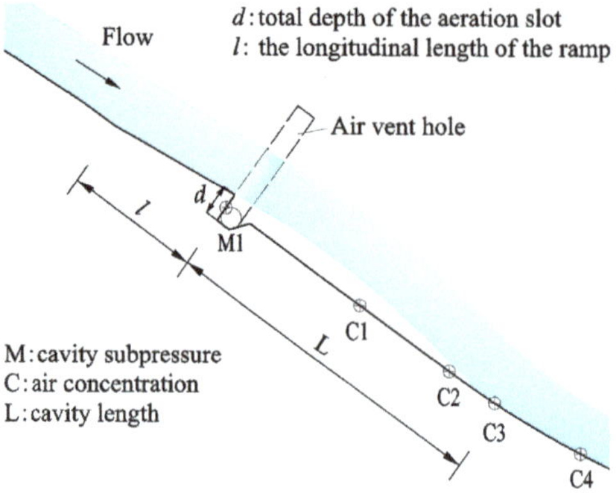

Figure 2. Aeration device size and measurement point arrangement.

Four experimental discharges Q (0.0740, 0.0528, 0.0354, 0.0199 m^3/s) and six different atmospheric pressure conditions (the pressure range inside the depressurization chamber varied from 26.3 to 101.3 kPa in steps of 15 kPa) were employed in this study, resulting in a total of twenty-four experimental conditions. Vacuum degree refers to the degree of gas rarity in a vacuum state, expressed by the formula $P_v = P_0 - P_a$, where P_a is the environmental pressure inside the vacuum chamber and P_0 is the atmospheric pressure in the external test laboratory of the vacuum chamber as shown in Table 1 for the vacuum degree inside the depressurization chamber. The test parameters are listed in Table 2.

Table 1. Vacuum degrees in the depressurization chamber under different atmospheric pressures.

Series	Atmospheric Pressure (kPa)	Simulating High Altitudes (m)	Depressurization Chamber Vacuum Degree (kPa)
1	101.3	0	0
2	86.3	1330	15.0
3	71.3	2900	30.0
4	56.3	4700	45.0
5	41.3	6950	60.0
6	26.3	10,000	75.0

Table 2. Test parameters.

Series	Discharge (m^3/s)	Incoming Flow Velocity (m/s)	Froude Number
1	0.0740	4.96	5.61
2	0.0528	4.63	5.99
3	0.0354	4.33	6.62
4	0.0199	4.19	8.41

Different measurement methods were employed to measure the ventilation hole air velocity, cavity length, cavity subpressure, and air concentration. The air velocity was measured using the KANOMAX multi-point air velocity measurement system (Andover, NJ, USA). The probe used was of the hot-wire type with a maximum range of 50 m/s and an accuracy of 2% full scale (F.S.). The probe with left and right ventilation holes was connected to a Model 1550 data acquisition system for automated measurements. The sampling frequency was 10 Hz, and the sampling duration was 120 s. The cavity length was determined through concentration measurements and observations. The downstream water air concentration under the aeration device was measured using a resistance-type air concentration meter. Four measurement points (C1, C2, C3, C4) were arranged on the bottom plate at distances of 24 cm, 40 cm, 48 cm, and 62 cm from the end of the aerator. The cavity subpressure was measured using an absolute pressure fluctuation sensor. Measurement point M1 was located in the middle of the aerator.

3. Forced Aeration Characteristics under Low Atmospheric Pressure

3.1. Ventilation Hole Air Velocity

The characteristic air velocity values of the ventilation holes under different atmospheric pressures are listed in Table 3 for typical discharges. Air velocity changed with atmospheric pressure under different discharges, as shown in Figure 3. The root mean square proportion of air velocity to the average air velocity under different pressures ranged from 13.3% to 15.3%, showing no significant trend with changes in atmospheric pressure. This suggested that changes in atmospheric pressure did not noticeably impact the turbulence of the air entrainment. There was a strong linear positive correlation between atmospheric pressure and the air velocity through the air vent holes. As shown in Figure 3, the air velocity through the air vent holes notably declined with decreased atmospheric pressure. With the use of a linear fit with a y-intercept of 0, the R^2 value approached 0.99. At a discharge of 0.0740 m^3/s, for every 15 kPa decrease in pressure, the air velocity decreased by approximately 17.2% to 29.2%, with an average reduction of about 24.1%. At a discharge of 0.0528 m^3/s, the air velocity decreased by around 16.9% to 32.8%, with an average reduction of about 25.0%. At a discharge of 0.0354 m^3/s, the air velocity decreased by approximately 17.5% to 35.7%, with an average reduction of about 25.4%. Finally, at a discharge of 0.0199 m^3/s, the air velocity decreased by about 17.9% to 34.8%, with an average reduction of approximately 23.9%.

Table 3. Air velocity characteristic values of ventilation holes in aeration device under different atmospheric pressures (the discharge was Q = 0.0740 m^3/s).

Atmospheric Pressure (kPa)	Feature Parameters (m/s)				Degree of Influence on the Average Value (%)
	Maximum Value	Minimum Value	RMS Value	Average Value	
101.3	1.63	0.60	0.14	1.00	
86.3	1.46	0.49	0.12	0.83	17.2
71.3	1.13	0.40	0.10	0.65	22.1
56.3	0.87	0.32	0.07	0.50	21.9
41.3	0.56	0.21	0.05	0.35	29.9
26.3	0.34	0.18	0.03	0.25	29.2

The ventilation volume of the flip bucket is equal to the sum of the air supply from the ventilation holes on both sides. If the ventilation volume is insufficient, the anti-cavitation protection effect is poor, while if the ventilation volume is saturated, it can lead to adverse flow conditions. Under various operating conditions, the ventilation volume decreased significantly with the decrease in atmospheric pressure. The probable reasons for this are as follows: (i) In low atmospheric pressure conditions, the gas density was lower compared to normal atmospheric pressure. This led to an increase in the average distance between air molecules and a reduction in the thermal motion of air molecules, resulting in inadequate air intake through the ventilation holes. (ii) There was a reduction in the air pressure gradient force. The horizontal air pressure gradient force served as the primary driving force behind wind formation and could be calculated using Equation (1).

$$G = -\frac{1}{\rho}\nabla P \quad (1)$$

where $-\nabla P$ is the pressure gradient, with its direction pointing from high pressure to low pressure.

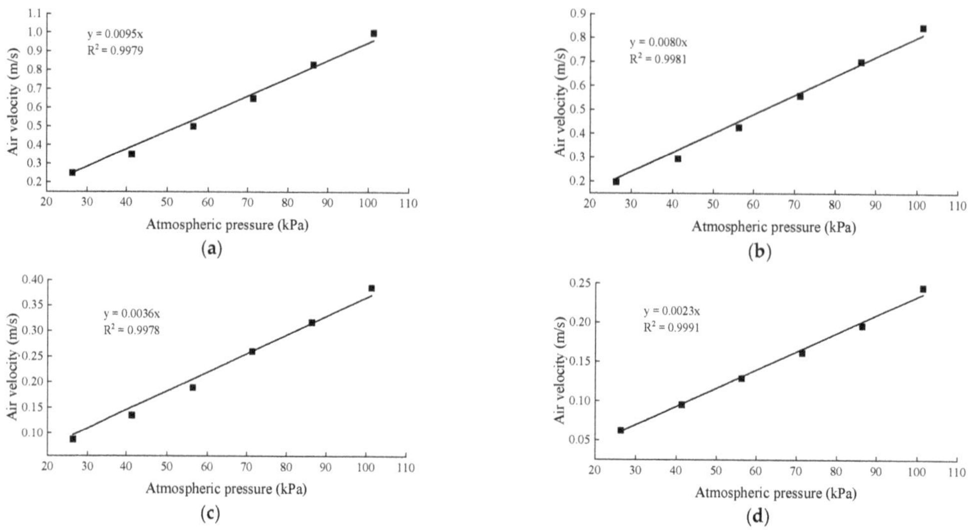

Figure 3. Variation trend of vent hole air velocity with changes in atmospheric pressure at different discharges. (**a**) $Q = 0.0740\ m^3/s$; (**b**) $Q = 0.0528\ m^3/s$; (**c**) $Q = 0.0354\ m^3/s$; (**d**) $Q = 0.0199\ m^3/s$.

For aeration devices, high pressure refers to the atmospheric pressure, while low pressure pertains to the pressure within the cavity. Air flows from the high-pressure region to the low-pressure region, creating airflow within the ventilation holes. Hence, its characteristics are closely linked to the cavity subpressure. The measurement results from point M1, the cavity subpressure, and the air pressure gradient force under different atmospheric pressures are listed in Table 4. The ventilation volume decreased with the reduction in discharge. The trend of changes in ventilation volume and cavity subpressure was essentially consistent with the variation in the incoming discharge. The larger the subpressure inside the cavity under the water jet, the stronger the intake capacity of the ventilation hole. Taking the air pressure gradient force from point M1 to the ventilation hole as an example, lower atmospheric pressure resulted in decreased air density and cavity subpressure. However, since the decrease in cavity subpressure was much greater than that in atmospheric air density, the air pressure gradient force noticeably diminished with decreased atmospheric pressure.

Table 4. Cavity subpressure values and pressure gradient forces under different atmospheric pressures.

Measurement Point		Atmospheric Pressure (kPa)					
		101.3	**86.3**	**71.3**	**56.3**	**41.3**	**26.3**
M1	Measurement Value (kPa)	101.108	86.166	71.230	56.259	41.275	26.294
	Subpressure Value(kPa)	0.192	0.134	0.070	0.041	0.025	0.006
Pressure gradient force (N/kg)		0.908	0.728	0.458	0.339	0.281	0.106

3.2. Cavity Length

The length of the cavity is an important parameter in the design of an aeration device. Both the ventilation volume and the effective protection length of the aeration device are influenced by it. The longer the cavity, the greater the ventilation volume, the more fully the water flow is aerated, and the longer the aeration protection distance. The variation in cavity length under different atmospheric pressures is shown in Figure 4.

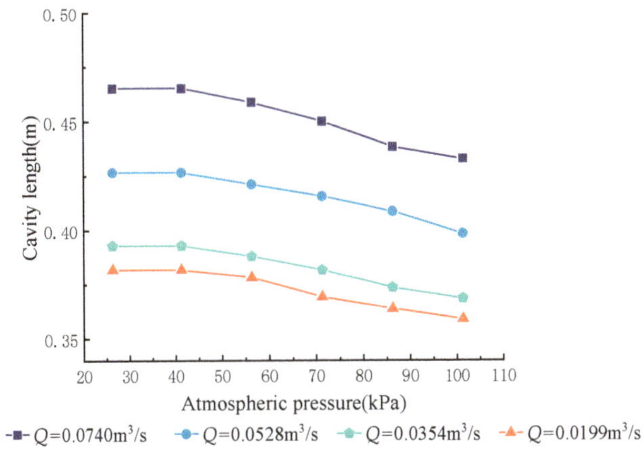

Figure 4. Cavity length under different atmospheric pressures.

The results indicated that, in the atmospheric pressure range of 101.3 to 41.3 kPa, the cavity length showed an increasing trend as the atmospheric pressure decreased. For every 15 kPa decrease in atmospheric pressure, the cavity length increased by an average of 1.68%. The growth amplitude was correlated with the change in the cavity negative pressure. As the atmospheric pressure decreased, the reduction in the cavity negative pressure was more significant, leading to a larger growth amplitude in the cavity length. The subpressure in the cavity exerts a downward drag force on the water flow, and the larger the subpressure, the shorter the cavity [31]. As seen from the previous section's results, the subpressure in the cavity showed a significant decrease with the decrease in atmospheric pressure. When the atmospheric pressure dropped below 41.3 kPa, the subpressure inside the cavity was almost equal to the ambient atmospheric pressure. Neglecting the effects of air resistance and surface tension, at this point, the ejected water could be considered to be only subject to the force of gravity. Therefore, the length of the cavity tended to increase as the atmospheric pressure dropped, and the extent of growth was related to the decrease in negative pressure within the cavity.

3.3. Correction of Aeration Coefficient under Low Atmospheric Pressure

The calculation of the ventilation volume in aeration devices is a crucial issue in the study of using air entrainment to alleviate cavitation. It also serves as the basis for the rational arrangement of aeration devices. Through dimensional analysis, the widely applicable formula for calculating ventilation volume in aeration devices takes the form of Equation (2). The empirical formulas proposed by Bruschin [12] and Pan et al. [32] were of the above form.

$$q_a = KvL \tag{2}$$

where q_a is the ventilation volume per unit width, K is the aeration coefficient, v is the average velocity at the flip bucket, and L is the length of the cavity.

The aeration coefficient K value is currently widely regarded as a constant related to the dimensions of the aeration device. Existing preliminary research indicated that the impact of atmospheric pressure on cavity length is relatively minor. Therefore, changes in atmospheric pressure have a significant effect on the aeration coefficient K value, necessitating adjustments; otherwise, significant errors will arise. The relationship between these two factors at different discharges is illustrated in Figure 5. Decreased atmospheric pressure resulted in lower air density, while K exhibited a linear decrease.

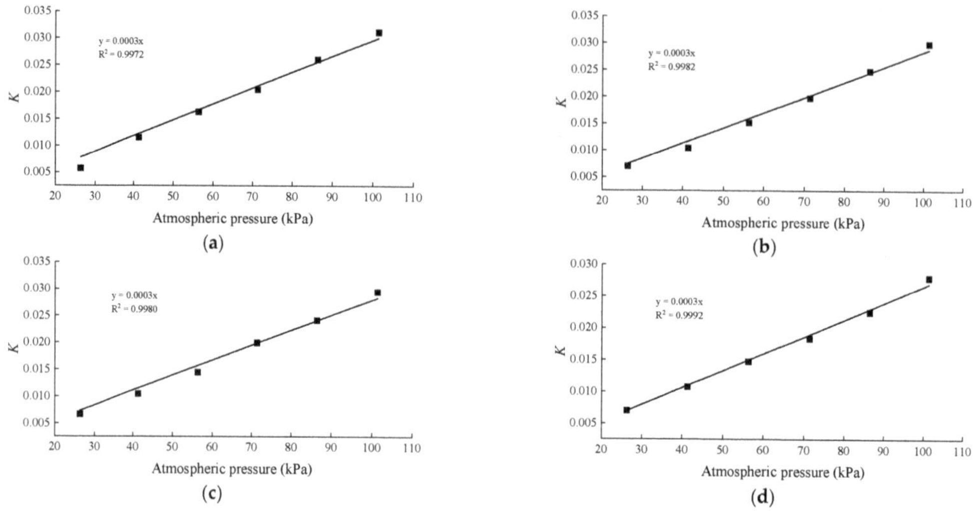

Figure 5. The correlation between the aeration coefficient K and changes in atmospheric pressure at different discharges. (**a**) $Q = 0.0740\ m^3/s$; (**b**) $Q = 0.0528\ m^3/s$; (**c**) $Q = 0.0354\ m^3/s$; (**d**) $Q = 0.0199\ m^3/s$.

Through regression analysis of 20 sets of test data with different discharges and atmospheric pressures, and incorporating corrections based on variations in cavity length, the modified calculation Equation (3) for the aeration coefficient $K\prime$ accounting for low atmospheric pressure was derived.

$$K\prime = 0.935\frac{p}{p_0}K_0 \tag{3}$$

where p is the atmospheric pressure; p_0 is the standard atmospheric pressure; K_0 is the aeration coefficient at standard pressure.

To verify the effect of the correction equation, experiments were conducted by lowering the flip bucket height by 1 cm from Figure 2; the correction Equation (3) was substituted into Equation (2), and the data comparison between the corrected value of the calculated ventilation volume and the test value is plotted in Figure 6. The maximum error did not exceed 10%, indicating the accuracy of the rapid estimation of the ventilation volume in aeration devices under low atmospheric pressure.

Currently, in high-altitude areas, especially those above 3000 m, there are relatively limited prototype observational data for hydraulic engineering. To verify Equation (3), model test data (scale 1:40) and prototype data from the Baihetan Hydropower Station spillway model were compared. The Baihetan Hydropower Station has a maximum dam height of 289 m, an altitude of 825 m, an environmental air pressure of approximately 91.6 kPa, a normal reservoir water level of 825 m, and a discharge of 3900 m^3/s. The calculated results are presented in Table 5. The error between the model test results of the single-width aeration flow and the prototype measured data under normal pressure was 19.60%. However, when using the corrected Equation (3), the error of the calculated single-width aeration flow was only 1.72%, demonstrating higher accuracy.

Table 5. Comparison of calculated values from equation (3) with experimental model values.

Upstream Flow Velocity (m/s)	Prototype q_a	Model Test q_a	Model Test Error (%)	Calculation of q_a Using Equation (3)	Calculation Error (%) Using Equation (3)
25	16.2	19.38	19.60	16.48	1.72

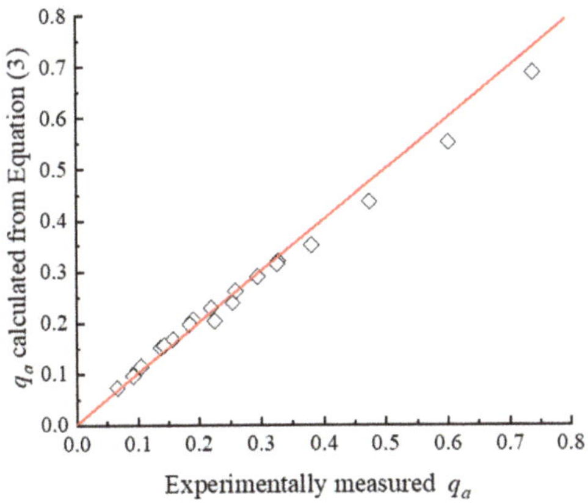

Figure 6. Comparison between the ventilation volume correction values calculated using Equation (3) and the experimental measurements.

4. The Effect of Atmospheric Pressure on the Air Concentration in Water

Air concentration is defined as the ratio of the volume of air to the volume of the air–water mixture in the aerated water flow. It is generally considered that when the air concentration in water is in the range of 1.5% to 2.5%, the damage caused by solid boundary cavitation is significantly reduced. When it reaches 3% to 5%, the damage from cavitation can be avoided. Experiments conducted by Peterka showed that when the air concentration in water close to the channel bottom was 2%, it significantly alleviated cavitation damage to solid boundary walls, and when the air concentration reached 7.4%, cavitation damage completely disappeared [33]. Russell and Sheehan conducted experimental research on the surface cavitation of concrete under different air concentrations. The results of the experiments indicated that an air concentration of 2.8% was sufficient to prevent cavitation damage to the concrete surface. Even in situations where severe air-entrained damage to the concrete surface was anticipated, an air concentration of 5% was enough to avoid cavitation damage [34]. Due to the decreased atmospheric pressure, there is a significant reduction in the ventilation volume of the aeration devices. This is bound to result in a weakening effect on the air concentration in the downstream water bodies, which is unfavorable for using air entrainment to alleviate cavitation.

The variation trend of the air concentration in characteristic downstream water bodies of the aeration device under different operating conditions with changes in atmospheric pressure is depicted in Figure 7.

The results indicated that the air concentration within the cavity remained constant regardless of changes in atmospheric pressure under different discharge conditions (Figure 7a). However, the air concentration at the other measurement points decreased with lower atmospheric pressure. When the atmospheric pressure decreased from 101.3 kPa to 26.3 kPa, the minimum reduction in air concentration in the cavity backwater area was 1.72%, and the maximum reduction was 14.9%. The minimum reduction in air concentration at the end of the cavity was 32.7%, and the maximum reduction could reach 38.5%. The minimum reduction in air concentration in the bubble escape section downstream of the cavity was 25.8%, and the maximum reduction could reach 38.3%.

Under various discharge conditions, with a uniform decrease in atmospheric pressure, there was a turning point in the extent of reduction in air concentration in the water at the end of the cavity's backwater region (Figure 7b). In the range of atmospheric pressure from 101.3 kPa to 71.3 kPa, for every 15 kPa decrease in atmospheric pressure, the air concentration only decreased by an average of 0.10%, showing a minor reduction. In the range of atmospheric pressure from 56.3 kPa to 26.3 kPa, for every 15 kPa decrease in atmospheric pressure, the air concentration decreased by 0.48% to 1.71%. Across different discharges, the average reduction was about 1.1%, indicating a relatively larger reduction.

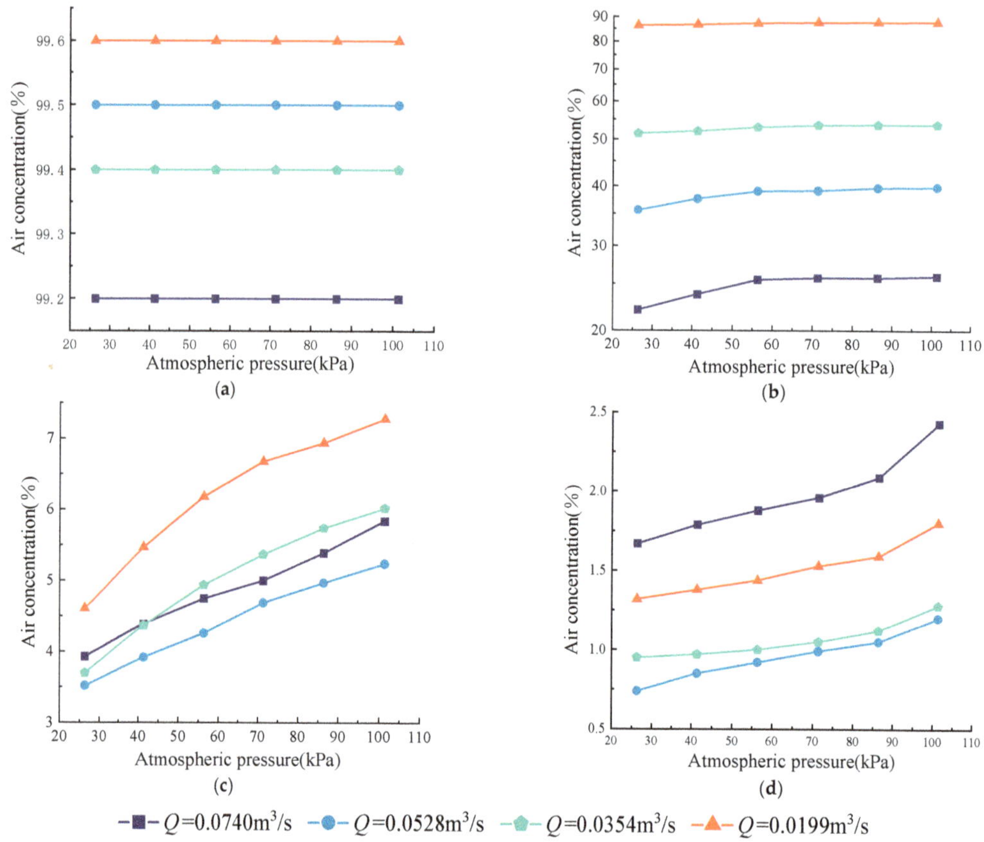

Figure 7. The relationship between air concentration in characteristic downstream water regions of an aeration device and atmospheric pressure under different operating conditions. (**a**) Inside the cavity; (**b**) cavity backwater area; (**c**) cavity end; (**d**) downstream bubble escape section of the cavity.

The air concentration at the end of the cavity exhibited a favorable linear relationship with atmospheric pressure (Figure 7c). With every 15 kPa decrease in atmospheric pressure, the air concentration decreased by approximately 0.34% to 0.54%, accounting for around 6.5% to 7.7% of the air concentration at normal atmospheric pressure.

Under different discharge conditions, with a uniform decrease in atmospheric pressure, there is a turning point in the extent of reduction in air concentration in the water within the cavity's downstream bubble escape section (Figure 7d). When atmospheric pressure was within the range of 101.3 kPa to 86.3 kPa, a significant reduction in the water air concentration was observed. With a 15 kPa decrease in atmospheric pressure, the air concentration decreased by 0.18% to 0.34% under various discharge conditions. This reduction accounted for about 11.7% to 15.3% of the air concentration at normal atmospheric pressure, with an average proportion of 13.7%. When atmospheric pressure was within the range of 86.3 kPa to 26.3 kPa, there was a favorable linear relationship between air concentration and atmospheric pressure. For every 15 kPa decrease in atmospheric pressure, the air concentration decreased by about 0.04% to 0.10% under different discharges. This reduction accounted for about 3.3% to 6.3% of the air concentration at normal atmospheric pressure, with an average proportion of 4.4%. And in this region, the rate of decline in the air concentration along the path was relatively fast under low air pressure. The decrease in atmospheric pressure may lead to a quicker reduction in the downstream along-path air concentration to below the minimum erosion-free air concentration, which is unfavorable for aeration protection. By assessing the correlation between the air concentration and atmospheric pressure in different characteristic regions of the water bodies, it can be inferred that air concentration decreases at lower atmospheric pressure, and the reduction trend varies depending on the initial air concentration. Specifically, when the water air concentration is higher, as the atmospheric pressure decreases, the reduction in air concentration initially shows a small decrease and then shows a larger decrease. Conversely, as the atmospheric pressure decreases, the reduction in air concentration exhibits an initial larger decrease followed by a smaller decrease when the water air concentration is lower. The current understanding of the air concentration decreasing with decreasing atmospheric pressure is as follows: Due to the decrease in atmospheric pressure, the surface tension of the water increases, making turbulent fragmentation more difficult. As a result, the exchange between the flowing water and the surrounding medium decreases during motion, resulting in a reduction in air entrainment. Turbulent diffusion is the fundamental reason for air entrainment. With decreased atmospheric pressure and increased surface tension, water particles require greater turbulent energy to overcome surface tension. Therefore, it is more challenging to achieve air entrainment under low atmospheric pressure.

The air concentration in the water of the downstream bubble escape region of the cavity is a crucial basis for analyzing the effectiveness of aeration device protection. As the air concentration in this area is often low, the impact of decreased atmospheric pressure on it is relatively significant. Based on regression analysis from depressurization generalized modeling experiments, considering the effect of atmospheric pressure, a correction formula for the air concentration C in the downstream bubble escape section of the aeration device cavity has been formulated as Equation (4).

$$C = \begin{cases} 0.0573(p_0 - p)C_0 & 86.3 \leq \mathrm{p} < 101.3 \\ [0.8978 - 0.0025(p_0 - p)]C_0 & 26.3 \leq \mathrm{p} < 86.3 \end{cases} \quad (4)$$

where C_0 is the air concentration in the water under normal atmospheric pressure, which can be obtained through model experiments, numerical simulations, or empirical formulas based on prototype observations; p is the atmospheric pressure; p_0 is the standard atmospheric pressure.

5. Conclusions

In this study, depressurized generalized model experiments comprising six different atmospheric pressure conditions and four discharges were conducted, resulting in a total of twenty-four operating conditions. This research aimed to investigate the variations in ventilation hole air velocity, ventilation volume, and air concentration under low atmospheric pressure conditions. Additionally, a low-pressure correction method for the aeration coefficient and water air concentration was proposed. The main conclusions are as follows:

(1) There is a favorable linear relationship between atmospheric pressure and the airflow velocity of the ventilation hole. With a decrease in atmospheric pressure, the airflow velocity and ventilation volume through the hole decreased. For every 15 kPa decrease in atmospheric pressure, the air velocity decreased by an average of 24.6%. Additionally, with the decrease in atmospheric pressure, the subpressure in the cavity significantly decreased, and there was a trend of increasing cavity length.

(2) When the atmospheric pressure decreased from 101.3 kPa to 26.3 kPa, the maximum reduction in air concentration in the cavity backwater area was 14.9%. The minimum reduction in air concentration at the end of the cavity was 32.7%, and the maximum reduction could reach 38.5%. The minimum reduction in air concentration in the

bubble escape section downstream of the cavity was 25.8%, and the maximum reduction could reach 38.3%. As the atmospheric pressure decreased, there was a trend of decreasing air concentration in the water, and the extent of this decrease depended on the initial air concentration. When the water air concentration was higher, the reduction was initially small, followed by a more significant decrease. Conversely, when the water air concentration was lower, the reduction showed an initially larger decrease followed by a smaller decrease.

(3) A low-atmospheric-pressure correction method for the aeration coefficient and water air concentration in the bubble escape section has been proposed. This method can be applied for the rapid estimation of ventilation volume and air concentration in aeration devices under low atmospheric pressure.

Author Contributions: Conceptualization, L.Z.; methodology, L.Z. and S.L.; formal analysis, L.Y. and Y.G.; data curation, C.L. and Y.L.; visualization, Y.G.; writing—original draft, Y.G.; writing—review and editing, L.Z. and L.Y. All authors have read and agreed to the published version of the manuscript.

Funding: This research was funded by the National Key R&D Program of China (2021YFC3090105) and the Special Research Fund of Nanjing Hydraulic Research Institute (Y123008).

Data Availability Statement: The data presented in this study are available on request from the corresponding author.

Conflicts of Interest: The authors declare no conflict of interest.

References

1. Qi, C.F. Intensive Study of Aeration and Ventilation Characteristics of Aeration Devices and Air Supply System for Hydraulic Structures. Ph.D. Thesis, Tianjin University, Tianjin, China, 2019.
2. Bhosekar, V.V.; Jothiprakash, V.; Deolalikar, P.B. Orifice spillway aerator: Hydraulic design. *J. Hydraul. Eng.* **2012**, *138*, 563–572. [CrossRef]
3. Falvey, H.T. *Cavitation in Chutes and Spillways*; Engineering Monograph 42; United States Department of the Interior, Bureau of Reclamation: Denver, CO, USA, 1990.
4. Guo, J.; Zhang, D.; Liu, Z.P.; Fan, L. Achievements on hydraulic problems in large spillway tunnel with a high head and large discharge flow and its risk analysis. *J. Hydraul. Eng.* **2006**, *10*, 1193–1198.
5. Lin, J.Y.; Zhang, S.R. *Hydraulic Structure*, 6th ed.; China Water & Power Press: Beijing, China, 2019; ISBN 978-7-5170-7791-6. (In Chinese)
6. Yang, J.; Li, S.C.; Lin, C. Plausible Differences between the Laboratory and Prototype Behaviors of Spillway Aerator Flows. *Water* **2022**, *14*, 3264. [CrossRef]
7. Wu, J.H.; Luo, C. Effects of Entrained Air Manner on Cavitation Damage. *J. Hydrodyn.* **2011**, *23*, 333–338. [CrossRef]
8. Wang, Y.J.; Zhang, L.C.; Luo, S.Z. Influence of Environmental Pressure Drop on Surface Tension Coefficient of Water. *Sci. Technol. Eng.* **2017**, *17*, 136–138. (In Chinese)
9. Pang, B.H. Influence of atmospheric pressure on cavitation characteristics of high velocity flow at high altitude. *J. Yangtze River Sci. Res. Inst.* **2019**, *36*, 64–67. (In Chinese) [CrossRef]
10. Liu, Z.C.; Wang, X.Q.; Fan, W.J.; Yv, K. Study on Some Properties of Hydrodynamic Cavitation Damage. In Proceedings of the 4th International Conference on Renewable Energy Technologies, Kuala Lumpur, Malaysia, 16–18 January 2018.
11. Zabaleta, F.; Bombardelli, F.A.; Toro, J.P. Towards an understanding of the mechanisms leading to air entrainment in the skimming flow over stepped spillways. *Environ. Fluid Mech.* **2020**, *20*, 375–392. [CrossRef]
12. Bruschin, J. Forced Aeration of High Velocity Flows. *J. Hydraul. Res.* **1987**, *25*, 5–14. [CrossRef]
13. Chanson, H. Study of air entrainment and aeration devices. *J. Hydraul. Res.* **1989**, *27*, 301–319. [CrossRef]
14. Luna-Bahena, J.C.; Pozos-Estrada, O.; Ortiz-Martínez, V.M.; Gracia-Sánchez, J. Experimental Investigation of Artificial Aeration on a Smooth Spillway with a Crest Pier. *Water* **2018**, *10*, 1383. [CrossRef]
15. Rutschmann, P.; Hager, W.H. Air entrainment by spillway aerators. *J. Hydraul. Eng. ASCE* **1990**, *116*, 765–782. [CrossRef]
16. Wu, J.H.; Fan, B.; Xu, W.L. Inverse problem of bottom slope design for aerator devices. *J. Hydrodyn.* **2013**, *25*, 805–808. [CrossRef]
17. Bai, R.D.; Liu, C.; Feng, B.Y.; Liu, S.J.; Zhang, F.X. Development of Bubble Characteristics on Chute Spillway Bottom. *Water* **2018**, *10*, 1129. [CrossRef]
18. Liu, X.C.; Bai, R.D.; Liu, S.J. Air entrainment scale effects in chute aerator flows. *Eur. J. Mech. B Fluids* **2023**, *99*, 163–172. [CrossRef]
19. Lian, J.J.; Qi, C.F.; Liu, F.; Gou, W.J.; Pan, S.Q.; Ouyang, Q.A. Air Entrainment and Air Demand in the Spillway Tunnel at the Jinping-I Dam. *Appl. Sci.* **2017**, *7*, 930. [CrossRef]

20. Zhang, L.C.; Luo, S.Z.; Wang, X. Experimental study of air entrainment to alleviate cavitations for chute with high-speed discharge. *Eng. Mech.* **2013**, *30*, 329–332. [CrossRef]
21. Yang, Y.S.; Chen, C.Z.; Yu, Q.Y. A mathematical model for self-aeration capacity of free jet on the aerator. *J. Hydraul. Eng.* **1996**, *3*, 13–21.
22. Pfister, M.; Hager, W.H.; Minor, H. Bottom Aeration of Stepped Spillways. *J. Hydraul. Eng.* **2006**, *132*, 850–853. [CrossRef]
23. Kells, J.A.; Smith, C.D. Reduction of cavitation on spillways by induced air entrainment. *Can. J. Civ. Eng.* **2011**, *18*, 358–377. [CrossRef]
24. Wei, L.; Yang, Q.; Dai, G.Q.; Du, Z.Y.; Zhong, J. Experimental Study on Ventilation Volume of the Flip Bucket. *Adv. Eng. Sci.* **2015**, *47*, 18–23.
25. Wang, Y.M.; Deng, J.; Wei, W.R. Investigation on the Cavity Backwater of Chute Aerators under Various Atmospheric Pressures. *Water* **2022**, *14*, 1513. [CrossRef]
26. Pang, B.H.; Zhang, L.C.; Lu, J.; Chi, F.D. Influence of Atmospheric Pressure on Distribution of Gas Nuclei in Water Body. *Water Resour. Power* **2018**, *36*, 30–33. (In Chinese)
27. Wu, J.H. Cavitation and its damage in hydro-electric engineering projects. In Proceedings of the 18th National Conference on Hydrodynamics (NCHD), Urumqi, China, 9–26 August 2004. (In Chinese).
28. Zhao, J.P.; Hu, Y.A. *Research on Pressure Reduction Model Testing for the Tail Section of the Spillway Tunnel during the Technical Implementation Phase of the Baihetan Hydropower Station*; Nanjing Hydraulic Research Institute: Nanjing, China, 2020. (In Chinese)
29. Pang, B.H.; Zhang, L.C.; Chi, F.D. Analysis of Ambient Air Pressure Effect on Cavity Length of Aeration Facility. *Water Resour. Power* **2020**, *38*, 114–116+121. (In Chinese)
30. Xue, B.; Zhao, Y.P.; Liao, W.L.; Li, Z.H. Numerical study on the influence of high altitude environment on hydraulic mechanical performance. In Proceedings of the IAHR's Symposium on Hydraulic Machinery and Systems, Beijing, China, 16–19 November 2017.
31. Wu, J.H.; Ruan, S.P. Cavity length below chute aerators. *Sci. China Ser. E-Technol. Sci.* **2008**, *51*, 170–178. [CrossRef]
32. Pan, S.B.; Shao, Y.Y.; Shi, Q.S.; Dong, X.L. The self-aeration capacity of the water jet over the aeration ramp. *J. Hydraul. Eng.* **1980**, *5*, 13–22. (In Chinese)
33. Peterka, A.J. The effect of entrained air on cavitation pitting. In Proceedings of the Minnesota International Hydraulic Convention, Minneapolis, MN, USA, 1–4 September 1953.
34. Russell, S.O.; Sheehan, G.J. Effect of Entrained Air on Cavitation Damage. *Can. J. Civ. Eng.* **1974**, *1*, 97–107. [CrossRef]

water

MDPI

Article

A Methodology for Water Resource Management and the Planning of the Coastal Reservoir in Indonesia

Indratmo Soekarno [1], Mohammad Bagus Adityawan [2,3,*], Calvin Sandi [4], Salma Izzatu Amatullah [3], Mohammad Farid [2,5], Yadi Suryadi [1], Bagus Pramono Yakti [2], Asrini Chrysanti [5], Arno Adi Kuntoro [2], Widyaningtias [2] and Muhammad Rizki Purnama [4]

1 Water Resources Engineering Research Group, Faculty of Civil and Environmental Engineering, Institut Teknologi Bandung, Jalan Ganesha 10, Bandung 40132, West Java, Indonesia; indratmosoekarno@yahoo.com (I.S.); yadisuryadi1991@gmail.com (Y.S.)

2 Center for Water Resources Development, Faculty of Civil and Environmental Engineering, Institut Teknologi Bandung, Jalan Ganesha 10, Bandung 40132, West Java, Indonesia; mfarid@ftsl.itb.ac.id (M.F.); bagusyakti@gmail.com (B.P.Y.); arnoak@ftsl.itb.ac.id (A.A.K.); widya@ftsl.itb.ac.id (W.)

3 Undergraduate School of Water Resources Engineering and Management, Faculty of Civil and Environmental Engineering, Institut Teknologi Bandung, Jalan Ganesha 10, Bandung 40132, West Java, Indonesia; izzatu.salma@gmail.com

4 Graduate School of Water Resources Management, Faculty of Civil and Environmental Engineering, Institut Teknologi Bandung, Jalan Ganesha 10, Bandung 40132, West Java, Indonesia; calvinsandi458@gmail.com (C.S.); mrizkipurnama@outlook.com (M.R.P.)

5 Center for Coastal and Marine Development, Faculty of Civil and Environmental Engineering, Institut Teknologi Bandung, Jalan Ganesha 10, Bandung 40132, West Java, Indonesia; asrini@ftsl.itb.ac.id

* Correspondence: mbagusadityawan@gmail.com

Abstract: Metropolitan areas in Indonesia, specifically Jabodetabek, are projected to continue experiencing population growth. This will result in increased water demand, which, if left unchecked, could lead to a water crisis. Additionally, the region faces persistent flooding issues. To mitigate these challenges, the Indonesian government proposes the creation of a coastal reservoir. However, a methodology for the pre-feasibility study of such reservoirs in Indonesia is currently lacking. This study addresses this gap by presenting a methodological framework encompassing hydrology and water allocation analyses as well as numerical modeling related to hydraulic and sedimentation aspects of coastal reservoirs. The methodology was designed to be adaptable based on data availability. Finally, this methodology was applied to a case study located at the Cisadane River estuary. The water allocation simulation showed that the reservoir can release an annual optimum discharge of 80.618 m^3/s in wet years and 22.731 m^3/s in dry years. Flood modeling demonstrated a modest impact on reducing early-hour floods, with a water level decrease of around 20–50 cm within 5 km of the reservoir. Furthermore, sediment modeling revealed varying sedimentation and erosion depths along the rivers, with approximately 62,300 m^3/year settling in the reservoir.

Keywords: coastal reservoir; methodology development; water balance; flood; sedimentation

1. Introduction

Jakarta, Bogor, Depok, Tangerang, and Bekasi (Jabodetabek) are metropolitan areas that continue to experience rapid population growth and industrial development. This has an impact on various aspects, including decreasing land availability, increasing demand for raw water, and continuously rising pollution and waste [1,2]. These factors significantly affect the condition of natural resources, particularly water resources. For instance, a study by [3] highlights the impact of coal mining on polluting water resources.

As a result, water resources will experience changes ranging from quantity to quality, which, if left unchecked, will cause a water crisis, especially during the dry season. A study conducted by the Indonesian government shows that the projected water demand for the

Jabodetabek area in 2030 will reach 110 m^3/s and in 2070 will reach 159 m^3/s. Therefore, adaptation is crucial as climate change intensifies and prolongs shortages [4]. Until now, piped water coverage in this area has remained at only 64% [5] and tends to decrease from year to year. Thus, massive groundwater extraction was conducted to meet increased water demands. This has caused land subsidence that has occurred for decades in this area [6–8]. A study by [9] shows that observed subsidence rates during the 1982–2010 period were about 1–15 cm/year and can be up to 20–28 cm/year at certain locations.

Apart from raw water problems, the Jabodetabek area also faces flooding issues. Over half of Jakarta's population resides in the coastal area in the northern part, which is vulnerable to floods from the riverside and seaside [10]. Floods have occurred not only in Jakarta but also in downstream areas around the Jakarta Bay area such as Tangerang and Bekasi. Various factors contribute to these floods, namely, river overflows and local rain; inadequate drainage systems; small water absorption areas, resulting in the potential for tidal flooding from the sea; and climate change, which increases flood inundation areas and depths [11,12]. Furthermore, land subsidence resulting from groundwater extraction amplifies vulnerability to tidal flooding [13,14].

Thus, to address these problems, the Indonesian government has planned a structural countermeasure, namely, a coastal reservoir, which is part of the National Capital Integrated Coastal Development Program (NCICD). Coastal reservoirs, successfully implemented in many countries [15], offer significant benefits for freshwater supply to coastal cities with growing populations [16,17]. This reservoir is planned to be built at the Cisadane River estuary, which will consist of a treated water reservoir (TWR) to address water shortage problems and an untreated water reservoir (UWR) to address flooding problems. In addition, another purpose of this infrastructure is to reduce groundwater extraction.

However, a methodology for the pre-feasibility study of the coastal reservoir in Indonesia has not been previously developed as this infrastructure is still rarely used in Indonesia. Therefore, this study addresses this gap by presenting a methodological framework for a pre-feasibility study of the coastal reservoir in Indonesia. The methodology involved an analysis of water needs and availability (water balance) as well as numerical modeling related to hydraulic and sedimentation modeling in the coastal reservoir. The analysis was carried out using several types of software, namely, WEAP 2023.0 and HEC-RAS 6.2. This methodology was then applied to a case study at the Cisadane River estuary to assess the feasibility and impact of the infrastructure through modeling since it is an essential step in the design process.

2. Methodology Development

2.1. Data Preparation

In the design of a coastal reservoir, the initial analysis focuses on the supply discharge from the river. This step is crucial for making decisions on water resource management [18]. Two types of discharge must be assessed: flood discharge and dependable discharge. Dependable discharge is the amount of discharge available to meet water needs with a calculated risk of failure [19]. This discharge can be calculated using two methods, depending on data availability. The first method involves using discharge observation data. If the discharge observation data are available, dependable discharge can be determined immediately. However, according to the Indonesia Hydrological Guidelines Standard [20], observed discharge data should consist of at least 20 years of data to accurately calculate dependable discharge. If this requirement is not met, a hydrological calculation (rainfall–runoff model) is needed to convert rainfall into runoff. The details of this process will be explained further in the next chapter.

Similarly, flood discharge can be calculated using the same method as that used for dependable discharge. The necessary discharge observation data are also required to cover a period of 20 years, or else a hydrological calculation is necessary. However, the methodology for calculating flood discharge is well-developed in the Indonesia Hydrological Guidelines Standard. Therefore, this study will not delve into detailed explanations regarding flood discharge calculation. To calculate the hydrological process, rainfall data with a minimum of 10 years of data are needed. However, in developing countries, including Indonesia, there are vast areas without rain gauges. Even in areas where rain gauges exist, the data are often incomplete. Hence, precipitation satellite products, namely, Tropical Rainfall Measuring Mission (TRMM), have been officially used in Indonesia.

However, the quality of TRMM data is less reliable in the tropical regions of Asia as a study conducted by [21] suggested that TRMM may not be suitable for applications without prior correction. Therefore, if the available rainfall data from rain gauges do not span 10 years, TRMM rainfall data can be utilized but must be calibrated to a rain gauge, as demonstrated by [22] in a hydrological analysis of upper Bengawan Solo River in Java Island. Additionally, other data required for hydrological calculations include land cover and climatology. The methodology framework is presented in Figure 1.

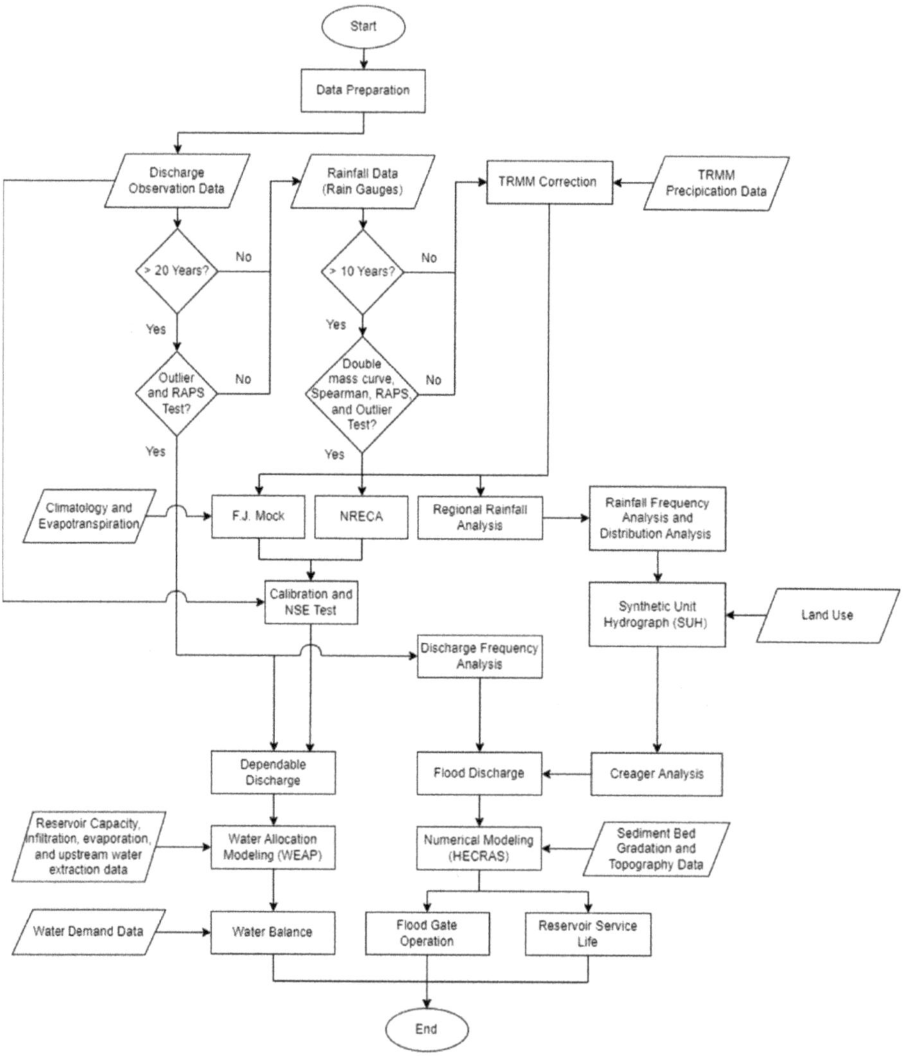

Figure 1. Methodology.

2.2. Hydrology Analysis

In a case where the discharge observation data fail to meet the requirement, the dependable discharge can be calculated using the F.J. Mock Model and National Rural Electric Cooperative Association (NRECA) rainfall data. The F.J. Mock Model is often chosen for its simplicity as a rainfall–runoff model with inflows, outflows, and groundwater storage considerations. The NRECA is a relatively simple rainfall–runoff model capable of calculating the percentage of surface runoff that enters a groundwater reservoir [18]. These two models have been widely used, especially in Indonesia [23–26].

Initially, rainfall data must be tested to ensure data quality with several methods, namely, Outlier, Stationer, the double mass curve test, and the Spearman trend test. If the data do not pass the test, correction using TRMM data is necessary. The regional rainfall is then calculated using the Polygon Thiessen Method for regional average rainfall data to calculate dependable discharge and the Polygon Thiessen method for regional maximum daily rainfall data to calculate flood discharge. In addition, land cover must be included in the analysis as it is an important factor in hydrological calculation that is used to calculate water losses due to infiltration. Research by [12,27] confirms that changes in land use result in increased flood discharge.

The dependable discharge model result is then calibrated using the observed discharge data. However, the observed discharge data should be tested with Outlier and a rescaled adjusted partial sums (RAPS) test before use. Calibration results for each model are then validated using Nash–Sutcliffe efficiency (NSE), with criteria provided by [28] (Table 1). The calibrated results should have a good interpretation or at least a satisfactory interpretation, and the best model, with the highest NSE value, should be chosen.

Table 1. NSE value interpretation.

NSE Value	Interpretation
NSE $\geq$ 0.75	Good
0.36 < NSE < 0.75	Satisfactory
NSE $\leq$ 0.36	Not Satisfied

For flood discharge, rainfall data are analyzed using a frequency analysis to obtain rainfall design for a certain return period. There are a lot of methods for frequency analysis. Gumbel, Normal, Log Normal, Pearson, and Log Pearson are the best frequency analysis methods according to [20]. Other methods not included in the standard can still be used if the results pass the Smirnov–Kolmogorov and chi-square tests. Multiple methods are recommended to obtain a distribution function with better confidence levels based on the lowest values of the Smirnov–Kolomogorov and chi-square test results. Thus, an optimum result can be obtained. The rainfall design is then distributed into hourly rainfall and used to calculate flood discharge using the synthetic unit hydrograph (SUH) method.

Currently, there are many referable SUH methods. Fortunately, the eligible SUH methods to be used in Indonesia have been determined in the Indonesia Hydrological Guidelines Standard [20], namely, the SCS (Soil Conservation Service), Snyder, Nakayasu, Limantara, GAMA-1, ITB-1, and ITB-2 unit hydrographs. Finally, to choose the appropriate SUH methods, the Creager curve is employed in the analysis. The Creager curve, developed by [29], serves as a tool to verify and select flood discharge from various SUH methods employed in an analysis. This curve is one of the most well-known methods and has been widely used in many hydrologic applications. Moreover, a study by [30] has conducted research that determines the Creager coefficient for Java Island.

2.3. Water Resource Allocation Modeling

The modeling of water resource allocation can be conducted using WEAP 2023.0 (water evaluation and planning) software. This software was made and developed by the Stockholm Environment Institute's U.S. Center. WEAP incorporates the necessary features to address water allocation needs. The algorithm integrates various elements, such as relative water flow and groundwater infiltration from rainfall, water conservation, water policy and allocation priorities, reservoir operations, hydropower generation, water quality, vulnerability assessment, and ecosystem requirements. Furthermore, this software has been extensively utilized in numerous studies [31–33]. Thus, this software is recommended for modeling water resource allocation.

The dependable discharge serves as an input for the model, along with other parameters such as the elevation–volume curve of reservoir capacity, infiltration, evaporation, etc. Additionally, upstream water extraction data must be incorporated into the model to prevent the overestimation of inflow discharge. The objective of this modeling is to determine a sustainable discharge that can be supplied and maintained throughout the year. This process may involve some trial and error to achieve the optimum discharge and sustain it for the following year. To become sustained, the water level elevation must be at the upper limit of effective storage at the beginning and end of the simulation, and it should not fall below the lower limit of effective storage throughout the year.

2.4. Numerical Model

Numerical models have been widely used in designing and analyzing the impact of hydraulic structures [29], including coastal reservoirs. In this study, numerical modeling was conducted to assess the reservoir's impact during periods of heavy rainfall, establish floodgate operation procedures, and calculate sediment rates to estimate the reservoir's service life. Numerical modeling software, such as HEC-RAS 6.2, MIKE Zero 2024, etc., can assist this process. However, despite the availability of several types of numerical modeling software, HEC-RAS 6.2 is preferred in Indonesia since it offers open-source software. For instance, studies by [34–38] used HEC-RAS 6.2 for their research in Indonesia. Moreover, HEC-RAS 6.2 has been widely used in numerous consultant projects as this software is well known among hydraulics practitioners in Indonesia, thus creating a large community of HEC-RAS 6.2 users in the country.

Given these considerations, we recommend the use of HEC-RAS software 6.2 in feasibility studies of coastal reservoirs in Indonesia. However, this does not imply restricting the use of other software. Alternative software, such as MIKE Zero 2024, are still eligible to use, provided that the results are thoroughly validated against the observation data. In this study, a HEC-RAS 6.2 one-dimensional model was used for the analysis. Developed by the US Army Corps of Engineers, HEC-RAS 6.2 is continuously improved and developed, standing as one of the most utilized modeling systems for analyzing channel flow and floodplain delineation [39]. Importantly, this software is capable of modeling sediment transport, which is notoriously difficult [40]. The flood discharge served as an input in the model, along with tidal, topography, and sediment bed gradation data.

3. Case Study

3.1. Research Location

The proposed coastal reservoir is located at the Cisadane River estuary in Tangerang Regency, Banten Province. The coastal reservoir scheme is given in Figure 2. The Cisadane River Basin covers a 1376 km^2 area with 126 km along the main river, as shown in Figure 3. The Cisadane River holds the potential to meet the water demand of Jabodetabek. However, the water quality of the Cisadane River is currently polluted due to industrial and domestic waste [41]. Therefore, waste water treatment is necessary before supplying water to the city. The coastal reservoir will consist of TWR, and a waste water treatment plan (WWTP) will be used to enhance water quality, resulting in raw water suitable for consumption. The treated raw water will then be distributed to surrounding residential and commercial areas.

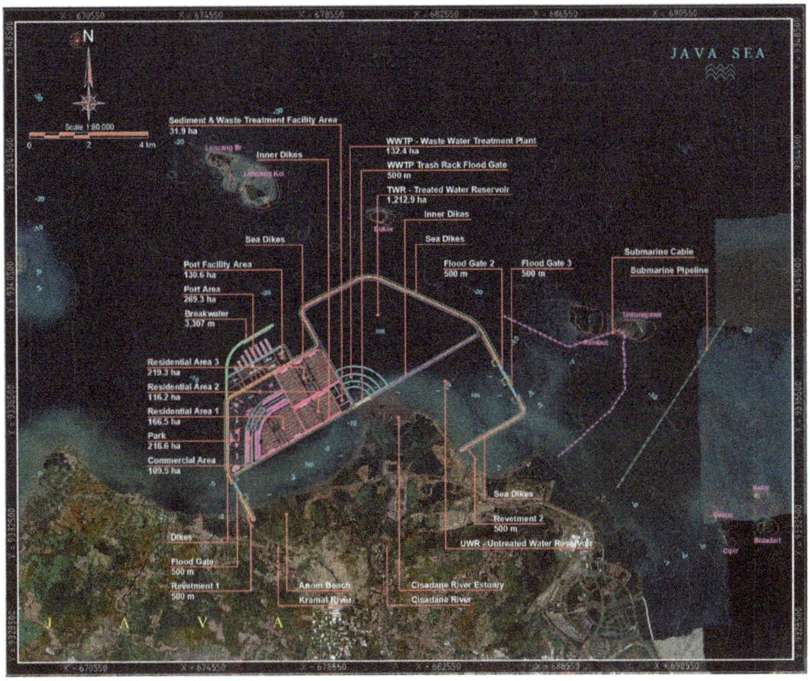

Figure 2. Coastal reservoir scheme.

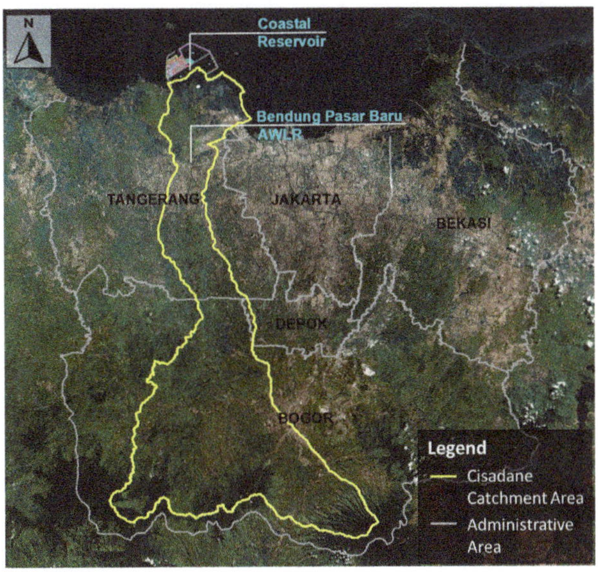

Figure 3. Catchment and administrative area of study location.

A flood control reservoir, referred to as the UWR, equipped with sea dikes and three identical flood gates is planned to mitigate flooding at the Cisadane River estuary. The total storage capacity of the UWR and TWR is 68.83 million m^3 and 202.78 million m^3, respectively, relative to +0 elevation. Sea dikes were built to prevent seawater contamination of the reservoir and mitigate tidal flooding. In addition, floodgates were built to prevent the overflow of UWR water. The operation of the flood gates is contingent on the reservoir and seawater levels, opening when the reservoir water level is above the seawater level. The capacity and area for each reservoir are shown in Table 2.

Table 2. Reservoir capacity and area.

	Volume (10^6 m^3)	Area (ha)
UWR	68.83	1739.69
TWR	202.78	1212.90

3.2. Hydrology Analysis

Hydrology analysis consists of dependable discharge and flood discharge calculations. Observed discharge data spanning 12 years were obtained from Bendung Pasar Baru. Due to limited observed discharge data, an empirical method was used to determine both dependable discharge and flood discharge, utilizing recorded rainfall data. The 10-year recorded rainfall data from Kracak, Cigudeg, Bendung Pasar Baru, Pasir Jaya, and Ranca Bungur Station were tested using Stationer and a double mass curve test. The Stationer test indicated that the entire recorded rainfall data were stationary. However, the double mass curve test indicated inconsistency for each recorded rainfall data station. Thus, a correction using TRMM data was necessary. Regional average and regional maximum daily rainfall data were then calculated using the Polygon Thiessen method.

The frequency analysis aimed to predict the maximum rainfall within a certain return period (rainfall design). Gumbel, Normal, Log Normal, and Log Pearson Type III methods were used for the frequency analysis in this research. The Smirnov–Kolmogorov test was conducted for every method, and all methods were found to be sufficient. Subsequently, a chi-square test was conducted to choose the most sufficient method. Based on the test results, the Gumbel method was selected with the lowest calculated chi-square value and was used in further analyses.

The flood discharge was calculated using several SUH methods, namely, SCS, Nakayasu, ITB-1, ITB-2, and GAMA-1 SUH. The peak discharge for each SUH method was then plotted on the Creager curve to verify and select the most appropriate flood discharge. The ITB-1 SUH was found to be the most appropriate according to the Creager curve, with the flood hydrograph results for the 100-year and 1000-year return periods presented in Figure 4.

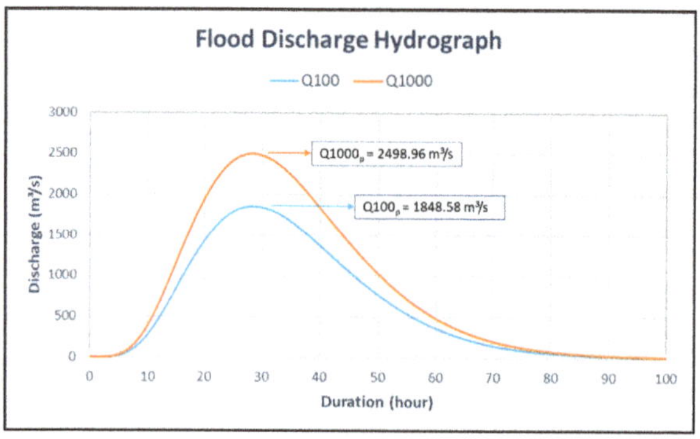

Figure 4. Flood discharge hydrograph.

The dependable discharge analysis was calculated using the F.J. Mock and NRECA models. Prior to analysis, evapotranspiration needed to be calculated as an input parameter for the model, referring to [42] and using the Penman–Monteith method. The model results were subsequently calibrated using recorded discharge data from Bendung Pasar Baru, with the NSE value as the calibrating parameter, as shown in Figure 5a. The NSE value for the F.J. Mock and NRECA models were 0.69 and 0.85, respectively. The NRECA model was chosen due to its higher NSE value with good interpretation. Consequently, the results from the NRECA model were used in water resource allocation modeling.

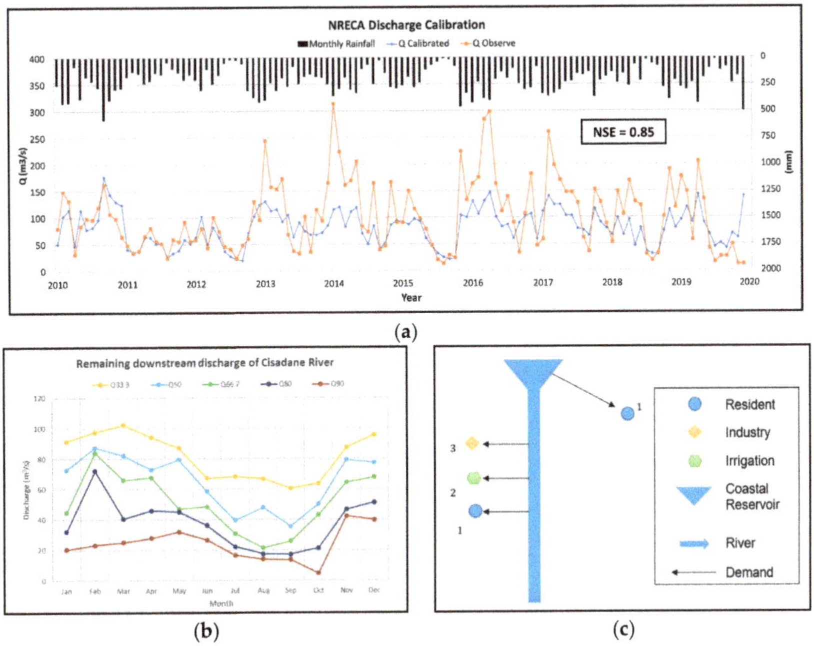

Figure 5. Dependable discharge analysis: (**a**) NRECA discharge calibration; (**b**) remaining downstream discharge of Cisadane River; (**c**) schematic model of WEAP.

3.3. Water Resource Allocation Modeling

After the calibrated discharge (existing condition) was obtained, an external factor, namely, water extraction, was added to the calculation. The water extraction data for upstream of the Cisadane River were from resident, irrigation, and industry data. Consequently, the dependable discharge was adjusted to account for upstream water extraction before being input into the WEAP model as the supply discharge for the coastal reservoir. The resultant downstream discharge of the Cisadane River is presented in Figure 5b. The schematic model of WEAP is illustrated in Figure 5c.

In this study, five dependable discharge scenarios were simulated based on their probabilities, namely, Q33.3, Q50, Q66.7, Q80, and Q90. The goal was to determine the optimal annual release discharge value from the TWR, ensuring that it maintained the upper limit of effective storage (normal water level) at the end of the simulation operation pattern and did not fall below the lower limit of effective storage throughout the year. The upper limit of effective storage for this coastal reservoir was at +0.00 m elevation, while the lower limit was at −18.20 m elevation.

Initially, the reservoir operation pattern was simulated with January as the initial simulation month. However, the results were not optimal, which was indicated by the large amount of water discharge wasted or overflowing into the sea through the floodgates. To mitigate this issue and reduce wasted water discharge, various simulations with different initial months were attempted. It was found that using May as the initial month yielded optimum results, characterized by the least wasted water discharge and the most supplied discharge. The WEAP simulation results of the reservoir operation pattern, along with the optimized discharge value, are presented in Figure 6, which were achieved through a trial and error process.

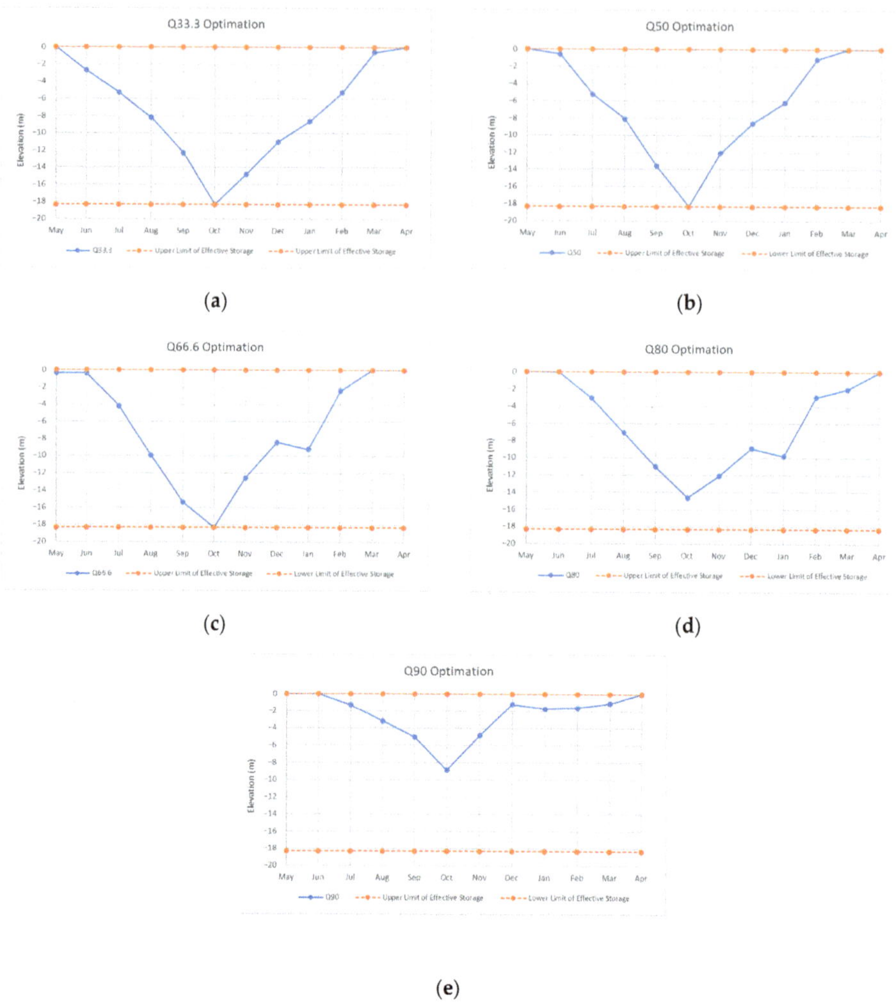

Figure 6. WEAP simulation results of reservoir operation pattern: (**a**) Q_{33} outflow = 80.618 m^3/s; (**b**) Q_{50} outflow = 61.660 m^3/s; (**c**) Q_{66} outflow = 48.817 m^3/s; (**d**) Q_{80} outflow = 36.313 m^3/s; (**e**) Q_{90} outflow = 22.731 m^3/s.

3.4. Numerical Modeling

The numerical modeling was divided into flood and sediment modeling. For flood modeling, floodgates were incorporated into the model, with five proposed sizes by the design team. All floodgates were required to be identical in size. The objective was to determine the most suitable yet cost-efficient size for the coastal reservoir. To evaluate each floodgate size's effectiveness, simulations were conducted under two conditions: flood design discharge with 100-year and 1000-year return periods, resulting in 10 scenarios, as shown in Table 3. In addition, an existing condition scenario (without the coastal reservoir) was simulated under flood design discharge of 100-year and 1000-year return periods to facilitate comparison and analyze the coastal reservoir's impact.

Table 3. Flood numerical modeling scenario.

Scenario	Q_{100}	Q_{1000}	Flood Gate Width (m)
1	v	-	500
2	-	v	
3	v	-	400
4	-	v	
5	v	-	300
6	-	v	
7	v	-	200
8	-	v	
9	v	-	100
10	-	v	

The modeling scope extended from Bendung Pasar Baru into the UWR. Cross-section data along the Cisadane River were obtained from a topographic survey. The modeled river's total length was approximately 22 km, with a 25 m distance between each cross section. The numerical model schematic set up is given in Figure 7. The simulation time was set at 100 h, with a time step of 10 s. The duration of the simulation was determined based on the overall time needed for the flood hydrograph. In our case, 100 h was the time needed for the flood hydrograph to reach its peak and return to its normal river discharge. Flood design discharge with 100-year and 1000-year return periods, derived from hydrology analysis, were used as the upstream boundary conditions. Tidal data for Jakarta Bay, obtained from two observation stations, were used as the downstream boundary condition. The model schematic setup is given in Figure 6. In addition, the floodgate operation rule was applied in the simulation to optimize the coastal reservoir's impact:

- When tide level > water level in the coastal reservoir, the floodgate is closed;
- When tide level ≤ water level in the coastal reservoir, the floodgate is opened.

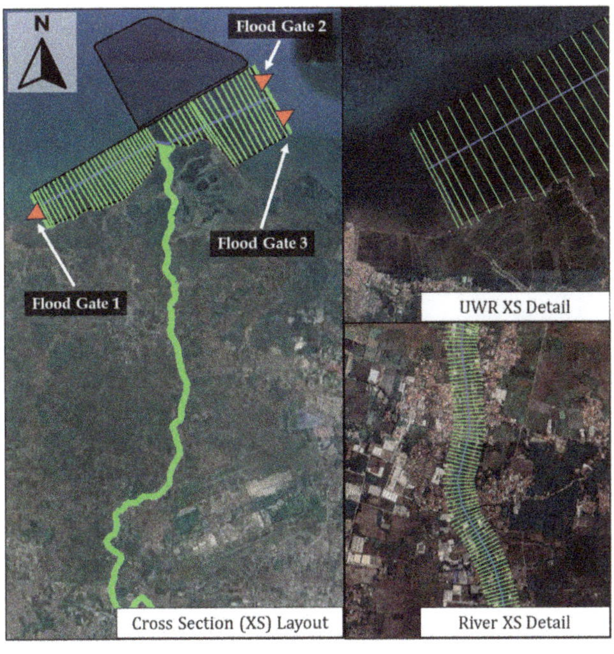

Figure 7. Numerical modeling schematic setup.

The numerical modeling results, comparing the existing scenario with the coastal reservoir scenario in a long-section view, are presented in Figure 8. In this figure, Scenario 1 is used as an example. The results indicated that the UWR had an impact on reducing early-hour flood levels, with a decrease in water levels of around 20–50 cm up to ±5 km upstream from the UWR, as shown in Figure 8a. However, under maximum Q100 conditions, the UWR had a minimal impact on reducing floods due to its small reservoir capacity compared to the flood volume of the Cisadane River, as shown in Figure 8b. Similar results were observed in each simulation scenario. The highest water level results for the UWR in each scenario are given in Figure 9. Figure 9a,b highlight a significant decrease in HWL between scenarios with 100 m, 200 m, and 300 m floodgates, with an up to 80 cm decrease in HWL between the 100 m and 300 m floodgate scenarios.

The decrease in HWL observed in the 400 m and 500 m floodgate scenarios was very insignificant. This indicated that the 300 m floodgate had reached its highest effectiveness. Thus, an increase in floodgate size beyond 300 m is unlikely to have a significant impact and tends to be cost-inefficient. In this case, the authors recommend using a 300 m floodgate for the coastal reservoir. A more detailed result is presented in Figure 10a, illustrating the comparison of water levels in the coastal reservoir and the sea, along with the total flow through the floodgates throughout the simulation. The negative value of the flow indicates that the discharge flowed into the coastal reservoir. Additionally, the modeling results were used to provide the floodgate operation, as shown in Figure 10b. At the beginning and end of the simulation, the floodgate was closed due to high tide levels and low river discharge. For the rest of the simulation, the floodgate was fully opened because of the large amount of flood discharge from the river.

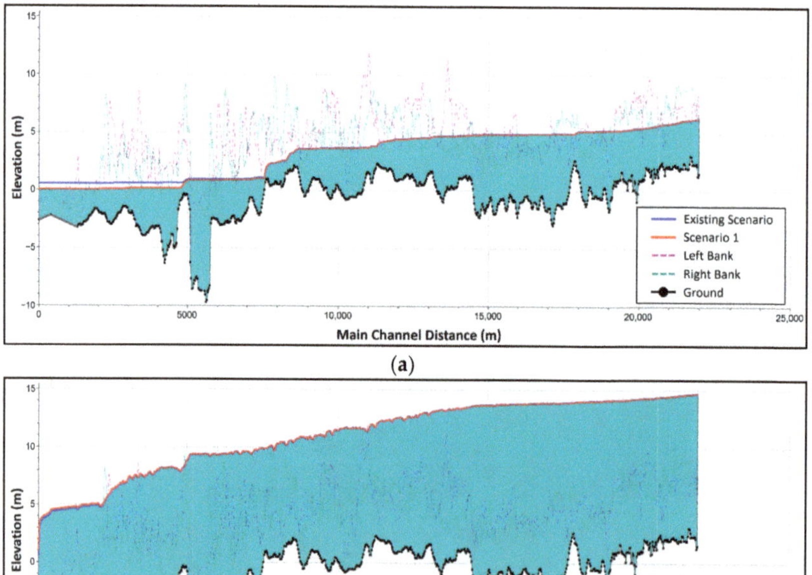

(**a**)

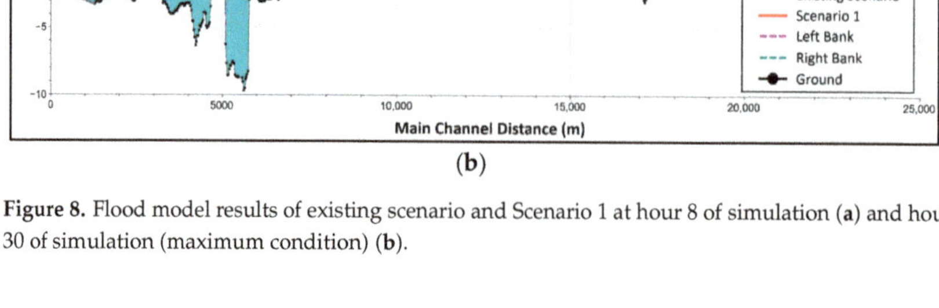

(**b**)

Figure 8. Flood model results of existing scenario and Scenario 1 at hour 8 of simulation (**a**) and hour 30 of simulation (maximum condition) (**b**).

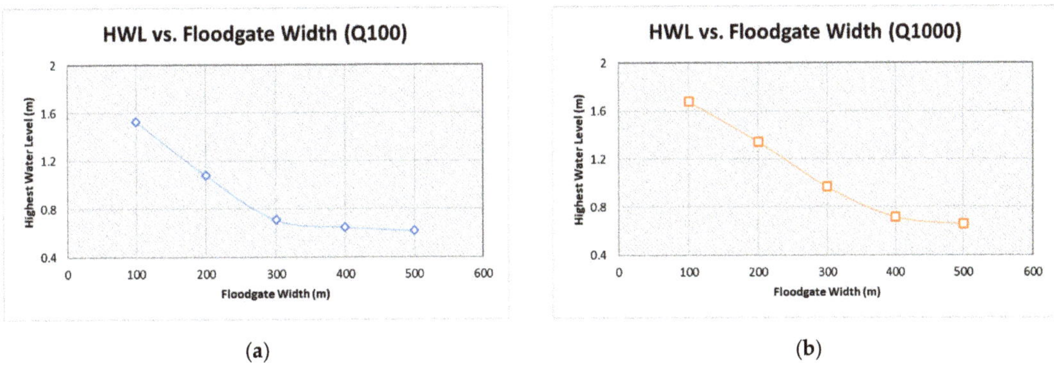

Figure 9. Highest water level and floodgate width correlation under two conditions: (**a**) Q100; (**b**) Q1000.

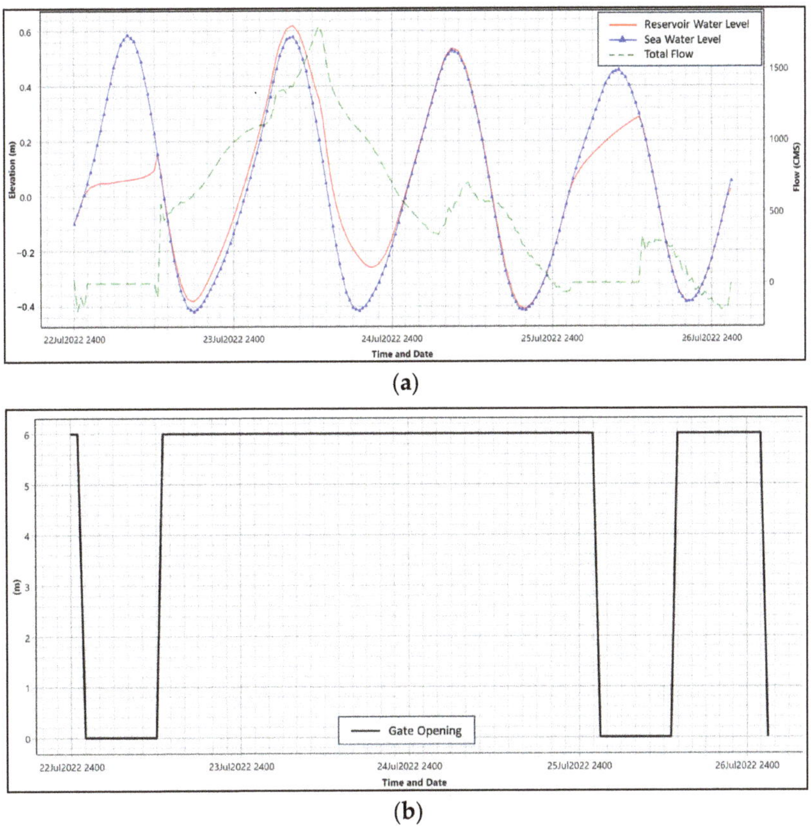

Figure 10. Result of flood simulation for Scenario 1: (**a**) reservoir water level and flow; (**b**) gate opening.

Finally, sediment modeling was conducted to simulate sediment transport in the UWR. The primary focus here was to analyze sediment deposition in the coastal reservoir. Thus, the floodgate was excluded from the model for simplicity, assuming that the floodgate would be continuously opened throughout the simulation. The model setup was the same as that for flood modeling, with dependable discharge at 50% probability used as the upstream boundary condition. Riverbed sediment gradation data were obtained from field measurements, as shown in Table 4, with three datasets classified as upstream, middle, and downstream segments. For the sediment boundary condition upstream, a rating curve of the flow–sediment load was utilized, which was also obtained from field measurements. To analyze the sediment behavior in the coastal reservoir, a long-term analysis was conducted, and the simulation time for sediment modeling was set at one year, with a time step of 1 min.

Table 4. River sediment gradation.

Segment	Percent Passing (%)			
	Gravel	Sand	Silt	Clay
Upstream	0	77.90	14.32	7.78
Middle	0	55.56	30.35	14.09
Downstream	0	49.72	33.75	16.53

The sediment modeling results, in a long-section view, are given in Figure 11. This figure shows the riverbed conditions at the beginning and end of the simulation. In general, there was substantial erosion along the Cisadane River and sedimentation in the UWR. Most parts of the Cisadane River experienced erosion, with approximately 90% of the riverbed parts eroded. The average erosion depth was 0.37 m, with a maximum erosion depth of 4.94 m. The volume of sediment deposited in the coastal reservoir was approximately 62,300 m^3/year or 0.35 cm/year of sediment layer, under the assumption that the sediment was evenly distributed along the reservoir.

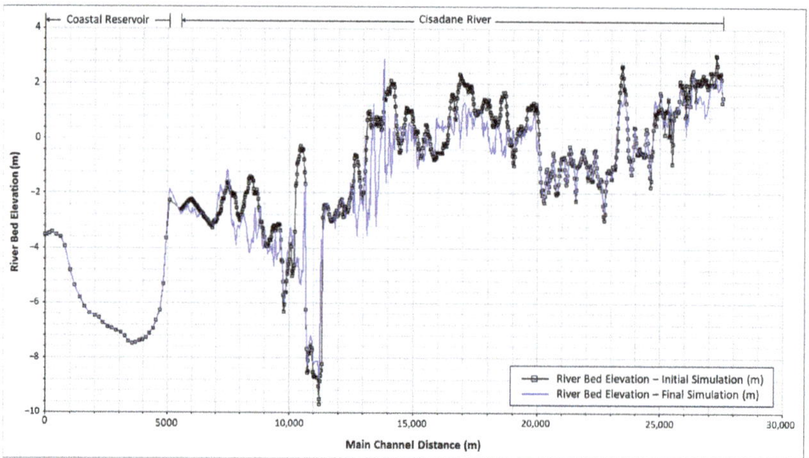

Figure 11. Cisadane riverbed changed for Q50%.

4. Conclusions

A methodology for the pre-feasibility study of coastal reservoirs in Indonesia has not been developed previously given the limited usage of such infrastructure in the country. Thus, this study addressed this gap by presenting a methodological framework for a pre-feasibility study and managing water resources of a coastal reservoir in Indonesia. The process encompassed hydrology analysis, water allocation modeling, and numerical modeling to determine dependable and flood discharge, conduct a water balance analysis, assess floodgate operation, and estimate reservoir service life. The methodology was designed to accommodate varying data availability, considering the common limitation of limited data in Indonesia.

Subsequently, we applied this methodology to a case study located at the Cisadane River estuary in Tangerang Regency, Banten Province. The water resource allocation simulation, considering different dependable discharge scenarios, revealed that the TWR could release an annual optimum discharge of 80.618 m^3/s during a wet year (Q33.3%) and 22.731 m^3/s during a dry year (Q90%). Flood modeling was simulated for both 100-year and 1000-year return periods, incorporating variations in floodgate size. The operation of the floodgate was based on the water levels of the UWR and the sea.

The flood modeling indicated that the UWR had a limited impact on reducing floods in the early hours of the simulation, with a decrease in water levels of around 20–50 cm up to $\pm$ 5 km upstream from the UWR. This impact was slight due to the small reservoir capacity compared to the flood volume of the Cisadane River. Similar results were observed in each simulation scenario. The water collected in the UWR is to supply the TWR during the dry season. Furthermore, sediment modeling was conducted for normal monthly discharge (Q50%) with a one-year simulation time, revealing sedimentation and erosion along the rivers with varying depths. Approximately 62,300 m^3/year of sediment volume or 0.35 cm/year of sediment layer is estimated to settle in the UWR.

To enhance the quality of the pre-feasibility study, the authors recommend using field measurement data rather than satellite projections, such as TRMM for precipitation data. Long-term and up-to-date time series data for discharge and precipitation are preferable for more credible results. In addition, the authors suggest installing new measuring instruments if the measurement instruments are inadequate or unavailable in the research location.

Author Contributions: Conceptualization, I.S. and M.B.A.; methodology, M.B.A., C.S., S.I.A. and M.F.; software, C.S. and S.I.A.; validation, C.S., S.I.A. and B.P.Y.; formal analysis, M.B.A. and M.F.; investigation, I.S. and M.F.; resources, Y.S., B.P.Y. and A.C.; writing—original draft preparation, C.S. and S.I.A.; writing—review and editing, M.B.A., Y.S., A.C., A.A.K. and W.; visualization, C.S. and S.I.A.; supervision, I.S., M.B.A., M.F., Y.S., A.A.K. and W.; project administration, M.R.P. All authors have read and agreed to the published version of the manuscript.

Funding: This research received no external funding.

Data Availability Statement: The data presented in this study are available upon request from the corresponding author.

Acknowledgments: This study was supported by Institut Teknologi Bandung Research (Riset ITB), P2MI ITB, Riset Unggulan 3P, and Program Riset Talenta Unggul LPIT ITB.

Conflicts of Interest: The authors declare no conflicts of interest.

References

1. Mellivera, A.; Zain, K.; Adityawan, M.B.; Harlan, D.; Farid, M.; Yakti, B.P. Development of FTCS Artificial Dissipation for Dam Break 2D Modelling. *J. Tek. Sipil* **2020**, *27*, 1–8. [CrossRef]
2. Younus, M.; Suswanta; Zaenuri, M.; Wildhani, A.M.; Rodriguez, M.J.D. From Crisis to Sustainability: Analyzing Fresh Water Shortage Crises and The Urgency for Government Intervention to Manage Resources. *J. Surv. Fish. Sci.* **2023**, *10*, 6228–6238.
3. Shi, X.; Lyu, G. Water-preserved coal mining in water-shortage mining areas: A case study in the Yonglong mining area of China. *Water Pract. Technol.* **2023**, *18*, 2123–2135. [CrossRef]
4. Sanchez, L.; Warziniack, T.; Knowles, M. The Inequitable Exposure of Socially Vulnerable Groups to Water Shortages across the United States. *Environ. Res. Lett.* **2023**, *18*, 044022. [CrossRef]
5. Taftazani, R.; Kazama, S.; Takizawa, S. Spatial Analysis of Groundwater Abstraction and Land Subsidence for Planning the Piped Water Supply in Jakarta, Indonesia. *Water* **2022**, *14*, 3197. [CrossRef]
6. Rahman, S.; Sumotarto, U.; Pramudito, H. Influence the condition land subsidence and groundwater impact of Jakarta coastal area. *IOP Conf. Ser. Earth Environ. Sci.* **2018**, *106*, 012006. [CrossRef]
7. Hendarto, H.; Standing, J.R. Influence of groundwater extraction on land subsidence in Jakarta. In Proceedings of the XVII ECSMGE, Reykjavík, Iceland, 1–6 September 2019; pp. 1–8.
8. Putri, R.F.; Rostika, M.D.; Abadi, A.W.; Rakhmatika, M. A Review Disaster Mitigation of Jakarta Land Subsidence Areas. *E3S Web Conf.* **2021**, *325*, 01002. [CrossRef]
9. Abidin, H.Z.; Andreas, H.; Gumilar, I. Land subsidence of Jakarta (Indonesia) and its relation with urban development. *Nat. Hazards* **2011**, *59*, 1753–1771. [CrossRef]

10. Suprayogi, H.; Rudyanto, A.; Bachtiar, H.; Limantara, L.M. Critical-phase sea dike construction of NCICD program in Jakarta as national capital city. *IOP Conf. Ser. Earth Environ. Sci.* **2018**, *162*, 012020. [CrossRef]
11. Stiawan, Y.A.; Adityawan, M.B.; Prasetyo, A. Flood Modeling on the Dadap River and Estuary, Banten Province. *J. Tek. Sipil* **2023**, *30*, 1–8. [CrossRef]
12. Mishra, B.K.; Emam, A.R.; Masago, Y.; Kumar, P.; Regmi, R.K.; Fukushi, K. Assessment of future flood inundations under climate and land use change scenarios in the Ciliwung River Basin, Jakarta. *J. Flood Risk Manag.* **2018**, *11*, S1105–S1115. [CrossRef]
13. Marwanza, I.; Anugrahadi, A.; Sumotarto, U.; Kurniawati, R.; Yudha, H.F.; Nugraheni, R.D. Land Subsidence and Geotechnical Impact of Jakarta Kota Area. *Indones. J. Urban Environ. Technol.* **2023**, *6*, 145–164. [CrossRef]
14. Batubara, B.; Kooy., M.; Zwarteveen, M. Politicising land subsidence in Jakarta: How land subsidence is the outcome of uneven sociospatial and socionatural processes of capitalist urbanization. *Geoforum* **2023**, *139*, 103689. [CrossRef]
15. Yang, S. Historical Review of Existing Coastal Reservoirs and Its Applications. In Proceedings of the 38th IAHR World Congress, Panama City, Panama, 1–6 September 2019; pp. 3957–3973.
16. Yang, P.; Liu, M.; Tang, K.M.; Yang, H.; Lai, D.Y.F.; Tong, C.; Chun, K.P.; Zhang, L.; Tang, C. Coastal reservoirs as a source of nitrous oxide: Spatio-temporal patterns and assessment strategy. *Sci. Total Environ.* **2021**, *790*, 147878. [CrossRef] [PubMed]
17. Mo, Y.; Jin, Q.; Zhang, C.; Xu, J.; Tang, H.; Shen, C.; Scheuermann, A.; Li, L. Combined effect of inland groundwater input and tides on flow and salinization in the coastal reservoir and adjacent aquifer. *J. Hydrol.* **2021**, *600*, 126575. [CrossRef]
18. Jayanti, M.; Sabar, A.; Ariesyady, H.D.; Marselina, M.; Qadafi, M. A comparison of three water discharge forecasting models for monsoon climate region: A case study in cimanuk-jatigede watershed Indonesia. *Water Cycle* **2023**, *4*, 17–25. [CrossRef]
19. Sebayang, I.S.D.; Fahmi, M. Dependable Flow Modeling In Upper Basin Citarum Using Multilayer Perceptron Backpropagation. *Int. J. Artif. Intell. Res.* **2020**, *4*, 75–85. [CrossRef]
20. *SNI 2415:2016*; Tata Cara Perhitungan Debit Banjir Rencana. Badan Standarisasi Nasional: Jakarta, Indonesia, 2016.
21. Suroso, S.; Santoso, P.; Birkinshaw, S.; Kilsby, C.; Bardossy, A.; Aldrian, E. Assessment of TRMM rainfall data for flood modelling in three contrasting catchments in Java, Indonesia. *J. Hydroinform.* **2023**, *25*, 797–814. [CrossRef]
22. Senjaya, T.; Yudianto, D.; Yuebo, X.; Adidarma, W. Application of TRMM in the Hydrological Analysis of Upper Bengawan Solo River Basin. *J. Civ. Eng. Forum* **2020**, *6*, 309–318. [CrossRef]
23. Komariah, I.; Motsumoto, T. Application of Hydrological Method for Sustainable Water Management in the Upper-Middle Ciliwung (UMC) River Basin, Indonesia. *J. Water Environ. Technol.* **2019**, *17*, 203–217. [CrossRef]
24. Rintis, H.; Suyanto; Setyoasri, Y.P. Rainfall-Discharge Simulation in Bah Bolon Catchment Area by Mock Method, NRECA Method, and GR2M Method. *Appl. Mech. Mater.* **2016**, *845*, 24–29.
25. Chandrasasi, D.; Limantara, L.M.; Juni, R.W. Analysis using the F. J. Mock Method for calculation of water balance in the Upper Konto Sub-Watershed. *IOP Conf. Ser. Earth Environ. Sci.* **2020**, *437*, 012019. [CrossRef]
26. Herman, R.; Andiesse, V.W.; Rahmi, O.S. Changes in Land Use Affect Dependable Discharge in the Miu River Basin. *MATEC Web Conf.* **2020**, *331*, 04004. [CrossRef]
27. Utamahadi, M.A.; Pandjaitan, N.H.; Rau, M.I. Land use change impacts on discharge analysis using SWAT model at Ciherang Pondok DAM catchment area. *IOP Conf. Ser. Earth Environ. Sci.* **2018**, *149*, 012015. [CrossRef]
28. Hamdi, A.; Abdulhameed, I.; Mawlood, I. Application of Weap Model for Managing Water Resources in Iraq: A Review. *IOP Conf. Ser. Earth Environ. Sci.* **2023**, *1222*, 012032. [CrossRef]
29. Creager, W.P.; Justin, J.D.; Hinds, J. *Engineering for Dams: Volume 1—General Design*; John Wiley and Sons: New York, NY, USA, 1945.
30. PLN (Perusahaan Listrik Negara). *Studi Inventarisasi Air*; PLN: Jakarta, Indonesia, 1997.
31. Sendrós, A.; Himi, M.; Rivero, L.; Lovera, R.; Urruela, A.; Tapias, J.C.; Casas, A. Enhanced Groundwater Protection and Management Using Gravity and Geoelectrical Data (Valls Basin, Spain). *Water* **2023**, *15*, 4130. [CrossRef]
32. Sheikha-BagemGhaleh, S.; Babazadeh, H.; Rezaie, H.; Sarai-Tabrizi, M. The effect of climate change on surface and groundwater resources using WEAP-MODFLOW models. *Appl. Water Sci.* **2023**, *13*, 121. [CrossRef]
33. Čož, N.; Ahmadian, R.; Falconer, R.A. Implementation of a Full Momentum Conservative Approach in Modelling Flow Through Tidal Structures. *Water* **2019**, *11*, 1917. [CrossRef]
34. Sandi, C.; Nugroho, E.O.; Cahyono, M. Dambreak Risk Analysis of Jenelata Dam and its Mitigation Plan. *Bull. Civ. Eng.* **2022**, *2*, 49–56. [CrossRef]
35. Kirana, P.H.; Farid, M.; Adityawan, M.B.; Kuntoro, A.A.; Widyaningtias. Study of Flood Risk Assessment on Banyumas and Cilacap District in Downstream Serayu River Basin, Indonesia. *J. Tek. Sipil* **2023**, *30*, 149–156. [CrossRef]
36. Burnama, N.S.; Rohmat, F.I.W.; Farid, M.; Kuntoro, A.A.; Kardhana, H.; Rohmat, F.I.W.; Wijayasari, W. The Utilization of Satellite Data and Machine Learning for Predicting the Inundation Height in the Majalaya Watershed. *Water* **2023**, *15*, 3026. [CrossRef]
37. Syarifudin, A.; Satyanaga, A.; Destania, H.R. Application of the HEC-RAS Program in the Simulation of the Streamflow Hydrograph for Air Lakitan Watershed. *Water* **2022**, *14*, 4094. [CrossRef]
38. Rohmat, F.I.W.; Sa'adi, Z.; Stamataki, I.; Kuntoro, A.A.; Farid, M.; Suwarman, R. Flood modeling and baseline study in urban and high population environment: A case study of Majalaya, Indonesia. *Urban Clim.* **2022**, *46*, 101332. [CrossRef]
39. Thapa, S.; Shrestha, A.; Lamichhane, S.; Adhikari, R.; Gautam, D. Catchment-scale flood hazard mapping and flood vulnerability analysis of residential buildings: The case of Khando River in eastern Nepal. *J. Hydrol. Reg. Stud.* **2020**, *30*, 100704. [CrossRef]
40. Erima, G.; Kabenge, I.; Gidudu, A.; Bamutaze, Y.; Egeru, A. Differentiated Spatial-Temporal Flood Vulnerability and Risk Assessment in Lowland Plains in Eastern Uganda. *Hydrology* **2022**, *9*, 201. [CrossRef]

41. Effendi, H.; Permatasari, P.A.; Muslimah, S.; Mursalin. Water quality of Cisadane River based on watershed segmentation. *IOP Conf. Ser. Earth Environ. Sci.* **2018**, *149*, 012023. [CrossRef]
42. *SNI 7745:2012*; Tata Cara Perhitungan Evapotranspirasi Tanaman Acuan dengan Metode Penman-Monteith. Badan Standarisasi Nasional: Jakarta, Indonesia, 2012.

water

Article

Study of a Tailings Dam Failure Pattern and Post-Failure Effects under Flooding Conditions

Zhong Gao [1,2], Jinpeng Liu [1], Wen He [1,3,*], Bokai Lu [1], Manman Wang [1] and Zikai Tang [1]

1 School of Resources and Environmental Engineering, Jiangxi University of Science and Technology, Ganzhou 341000, China; littleaction2023@gmail.com (Z.G.)
2 Zijin (Changsha) Engineering Technology Co., Ltd., Changsha 410006, China
3 Jiangxi Province Key Laboratory of Mining Engineering, Jiangxi University of Science and Technology, Ganzhou 341000, China
* Correspondence: hewen@jxust.edu.cn

Abstract: Tailings dams are structures that store both tailings and water, so almost all tailings dam accidents are water related. This paper investigates a tailings dam's failure pattern and damage development under flood conditions by conducting a 1:100 large-scale tailings dam failure model test. It also simulates the tailings dam breach discharge process based on the breach mode using FLOW-3D software, and the extent of the impact of the dam failure debris flow downstream was derived. Dam failure tests show that the form of dam failure under flood conditions is seepage failure. The damage manifests itself in the form of flowing soil, which is broadly divided into two processes: the seepage stabilization phase and the flowing soil development damage phase. The dam failure test shows that the rate of rise in the height of the dam saturation line is faster and then slower. The order of the saturation line at the dam face is second-level sub-dam, third-level sub-dam, first-level sub-dam, and fourth-level sub-dam. The final failure of the tailings dam is the production of a breach at the top of the dam due to the development of the dam's fluid damage zone to the dam top. The simulated dam breach release results show that by the time the dam breach fluid is released at 300 s, the area of over mud has reached 95,250 square meters. Local farmland and roads were submerged, and other facilities and buildings would be damaged to varying degrees. Based on the data from these studies, targeted measures for rectifying hidden dangers and preventing dam breaks from both technical and management aspects can be proposed for tailings dams.

Keywords: tailings dam; flood; model experiment; dam break; FLOW-3D

1. Introduction

1.1. Research Status

The mud wastewater containing tailings will be discharged after metal and non-metal mine beneficiation. Tailings slurry contains mercury, arsenic, and other heavy metal ions, both resources and pollution sources [1]. The tailings dam is a dam body formed by the accumulation and rolling of the tailings after the mine selects the useful components [2]. It is of great significance to research the dynamic stability of tailings reservoirs for mine safety production, protection of downstream life and property safety, and the surrounding environment [3]. Tailings dams, an important source of danger if an accident, are bound to people's lives and property [4]. In 2008, a dam break accident occurred in the 980 ditch tailings pond of Shanxi Xinta Mining Co., Ltd., Yuncheng, China, resulting in 281 deaths and 33 injuries. The direct economic loss was as high as CNY 96,192,000 [5]. On the afternoon of 30 April 2006, the tailings dam of Zhen'an Gold Mine in Shaanxi Province was constructed. The accident caused 17 people to disappear, five people were injured, and 76 houses were destroyed [6,7].

In many tailings reservoir accidents, due to the lack of flood discharge capacity of flood discharge structures in the reservoir area, flood overtopping, tailings dam break, and other phenomena occur occasionally [8,9]. In this regard, scholars in related fields have performed much research and achieved certain results. Chen Zhang et al. [10] established three-dimensional and two-dimensional finite element models. The seepage field of the project under different operating conditions was simulated, and the safety factor under different operating conditions was obtained by combining the seepage field with the stable surface. The influence of the length of the dry beach and the upstream slope ratio on the seepage and stability of the tailings dam was determined. Sánchez-Peralta et al. [11] took a dry tailings pond in Colombia as the research object, studied the movement characteristics of dam break debris flow with different water contents, and obtained the relationship between the length and width of dam break debris flow movement. Changbo Du et al. [12] studied and analyzed the influence of reinforcement on tailings dam and the change law of pore water pressure and internal pressure of the dam body after mud discharge. The pore water pressure and internal earth pressure of the accumulation dam after grouting gradually increased with time. Reinforcement can greatly reduce the pore water pressure and internal pressure of reinforced dams. Gregor Petkovšek et al. [13] proposed a dam break model EMBREA-MUD to calculate the water and tailings outflow of the tailings reservoir and the corresponding break growth. Weile Geng et al. [14] conducted experimental research on the settlement deformation and mesostructure evolution of unsaturated tailings under continuous load. The results showed that the mesostructure deformation of unsaturated tailings with different moisture contents under load was the same and could be divided into four stages: pore compression, elastic deformation, structure change, and further compaction. Alan Lolaev et al. [15] developed a method to determine the tailings filtration and secondary consolidation coefficient in the process of alluvial according to the physical conditions, density, and water phreatic, and a mathematical model to calculate the consolidation time. Kun Wang et al. [16] proposed a multidisciplinary program to simulate the dam break runoff of hypothetical tailings reservoirs on the downstream complex terrain using UAV photogrammetry and smooth particle hydrodynamics (SPH) numerical method. Rawya M. Kansoh [17] studied the influence of the earth-rock dam's structural parameters on the dam failure process. Kehui Liu et al. [18] studied the microscopic characteristics of hydraulic erosion of reinforced tailings dams and revealed the influence of different reinforcement spacing on the critical start-up speed of tailings particles. It shows that the smaller the reinforcement spacing, the greater the critical start-up speed of the reinforced tailings samples. Luca Piciullo et al. [19] proposed a regression analysis that considers the functional relationship between the release amount and the characteristics of the tailings dam, such as height and water storage (i.e., dam factor). The effects of construction type, filling material, and failure mode on the release amount were also evaluated, as well as the failure frequency of the tailings dam as a function of the construction method. Tailing dams built using upstream construction methods are more prone to failure and are more susceptible to static and dynamic liquefaction. Chunhui Ma et al. [20] pointed out that a reasonable construction schedule and flexible waterproof material are key features of impervious bodies for dams with significant deformation. When the dam deformation becomes stable, consideration should be given to secondary treatment of the impervious body to enhance dam safety. Fukumoto et al. [21] used finite element software to simulate the seepage failure process caused by seepage. Alibek Issakhov et al. [22] combined the k-ω turbulence model to study this process numerically. The VOF (volume of fluid) method was used to simulate the fluid movement behind the tailings dam during the break-up of the fluid and the riverbed landscape. Yonas B. Dibike et al. [23] A two-dimensional hydrodynamic and component transport model was used to study the effect of OS tailings release on the water quality and sediment quality of LAR by simulating sediments and related chemicals. It was concluded that the tailings release location was different; 40% to 70% of the sediments and related chemicals were deposited on the riverbed of the 160 km study section, while the remaining sediments and related chemicals left the study area in the first three days after the release event. Research conducted by Xiaofei Jing et al. [24] investigated the overflow characteristics of tailings dams reinforced with steel bars. During the overflow process, they measured dam displacements, saturation lines, and internal stresses. The study demonstrated that the erosion resistance of tailings dams significantly improves with an increase in the number of reinforcement layers. Abdellah Mahdi et al. [25] studied the potential consequences of a hypothetical oil sand tailings dam failure. For this reason, a non-Newtonian dam–dam model with a viscoplastic rheological relationship is used. The model can reproduce the flood and water level changes in downstream lakes (due to destructive waves). The simulation study of oil sand tailings overflow proves the importance of considering the non-Newtonian characteristics of tailings. Naeini et al. [26] used SIGMA/W and QUAKE/W software to analyze the high-middle line tailings dam's dynamic response and permanent deformation and evaluated the dam's performance. Mohammad Reza Boroomand [27] used the numerical analysis method to analyze the earth dam's seepage under the uncertainty of geotechnical parameters and analyzed the seepage of the earth dam under the condition of uncertainty of geotechnical parameters. Sumin Li et al. [28] simulated and analyzed the hazard range, degree, and spatial state of sediment flow after the dam break and obtained the influence of sand flow velocity, flow depth, and impact on the downstream villages in the disaster area. The feasibility of the expansion and heightening of the tailings dam project was demonstrated, and the disaster risk levels of different spatial locations in the downstream villages were obtained through simulation. Through experiments, Kong et al. [29] studied the influence parameters of tailings dams under seepage. They concluded that the particle size gradation, non-uniformity coefficient, and water content of tailings sand were the main factors affecting the critical hydraulic gradient. It is concluded that the seepage failure gradient with suitable gradation, uniform particles, and suitable water content is significantly higher than that with poor gradation, uneven particles, and poor water content.

Flood overtopping and seepage failure account for 80% of the total accidents, and these two failure forms mainly occur in flood season and are closely related to water. Therefore, it is necessary to explore the tailings dam failure mode, development process, and the impact on the downstream after dam failure under flood conditions to ensure its safe operation. Based on the engineering background of a tungsten mine tailings dam in Ganzhou City, Jiangxi Province, a 1:100 physical model test was carried out to explore the dam break form and failure development process of the tailings dam under flood conditions. The FLOW-3D fluid simulation software was used to solve the influence of the tailings dam on the downstream after the dam break, and the change law of the flow area, velocity development, and submerged depth of the dam break fluid during the flood discharge process was analyzed. Finally, reasonable prevention and remediation suggestions are proposed for the hidden dangers of tailings dams.

The innovation of this paper is to determine the dam break mode and dam break position of the tailings dam under flood conditions by constructing a physical model, which provides a basis for simulating the influence of dam break on the downstream of the dam body.

1.2. Research Flowchart

The research flowchart of this paper is shown in Figure 1.

2. Design and Construction of Large Physical Models

2.1. Overview of the Prototype Tailings Dam

The prototype tailings dam is a tungsten tailings dam located in a narrow valley running north–south in Ganzhou City, Jiangxi Province, China. The downstream of the tailings accumulation dam is farmland, dormitory buildings, mountain roads, etc., and the valleys in the downstream are relatively open. Figure 2 shows an overhead view of the prototype tailings dam. The tailings dam is built using the upstream damming method. The initial dam is a clay core wall weathering material dam located at the mouth of the northern valley. The bottom elevation is 262.0 m, the top elevation is 284.0 m, the dam height is 22 m, the upstream and downstream slope ratio is 1:2.5. The design average external slope ratio of the tailings accumulation dam is 1:5, the average slope of the tailings deposition beach is 5%, the design final tailings accumulation dam elevation is 368.00 m, the total dam height 106.00 m, with a complete storage capacity of 1550×10^4 m^3 and a service life of 65 years. The present top elevation of the stacked dam is about 315 m, the height of the dam is 53 m, the accumulated storage capacity is about 559×10^4 m^3, and the average external slope ratio of the stacked dam is 1:4.9. The reservoir is currently a fourth-class reservoir, with a flood protection standard of one in 200 years. At a later stage, it will be a second-class reservoir with a flood protection standard of 1000 years.

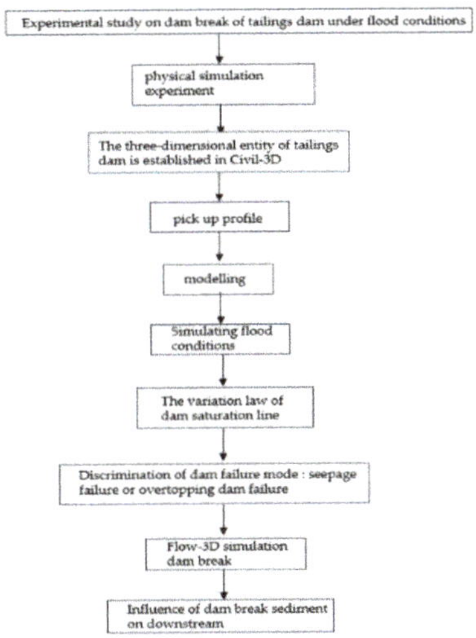

Figure 1. Research flowchart.

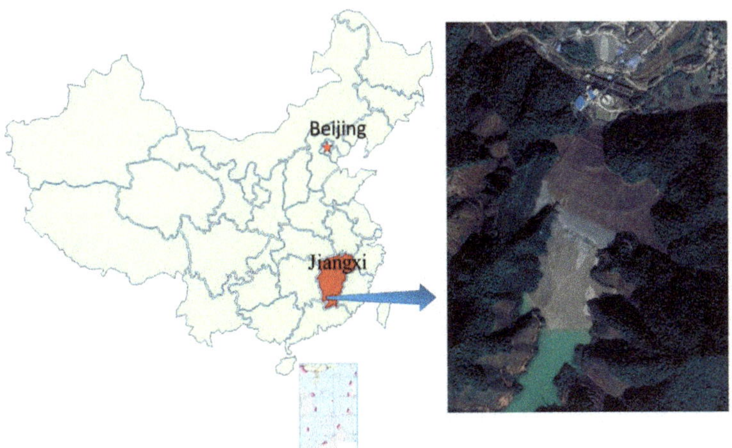

Figure 2. Top view of a tailings dam.

2.2. Selection of Model Sand

To ensure the relative reliability of the test results, the selected dam materials are properly relaxed to meet the primary conditions of similar main influencing factors. The model test focuses on the agglomeration effect of particle movement during the deformation of the dam body. For the selection of model sand, the initial dam is built with silty clay, and the accumulation dam adopts the mine prototype tailings. Figure 3 is the particle size distribution curve of the model sand. According to the particle size distribution curve, two quantitative indexes of soil particles can be determined: non-uniformity coefficient C_u and curvature coefficient C_c. C_u and C_c can jointly determine the gradation of soil. The expressions of the two are:

$$C_u = \frac{d_{60}}{d_{10}}, \tag{1}$$

$$C_c = \frac{d_{30}^2}{d_{10} \cdot d_{60}}, \tag{2}$$

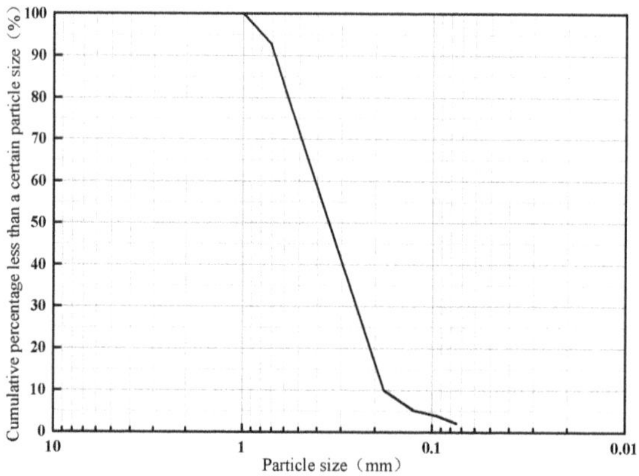

Figure 3. Cumulative distribution curve of particle size.

The calculated C_u and C_c of the model sand are 2.29 and 0.84, respectively. It is generally believed that the sand soil with $C_u < 5$ or C_c outside 1~3 belongs to the poorly graded soil, so the model sand is determined to be the poorly graded soil. If the seepage failure occurs in the dam, the development mode of seepage failure can be predicted by some parameters of the soil, that is, whether the soil is piping or flowing soil. According to the non-uniform coefficient discrimination method proposed by the former Soviet Union scholar Istomina, it is preliminarily judged that the model sand is a flowing soil-type soil.

2.3. Construction of the Dam Failure Model

The dam break process of a tailings reservoir involves many aspects, such as hydraulics, mud and sand dynamics, and soil mechanics. It involves many disciplines and is highly complex, which leads to the similarity relationship of model tests.

Therefore, we must put aside the generality of similarity and focus on the similarity of critical elements. This experiment uses the engineering background of a tungsten mine tailings dam in Jiangxi Province, China. The similarity criterion is appropriately relaxed, and the accumulation effect of particle movement during the deformation of the dam body is emphasized to construct the physical model.

Under the condition of geometric similarity, the physical model test of the 1:100 large-scale tailings dam is carried out according to the level of the second-class reservoir of the prototype tailings dam. The prototype range of the tailings dam is 1200 m × 700 m, and the model size is 12 m × 7 m. The model mainly comprises bedrock, a dam body, an observation system, and a water supply circulation system. The specific steps are as follows: According to the topographic map data provided by the mine, the three-dimensional model of the prototype tailings dam is established using Civil-3D modeling software (ver.2018) according to the size of the actual tailings dam (Figure 4). Then, several vertical sections are cut out in the model with the east–west direction as the standard line, and the points on each vertical section are taken equidistantly to extract the elevation value of each point on each vertical section. The model is intended to build a model with an elevation of 230 m in the actual terrain. The steel frame structure of the bedrock is made based on the elevation of each point on the vertical section. The square steel pipe is used as the bedrock support. Each steel pipe corresponds to the elevation of its relative point in proportion. Finally, the waterproof cloth is covered on the steel frame group and fixed to obtain a complete view of the bedrock terrain. Figure 5 is the completed mountain steel frame group, and Figure 6 is the complete bedrock after laying waterproof cloth.

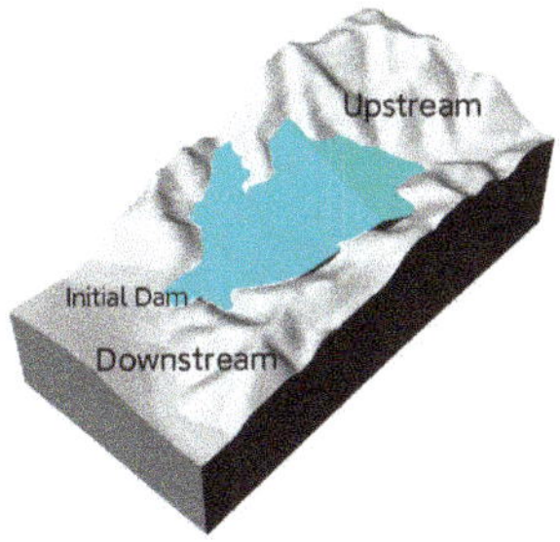

Figure 4. Three-dimensional model of the tailings dam.

Figure 5. Mountain support structures.

Figure 6. A complete view of bedrock.

The initial dam is piled up with silty clay. In the process of stacking the initial dam, two PVC pipes with holes in the wall body and tightly wrapped with permeable geotextiles are symmetrically buried at the bottom of the initial dam to simulate the drainage pipe. A valve is installed at the outlet end of the two drainage pipes to control the drainage speed. The sub-dam uses the pipeline method commonly used in the mine to simulate the ore drawing. An ore drawing main pipe is introduced from the slurry pool to start the ore drawing from the model's right side, and a valve is set in the main pipe to control the flow rate of the square ore. When the pulp flows into the tailings pond, the tailings will be layered and precipitated under hydraulic screening. After precipitation, the ore is suspended when the tailings reach the target dam height. Start to build the next sub-dam, use the layered filling method to build the dam body to the design elevation, and use the vertical line method to control the elevation when building the dam. (Figure 7) is the construction of the second sub-dam. Four pore water pressure gauges are buried in the dam construction process to monitor the position of the saturation line of the dam body. The four pore water pressure gauges' positions are arranged along the dam body's central axis. They are located directly below the dam crest of the first-, second-, third-, and fourth-level accumulation dams. They are named as site 1, site 2, site 3, and site 4 (Figure 8).

Figure 7. Second-stage sub-dam stacking.

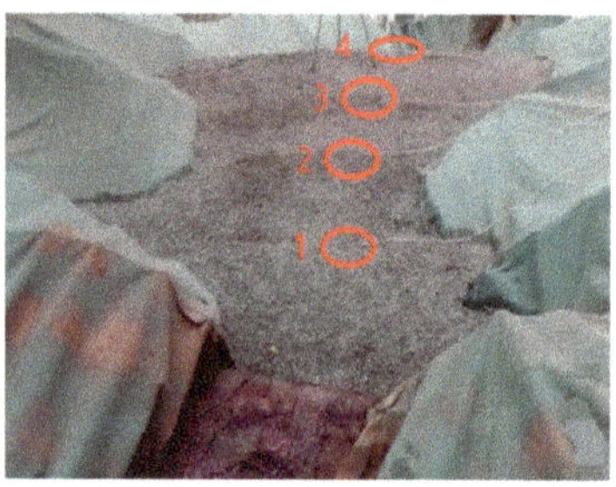

Figure 8. Buried pore water pressure gauges.

Due to the need to supply a large amount of water for the test, a water tower was placed on the site (Figure 9), and a return water collection system was designed to achieve a water supply cycle (Figure 10). The observation equipment of the test (Figure 11) uses a trinocular camera and a high-definition camera to record the dam break process of the tailings dam.

Figure 9. Water tower.

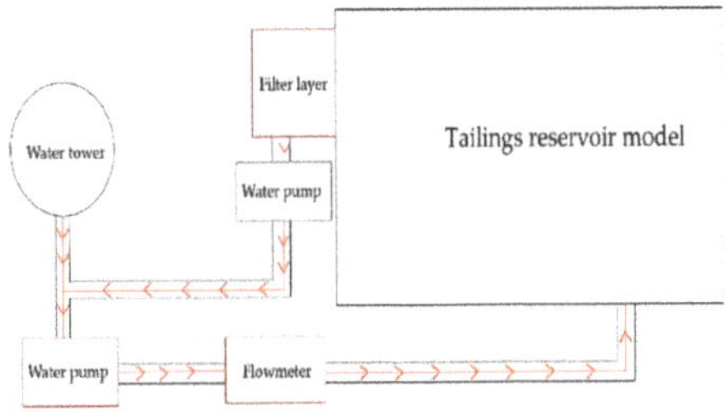

Figure 10. Water supply circulation system.

Figure 11. Experimental observation system.

3. Tailings Dam Break Model Experiment

The dam failure mode under specific flood conditions is characterized by permeation damage, manifested as soil erosion. Through analysis of experimental phenomena and data, the development process of dam failure is elucidated, revealing the variation patterns of pore water pressure at different locations and the saturation line of the dam body.

3.1. Dam Failure Experiment

The test was carried out by intermittently injecting water into the reservoir to simulate flood conditions, keeping the flow rate stable during the injection, and keeping the drainage pipe open during the whole test. The beginning of the water injection was taken as the beginning of the test, and the entire dam break test lasted 448 min. It can be roughly divided into two stages, each accounting for one-half of the total length. Figure 12 shows a typical picture of the damage to the dam during the test. Figure 13 shows a timeline of the test damage development. The specific tailings dam damage development process is as follows:

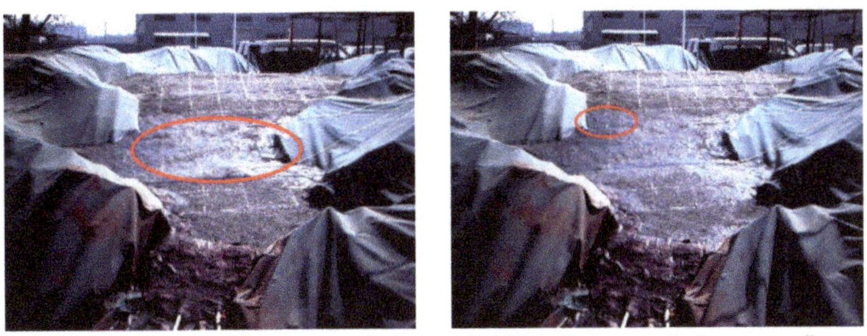

(**a**) Second-stage sub-dam overflow (**b**) Third-stage sub-dam overflow

Figure 12. *Cont.*

(c) First-stage sub-dam overflow

(d) First flow soil damage

(e) Second flow soil damage

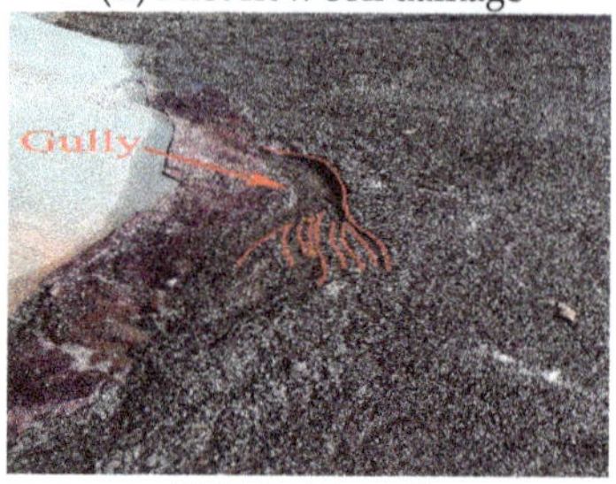

(f) Left-side gully

(g) Right-side gully

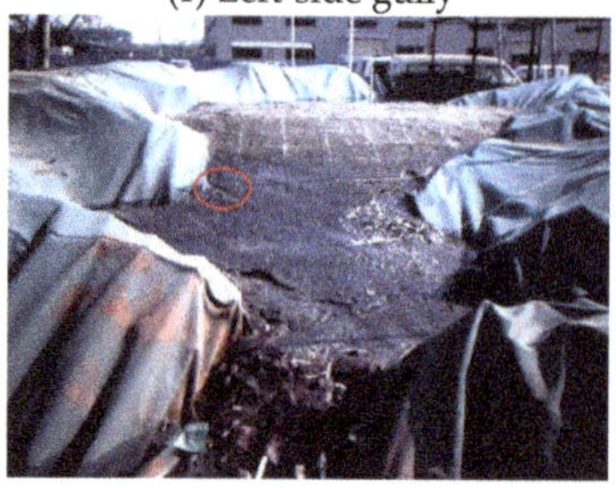

(h) Third flow soil damage

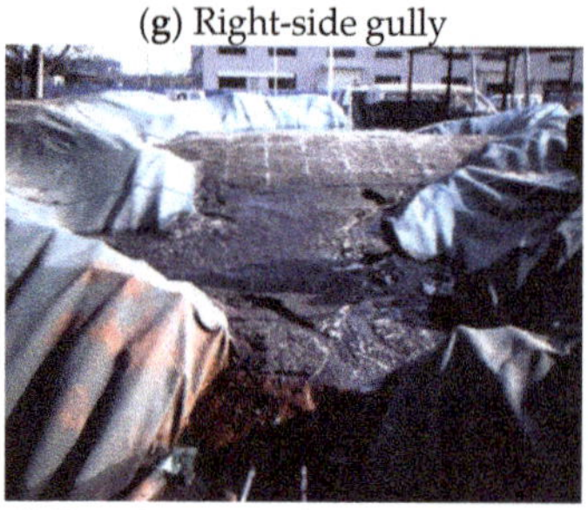

(i) Fourth-stage sub-dam overflows

(j) Final damage pattern of the dam

Figure 12. Dam failure model tests.

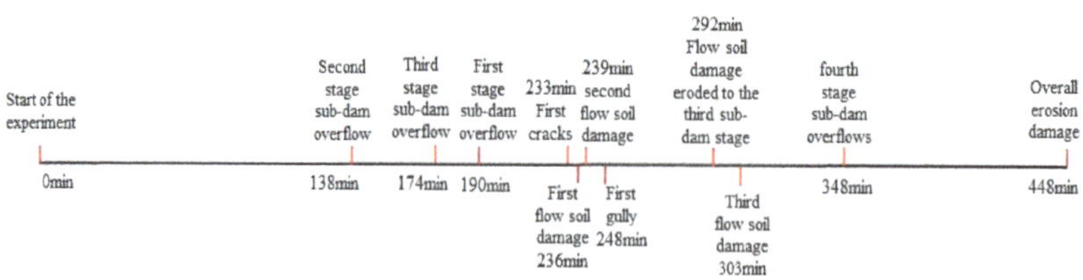

Figure 13. Timeline of the development of the flowing soil destruction.

The first stage is seepage stabilization: the overflow water is clear, the dam's surface is stable, and there is no movement of particles. At 138 min of the test, the contact zone between the right end of the second sub-dam and the bedrock began to seep first (Figure 12a). The seepage water flows along the contact zone between the dam body and the bedrock and overflows down the dam face. There are two reasons for the seepage here. The first is fine cracks in the contact area between the soil and the bedrock, which provides a breakthrough for the seepage water. The second is based on calculating the data collected by the pore water pressure gauge. It can be seen that the saturation line at this time escapes on the slope of the second sub-dam, where the dam surface overflows. Subsequently, the second-stage sub-dam continued to seep, and the overflow area gradually expanded and merged with the second-stage sub-dam dam surface. At 174 min, the third-stage sub-dam began to overflow on the left side (Figure 12b). At this time, according to the collected data, it can be calculated that the buried depth of the saturation line has been exposed to the third-level sub-dam. At 190 min, the first sub-dam also overflowed (Figure 12c). Then, the sand boiling point appears at the right end of the first-stage sub-dam, and the soil particles fluctuate obviously with the overflow water. The sand boiling causes the soil particles to be continuously taken out of the soil body. At 203 min, the dam surfaces of the first, second, and third sub-dams have all become swampy.

The second stage is the development and failure stage of the flowing soil: seepage deformation occurs continuously, and more earthwork is lost. At 236 min, the first flow soil damage happened at the right end of the second sub-dam (Figure 12d). The failure form is flow slip. At 239 min, a second flow soil damage occurred on the left side of the first sub-dam (Figure 12e). The flow-slipping soil will form a pit that evolves into a breach, making the seepage velocity and seepage flow faster and larger. Then, the pit part of the soil slides, and the seepage water erodes the downstream dam surface. At 248 min, two erosion ditches have been formed in the flow soil failure area on both sides of the dam (Figure 12f,g).

The erosion gully on the left side is located at the junction of the right side of the first-order sub-dam and the bedrock. The critical hydraulic gradient is lower, the dominant flow develops more rapidly, the sand is wrapped violently, and the subsequent seepage damage is more likely to occur. The erosion gully produces more water flow to scour multiple branches on the dam's surface. The right erosion ditch is located at the junction of the secondary dam and the bedrock. At this time, the erosion ditch has developed to a certain depth, and the sand boiling point has reached 6. The flowing soil migrates downward under the action of overflow water. The flowing water will bring the fine particles to the downstream area. The coarse particles will be accumulated to form a 'filter layer' to block the overflow water channel. The seepage pressure on both sides of the filter layer gradually increases. A new seepage channel will be formed when the seepage pressure on one side reaches the critical value. At 292 min, the flow soil damage eroded to the third sub-dam and further developed upstream along the boundary. Part of the erosion gully's inner wall soil is washed away underwater, and the internal wall forms holes and expands upward until the upper part forms a suspended surface. When the shear strength of the upper soil is greater than the shear strength of the soil, it will collapse and continue to repeat the next round of erosion. At this time, the left-flowing soil does not develop to the upstream failure but to the proper lateral erosion, and the right side flushes out a new channel due to the obstruction of the 'filter layer'. At 303 min, the third flow soil failure occurred on the left side of the third sub-dam (Figure 12h). Because of the increase in overflow water and the acceleration of water flow, the right scouring area opens the downstream channel at the particle deposition, and fine particles are continuously taken out of the dam by seepage water. The overflow water also washes away the 'filter layer' on the left side. After that, the first sub-dam eroded to the deep, and the dam surface failure area did not expand. There is a hydraulic–gravity erosion cycle in the flow soil damage area of the second-stage sub-dam, which extends to the upstream and the middle of the dam body. With the increase in the erosion damage area of the water flow, the more the sand boiling point, the faster the seepage damage, and the erosion area of the lower section continues to expand, and the water flow in the erosion gully is large and fast. The flowing soil failure zone of the third-stage sub-dam has not yet formed a penetrating failure path and is in the initial stage of erosion. At 348 min, the fourth-stage sub-dam overflowed (Figure 12i). At 448 min, the flow soil was eroded to the fourth sub-dam (Figure 12j).

The flow soil damage area is eroded to the fourth sub-dam, which is regarded as the whole dam damage. It is measured that the depth of the collapse area is about 12 cm, and the width is about 80 cm. It should be noted that although the dam body has undergone a large area of seepage failure, the dam body has not yet experienced an unstable landslide. The tailings dam finally broke because the dam body soil damage zone developed to the top of the dam to produce a breach.

3.2. The Change Rule of the Saturation Line

Figure 14 is about the change curve of the saturation line. At the beginning of the test, as the upstream water level rose, the saturation line rose rapidly despite the drain being in a normal discharge condition. After the lifting of the head has ceased, the rate of the upward lifting of the saturation line becomes significantly slower due to the hysteresis effect. Then, a certain depth of burial is maintained. In the middle and late stages of the test, most of the dam had become saturated, and the soil matrix suction had weakened. When water is again stored in the reservoir, the saturation line will again lift, but at a reduced rate compared to the initial period. If the reservoir level is no longer raised, the saturation line tends to fall after a period of time. By approximately 270 min into the test, the dam face had already developed a certain size of the flow damage zone, and it was no longer meaningful to discuss the depth of the saturation line.

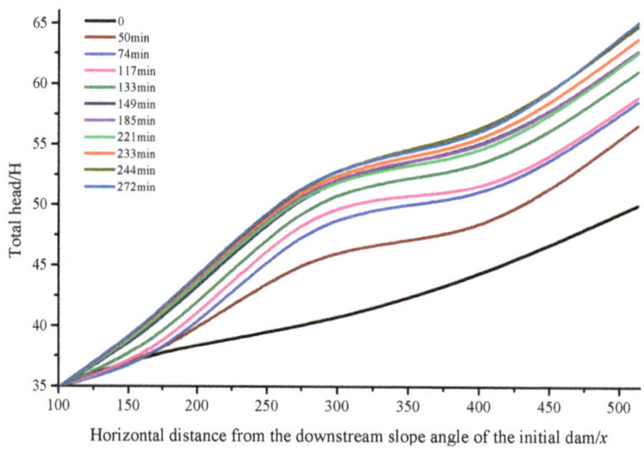

Figure 14. Variation curve of saturation line.

In the previous study [30], a two-dimensional finite element model of the tailings dam, chosen from the central axis of the three-dimensional tailings dam model, was used to analyze the distribution of saturation line in the tailings dam under flood conditions. The numerical simulation results show that when the upstream water head rose to 125 m (Figure 15), the saturation line intersected with the first and fourth-level accumulation dams and was exposed throughout the dam surface. The variation law of the saturation line obtained by the numerical simulation is consistent with the experimental phenomenon; that is, the saturation line increases with the rise of the reservoir water level, and the order of the dam surface exposure is the second-stage sub-dam, the third-stage sub-dam, the first-stage sub-dam, and the fourth-stage sub-dam. According to the simulation results, the displacement of the dam body does not change greatly, and the plastic strain zone does not appear on the slope and crest of the dam body, and there is no penetration. It can be judged that the tailings dam model does not have deep slip when the water level is about to overflow; that is, the skeleton structure of the dam body is stable. Combined with the physical model test, before the saturation line of the dam body reaches the dam surface of the fourth-level sub-dam, the tailings dam has undergone seepage failure, but the dam body has not undergone structural instability. The results of numerical simulation are consistent with the phenomenon of physical model test. After that, with the development of flowing soil, the damaged area of the dam body continues to extend to the top of the dam, which will eventually cause the breach of the dam top and cause the flood discharge of the dam.

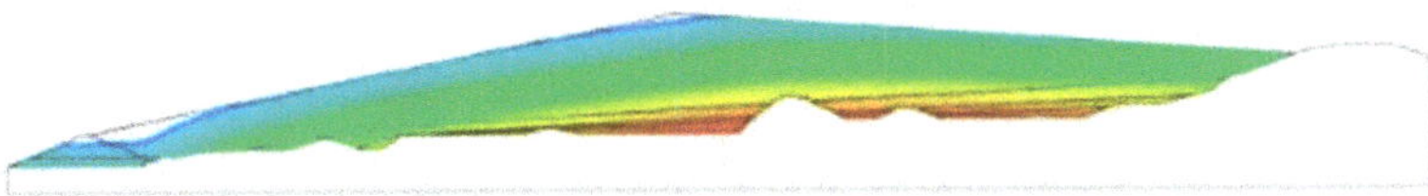

Figure 15. Saturation line distribution of the tailings dam under 125 m water level.

4. Impact Analysis after Dam Break and Prevention Suggestions

Based on the results of the physical model experiment, it can be inferred that the tailings dam failure was triggered by seepage failure. This means the area of flowing soil gradually eroded upstream until a breach was created at the top of the dam, and the reservoir fluid poured downstream. Therefore, an erosion damage trench was set up on the model for the dam breach calculation in FLOW-3D (ver. 9.3), extending from the top of the initial dam to the top of the dam, and the shape was simplified to a semi-cylinder. Figure 16 shows a model of the tailings dam after completion of the pre-treatment.

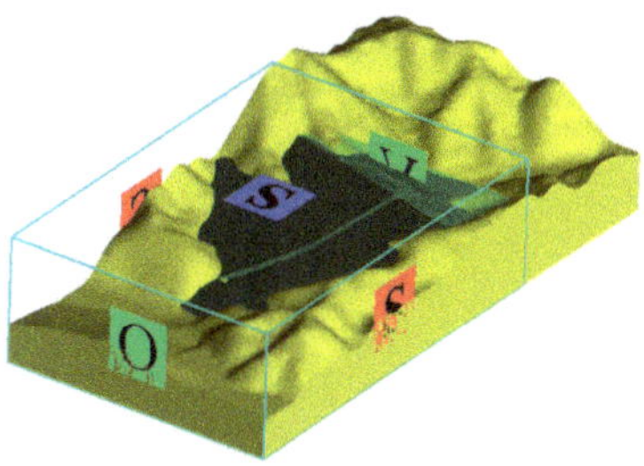

Figure 16. FLOW-3D 3D calculation model.

4.1. Dam Failure Test Results

An overview of the area downstream of the tailings dam is shown in Figure 17. The downstream area is dominated by the production facilities (red and yellow line areas in the figure), staff accommodation buildings (pink line area), the road around the mountain (blue curve), villages (green line area), and scattered agricultural land.

Figure 17. Aerial view downstream of tailings dam.

4.1.1. Overflow Area

Figure 18 shows the change in the extent of fluid inundation at 60 s, 120 s, 180 s, 240 s, and 300 s as calculated by the software, with the fluid in blue in the figure. As can be seen from the diagram, the breached fluid was rapidly released downstream in a short period and, by 300 s, covered the entire flat area downstream, with an overflow area of approximately 95,250,000 square meters. Farmland and roads in the area will be flooded, and production facilities and residential buildings will also be affected. In addition, emergency escape plans can be challenging to implement successfully at short notice. It is thus clear that in the event of a breach of this tailings dam, it would be a major accidental disaster.

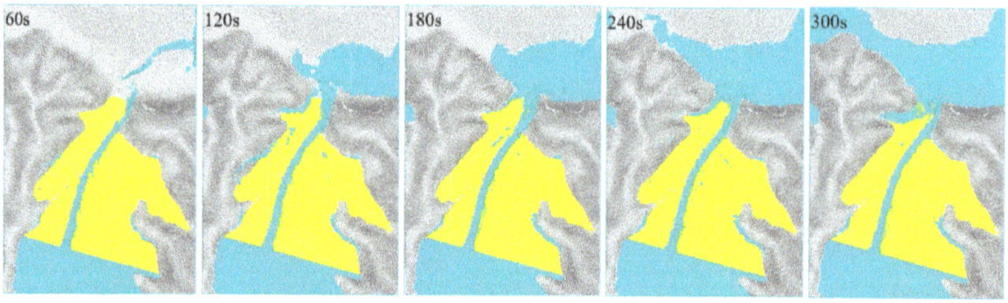

Figure 18. Time-course diagram of mud area.

4.1.2. Flooding Depth

Figure 19 shows a cloud of the distribution of flooding depth at 60 s, 120 s, 180 s, 240 s, and 300 s. Due to the lower topography in the eastern part of the downstream area, the fluids that wash down first collect in the east and then spread westwards. As can be seen from the graph, the maximum inundation depth is always located in the eastern part of the lower reaches near the initial dam. The mudslide did not affect the northern area due to the terrain's advantage; when the situation was urgent, people could be evacuated along the northwest-facing road to the north.

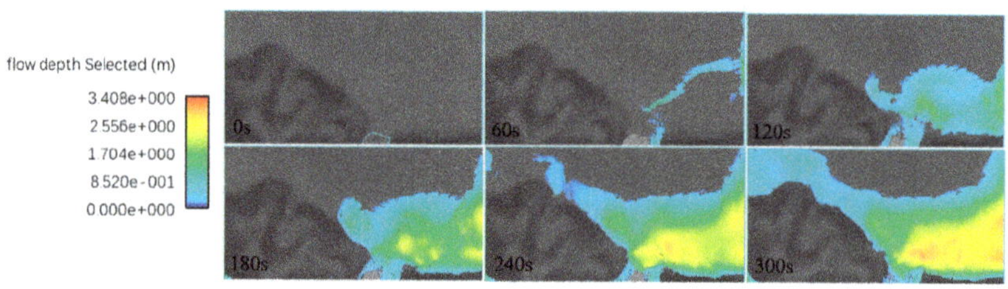

Figure 19. Cloud map of flooding depth.

4.1.3. Flow Rate Analysis

The flow velocity during the release process reflects the magnitude of the fluid impact. Figure 20 shows the flow velocity clouds during the dam breach release process at 60 s, 120 s, 180 s, 240 s, and 300 s. Due to inertia, the fluid emerges from the breach. It rapidly completes the transformation from potential energy to kinetic energy in the trench eroded by the flowing soil, with the flow velocity reaching a maximum. In addition, there is some leakage around the dam at the junction of the tailings dam and the mountain. After the fluid is flushed off the tailings dam, the average flow velocity decreases due to the diffusion principle and frictional forces. In general, the flow of emissions increases and then falls.

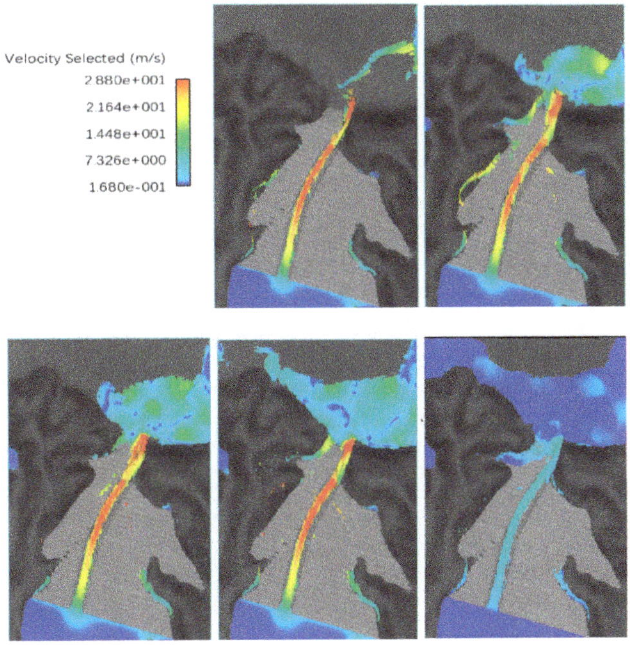

Figure 20. Cloud map of flow rate.

Three points, A, B, and C, are selected in the flow direction of the release to analyze the fluid's flow velocity characteristics, specifically during the dam failure process. The three points are located at the top of the initial dam, the foot of the initial dam, and the downstream area adjacent to the tailings dam (Figure 21). Figure 22 shows the variation in flow rate over time at three points. Overall, the flow velocities at points A, B, and C are successively reduced as the flow path develops. From the point of view of the flow velocity at a single point, it does not increase to a peak all at once but has an undulating, phased variation. At about 30 s, the overflow velocity starts to appear at the three points, after which the trend is a cyclic process of "increase-smooth or decrease" because the increase in flow velocity does not coincide with the expansion of the breach, which, in turn, determines the flow velocity of the discharge. The flow rate increases accordingly when the breach expands and becomes deeper again. After several cycles of this until the breach is no longer extended, the flow rate at points A, B, and C all fall during the last 30 s of the figures and will return to zero as the flooding stops.

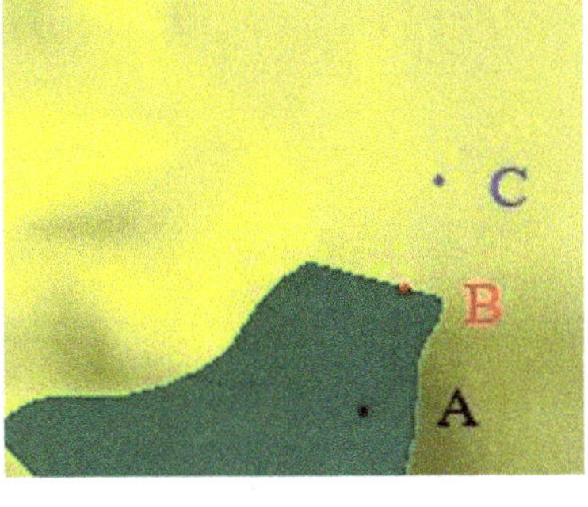

Figure 21. Flow rate reference points.

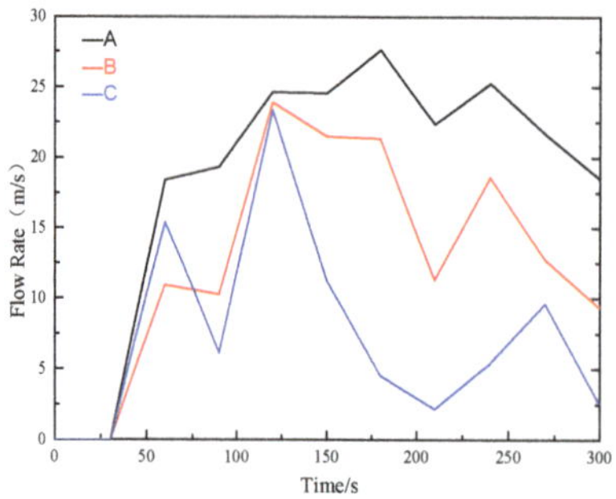

Figure 22. Flow rate time history diagram.

The analysis of the variation in the flow rate of the release shows that the debris flow impacts downstream in a segmented manner. Therefore, the decrease in flow velocity should not be regarded as the end of the entire dam break, nor should blindly carry out the aftermath of the accident at this stage, but should wait for a longer period to observe and confirm so as not to cause more damage.

4.2. Recommendations for Prevention and Management

4.2.1. Technical Measures

Dam surface treatment: According to the seepage characteristics of the tailings dam, to prevent overflow water and rainwater from scouring the shoulder and face of the dam and to collect the seepage water, a shoulder drainage ditch should be installed along the junction of the dam with the slopes of the two banks, and a face drainage ditch should be installed on the face of the dam. Moreover, the downstream slope of the dam can be mulched, turfed, and, if necessary, reinforced by stone pitching at the foot of the dam.

Additional seepage facilities: Combined with the model test results, it is clear that control of the saturation line of the tailings dam should be a top priority for safety management. To effectively control the depth of the saturation line, additional drainage facilities can be provided in the form of a combined horizontal drainage pipe and a vertical shaft connected to the end of the horizontal drainage pipe. In addition, the vertical drainage pipe should be raised with the height of the stockpile dam and pumped out periodically.

4.2.2. Management Measures

Routine inspection and maintenance: Besides monitoring various safety indicators such as the tailings dam saturation line and dry beach length, the person responsible for safety should regularly inspect the dam body for cracks, collapses, and surface erosion. They should also ensure that the slope protection is intact and that the drainage facilities are clear of blockages, siltation, or waterlogging. Check for seepage, pipe surges, or flowing soil, focusing on the junction between the dam and the hills on either side, and be vigilant for changes in seepage flow and turbidity. If a potential problem is identified, the cause must be immediately determined, and remedial action must be taken to prevent it.

Ensure excellence in flood management, including pre-flood preparation, response during flooding, and post-flood rescue work.

5. Conclusions

(1) The reservoir's water level had not yet crested before the dam was damaged. In other words, the cause of dam failure under flood conditions is seepage failure, which manifests itself in the form of flowing soil. Before the flow soil is destroyed, the dam surface will produce overflow, water accumulation, sand boiling, and other phenomena. The phenomenon of the dam failure test shows that the flow soil damage starts at the weak point of the dam at the junction with the bedrock. These areas have a high saturation gradient and are more prone to local damage. In the early stage

of soil flow failure, multiple sand boiling points were generated on the dam surface. With the development of seepage, collapsible cracks appeared on the dam surface one after another, forming erosion ditches. In the middle stage of soil failure, the failure area is widened. The soil cycle undergoes the process of erosion–gravity erosion, and the 'filter layer' will slow down the failure rate to a certain extent. In the later stage of flow soil damage, the flow water damage area began to penetrate, and the erosion intensified until the whole dam body was damaged. Therefore, when the sand boiling point is generated, and the collapsible cracks appear on the dam surface, these can be used as a warning sign of seepage failure.

(2) The buried depth of the saturation line becomes shallow with the increase in the upstream water head. And, the rate of increase is first fast and then slow. After the lifting head is stopped, the saturation line will still rise slightly for a period of time due to the lag effect. If the reservoir water level is not replenished for a long time, the saturation line will be reduced under normal drainage. The order of the saturation line escaping from the dam surface is the second sub-dam, the third sub-dam, the first sub-dam, and the fourth sub-dam. It can be seen that before the flood, it is necessary to check and repair the drainage facilities to ensure their suitable operation. During the flood season, all measures should be taken to enhance the flood discharge, reduce the saturation line, and avoid the seepage damage of the tailings dam.

(3) The results of the FLOW-3D hydrodynamic simulation software show that the breach fluid was rapidly discharged within a short period, covering the entire flat area downstream by 300 s. The local farmland and roads were submerged, and the rest of the construction facilities were also damaged to a certain extent. Therefore, it will be a major disaster once the tailings dam breaks. The rapid development of the dam breach mudslide and the short release time make it impractical to organize the evacuation of people when the release occurs. Therefore, in combination with the mechanism of tailings dam failure, targeted measures for potential remediation and dam failure prevention can be proposed from both technical and management aspects.

(4) The innovation point is to use a large-scale physical model test to study the dam break mode of tailings dam under flood conditions. By monitoring the internal changes of the tailings reservoir under flood conditions, the stage of seepage failure of the dam body can be judged, which can serve as an early warning for the subsequent break of the tailings dam. The experimental process and experimental results of the model can provide a reference for the changes in tailings reservoir under flood conditions under real working conditions so as to correspond to the changes of tailings reservoir fluid under flood conditions under real working conditions. Provide guidance for staff to monitor changes in tailings ponds. The determination of dam break position and dam break mode by model test provides a basis for simulating the influence of tailings dam break on the downstream. The use of a steel frame structure to build a tailings dam model can cover the entire tailings dam terrain more comprehensively and economically and can more comprehensively analyze the entire dam break process of the tailings dam. Compared with the local tailings dam similarity simulation and on-site exploration, it is more profound and comprehensive, which has practical significance for the safety of the tailings reservoir. The defect is that there is a prototype of the model, and it cannot be used for all tailings mines. The actual situation needs to be analyzed in detail. In addition, according to the tailings pond model test, it can be expected that the tailings pond model can be used to study the useful mineral components in the recovery reservoir, which has practical significance for environmental protection and resource recovery.

Author Contributions: Software, J.L.; Data curation, B.L.; Writing—original draft, Z.G. and Z.T.; Visualization, M.W.; Supervision, W.H. All authors have read and agreed to the published version of the manuscript.

Funding: This study is supported by the National Natural Science Foundation of China under the project (No. 51604127 and No. 51874268) and the Science and Technology Innovation Talent Project of Ganzhou City (No. 202101094905).

Data Availability Statement: The data presented in this study are available on request from the corresponding author.

Conflicts of Interest: Author Zhong Gao was employed by the company Zijin (Changsha) Engineering Technology Co., Ltd. The remaining authors declare that the research was conducted in the absence of any commercial or financial relationships that could be construed as a potential conflict of interest.

References

1. Yue, T.; Wu, X. Effect of magnetic starch on the clarification of hematite tailings wastewater. *IOP Conf. Ser. Earth Environ. Sci.* **2018**, *121*, 032051. [CrossRef]
2. Jin, W.; Wei, Z. Research progress on land reclamation and environmental management benefit estimation of non-ferrous metal tailings reservoir. *Environ. Sci. Res.* **2019**, *8*, 1304–1313. (In Chinese)
3. Kang, Z.; Han, Q.; Wang, S. umerical model of dynamic stability of a proposed tailings dam To be studied. *Min. Res. Dev.* **2015**, *35*, 59–62.
4. Zhou, K.; Liu, F.; Hu, J.; Gao, F. Research on dam-break disaster chain and chain-breaking disaster reduction control technology of tailings reservoir. *Disasterology* **2013**, *28*, 24–29. (In Chinese)
5. Chen, C.; Zhao, Y.; Jiang, L. Review on the current research of dam break of tailings pond. *Min. Res. Dev.* **2019**, *39*, 103–108. (In Chinese)
6. Zhou, Z.; Li, X. Causes and ecological environment impact assessment of tailings dam break accident in China. *Met. Mine* **2012**, *11*, 121–124. (In Chinese)
7. Zhou, L.; Huang, Y. Discussion on the causes and countermeasures of tailings reservoir accident. *China Manganese Ind.* **2016**, *34*, 37–39. (In Chinese)
8. Yang, M.; Song, Z. A tailings reservoir based on flood calculation and flood routing analysis. *China Met. Bull.* **2020**, *5*, 263–264. (In Chinese)
9. Zhang, X.; Song, Z.; Zhang, Z. Example analysis of a tailings pond flooded with a dam failure. *Eng. Constr.* **2020**, *52*, 44–48. (In Chinese)
10. Zhang, C.; Chai, J.; Cao, J.; Xu, Z.; Qin, Y.; Lv, Z. Numerical simulation of seepage and stability of tailings dams: A case study in Lixi, China. *Water* **2020**, *12*, 742. [CrossRef]
11. Sánchez-Peralta, J.A.; Beltrán-Rodríguez, L.N.; Trujillo-Vela, M.G.; Larrahondo, J.M. Flows of liquefied filtered tailings: Laboratory-scale physical and numerical modeling. *Int. J. Civ. Eng.* **2020**, *18*, 393–404. [CrossRef]
12. Du, C.; Liang, L.; Yi, F.; Niu, B. Effects of geosynthetic reinforcement on tailings accumulation dams. *Water* **2021**, *13*, 2986. [CrossRef]
13. Petkovšek, G.; Hassan, M.A.A.M.; Lumbroso, D.; Roca, M. A two-fluid simulation of tailings dam breaching. *Mine Water Environ.* **2021**, *40*, 151–165. [CrossRef]
14. Geng, W.; Wang, W.; Wei, Z.; Huang, G.; Jing, X.; Jiang, C.; Tian, S. Experimental study of mesostructure deformation characteristics of unsaturated tailings with different moisture content. *Water* **2021**, *13*, 15. [CrossRef]
15. Lolaev, A.; Oganesyan, A.; Badoev, A.; Oganesyan, E. Tailing dams formation algorithm. *Arab. J. Geosci.* **2020**, *13*, 974. [CrossRef]
16. Wang, K.; Yang, P.; Yu, G.; Yang, C.; Zhu, L. 3D numerical modelling of tailings dam breach run out flow over complex terrain: A multidisciplinary procedure. *Water* **2020**, *12*, 2538. [CrossRef]
17. Kansoh, R.M.; Elkholy, M.; Abo-Zaid, G. Effect of shape parameters on failure of earthen embankment due to overtopping. *KSCE J. Civ. Eng.* **2020**, *24*, 1476–1485. [CrossRef]
18. Liu, K.; Cai, H.; Jing, X.; Chen, Y.; Li, L.; Wu, S.; Wang, W. Study on hydraulic incipient motion model of reinforced tailings. *Water* **2021**, *13*, 2033. [CrossRef]
19. Piciullo, L.; Storrøsten, E.B.; Liu, Z.; Nadim, F.; Lacasse, S. A new look at the statistics of tailings dam failures. *Eng. Geol.* **2022**, *303*, 106–657. [CrossRef]
20. Ma, C.; Gao, Z.; Yang, J.; Cheng, L.; Chen, L. Operation performance and seepage flow of impervious body in blast-fill dams using discrete element method and measured data. *Water* **2022**, *14*, 1443. [CrossRef]
21. Fukumoto, Y.; Ohtsuka, S. 3-D direct numerical model for failure of non-cohesive granular soils with upward seepage flow. *Comput. Part. Mech.* **2018**, *5*, 443–454. [CrossRef]
22. Issakhov, A.; Borsikbayeva, A. The impact of a multilevel protection column on the propagation of a water wave and pressure distribution during a dam break: Numerical simulation. *J. Hydrol.* **2021**, *598*, 126–212. [CrossRef]
23. Dibike, Y.B.; Shakibaeinia, A.; Droppo, I.G.; Caron, E. Modeling the potential effects of Oil-Sands tailings pond breach on the water and sediment quality of the Lower Athabasca River. *Sci. Total Environ.* **2018**, *642*, 1263–1281. [CrossRef] [PubMed]
24. Jing, X.; Chen, Y.; Williams, D.J.; Serna, M.L.; Zheng, H. Overtopping failure of a reinforced tailings dam: Laboratory investigation and forecasting model of dam failure. *Water* **2019**, *11*, 315. [CrossRef]

25. Mahdi, A.; Shakibaeinia, A.; Dibike, B.Y. Numerical modelling of oil-sands tailings dam breach runout and overland flow. *Sci. Total Environ.* **2020**, *703*, 134–568. [CrossRef] [PubMed]
26. Naeini, M.; Akhtarpour, A. Numerical analysis of seismic stability of a high centerline tailings dam. *Soil Dyn. Earthq. Eng.* **2018**, *107*, 179–194. [CrossRef]
27. Boroomand, M.R.; Mohammadi, A. Evaluation of earth dam leakage considering the uncertainty in soil hydraulic parameters. *Civ. Eng. J.* **2019**, *5*, 1543–1556. [CrossRef]
28. Li, S.; Yuan, L.; Yang, H.; An, H.; Wang, G. Tailings dam safety monitoring and early warning based on spatial evolution process of mud-sand flow. *Saf. Sci.* **2020**, *124*, 104–579. [CrossRef]
29. Kong, X.; Liu, Y.; Li, X.; Li, R.; He, Y. Analysis and experimental study of mechanical factors of permeation damage of a uranium tailings pond dam. *Eng. Saf. Environ. Prot.* **2019**, *45*, 30–36. (In Chinese)
30. He, W.; Chen, H.; Zheng, C. Experimental study on dam break of a tailings dam under flood conditions. *J. Saf. Environ.* **2022**, *22*, 3126–3134. (In Chinese)

water

Article

Study on the Hydraulic Response of an Open-Channel Water Transmission Project after Flow Switching

Naifeng Zhang [1], Honglei Ren [2] and Fei Lin [3,4,*]

1 Key Laboratory of Water Conservancy and Water Resources of Anhui Province, Water Resources Research Institute of Anhui Province and Huai River Water Resources Commission, Ministry of Water Resources, Bengbu 233000, China; zhangnf0901@126.com
2 College of Civil Engineering, Hefei University of Technology, Hefei 230009, China; renhonglei2021@163.com
3 Institute of Intelligent Machines, Hefei Institutes of Physical Science, Chinese Academy of Sciences, Hefei 230031, China
4 Hefei Institutes of Collaborative Innovation for Intelligent Agriculture, Hefei 231131, China
* Correspondence: feilin@iim.ac.cn

Abstract: For the complex and changing water transfer conditions along the nullah water transfer project, it is of great significance to analyze the hydraulic response characteristics of the channel pool to ensure the safe and efficient operation of the project. In this paper, using the adjustable reservoir capacity of the channel pool, combined with the earliest and latest regulation time, a method is proposed to determine the optimal regulation time of water conservancy engineering structures through the hydraulic response characteristics. In order to validate the method, the hydraulic response characteristics of the channel pool during the whole regulation period were investigated using an actual river diversion project as an example. By using the Sobol global sensitivity analysis method, the sensitivity analysis of three variables (water transfer flow, upstream initial flow, and downstream water level) was conducted, and the results showed that the change of water transfer flow was the most-critical factor affecting the optimal regulation time. In order to quantitatively study the correlation between the water transfer flow and the optimal regulation time, the water transfer flow factor was coupled to the existing prediction model, and the results showed that the coupled model can efficiently predict the optimal regulation time, and the NSE coefficient can reach 0.98. In addition, the coupled model can be efficiently used to serve on-site dispatchers for accurately predicting the optimal regulation time and making the corresponding safety decisions.

Keywords: open channel; water transfer project; hydraulic response characteristics; regulation time

1. Introduction

The spatial and temporal distribution of water resources in China is uneven, with more water resources in the south and fewer in the north, and regional water resources' supply and demand conflicts are prominent. To optimize the allocation of water resources, many large open-channel water diversion projects have been constructed in various regions [1]. However, the open-channel water transfer project has a long water transmission line, involves a wide range, and is difficult to control. During the operation process, various external factors such as weather will affect the normal operation of the project, which further increases the work intensity of dispatchers. Analyzing the hydraulic response characteristics of canals and ponds can provide important technical support for the actual operation of the project; thus, many scholars have conducted in-depth research in this direction by constructing hydrodynamic models. In open-channel water transfer projects, the length of the channel is usually much larger than the width and depth of the water transfer section, so the hydrodynamic simulation model of the whole system can be approximated as a one-dimensional problem [2]. Qi Jiashuo, Li Kangbin [3,4], and others simulated the variation trend of the channel water level under different gating control schemes through a one-dimensional hydrodynamic model and analyzed various hydraulic response characteristics to provide a data reference for managers to control the gating. Schuurmans, Bautista, and Paris [5–7] all used the hydraulic response characteristics of the channel to estimate the feedforward control time of the regulating building. Zheng Hezhen and Wahlin [8,9] built a one-dimensional hydrodynamic model to analyze the internal relationship between the water level, flow rate, head, regulation time, and storage capacity between the canals and ponds of each pumping station and optimized the predictive control method of the water delivery system and the pumping station control strategy. Lu Longbin, Li Mengxuan, Liu Zhiyong [10–12], and others have all deeply described the evolution of water conditions in separate pumping stations under emergency conditions by analyzing the trend of water level changes before and after the pumping station, which can provide support for engineering emergency treatment. Zhang et al. [13] analyzed the hydraulic control conditions for the operation of the dumping station in the open-channel water transfer project based on the operation case of the dumping station of the pipeline water delivery system and proposed the optimal operation mode of the dumping station in the open channel. Yan et al. [14] constructed a single-canal pond model, assuming that the water level in front of the downstream pumping station was a fixed value, reversed the flow change of each section, and judged the optimal regulation time of the pumping station based on the flow change trend of the downstream pumping station. By coupling a one-dimensional hydrodynamic model with a multi-objective genetic algorithm and taking the gate opening as a decision variable, they generated a cascade gate group adjustment scheme [15]. In other studies, the hydraulic response relationship has been extended to the field of channel control. Among them, the lag time in feedforward control is another manifestation of the hydraulic response relationship, and many scholars [16–18] use known perturbations to calculate the feedforward control strategy of the throttle gate based on the lag time. Subsequent unpredictable disturbances within the channel occur from time to time, decreasing the ability to perceive water conditions within the channel. Thus, predictive control models have emerged to solve such problems. The integral delay (ID) model is currently used as a predictive model. The principle of the ID model is to divide the channel into uniform and backwater zones according to the hydraulic response relationship and use the hydraulic response relationship to predict the evolution of the water level change and flow rate in the channel. This control method has been widely used in water transfer and irrigation projects [19–22]. However, all of the above articles focus on the accurate control of the water level, ignoring the role of its storage capacity in the regulation process. Therefore, this paper adopted the method of interval water level control to give full play to the application of its storage capacity in water transfer projects. Through the fast-fitting equation, the feasible region of the safe control time of the channel pool is calculated quickly.

To make the feasible region calculation results better applicable to practical engineering, it is also necessary to find a simple and quick calculation method. Based on historical data or the calculation results of the mechanism model, the simplified method of fitting the objective function's fast calculation equation has been widely used in many industries of water conservancy. Wang Jianping [23] deduced the calculation equation of the equilibrium scour depth by analyzing the influencing factors of the scour depth of the open channel under different working conditions using dimensional analysis and regression analysis methods. Xia et al. [24], based on the measured data of hydrology and sediment in the Jiaojiang River Estuary for many years, combined with the comparative relationship between the turbulent action of water flow and the action of sediment gravity, obtained an empirical equation for the sediment-carrying capacity suitable for the study area. By analyzing the relationship between the upstream initial flow and the optimal regulation time interval, Li Guhan et al. [25] used the least-squares method for regression analysis to obtain an empirical equation that can quickly generate the hydraulic regulation scheme of the Shaping secondary power station. The above research analyzes the relationship between the parameters and selects the key influencing factors to formulate the coupling of the independent variables and dependent variables according to the level of correlation, to achieve the rapid calculation of the target values.

Based on the research on the hydraulic response characteristics of conventional canal tanks, this study proposes a quick calculation method for safe regulation intervals using the relationship between regulation volume and hydraulic response. By analyzing the earliest and latest regulating times of each load type within a project case study, we calculate the feasible safe regulating time range for the pumping station and determine the regulation scheme based on the quick regulating formula. This computational method is compared with the traditional method for controlling the water level, with the main differences being as follows. First, the regulating method can provide full play to the buffer function of the channel-water-storage process during scheduling, in line with actual needs, and can meet various scheduling requirements in water diversion projects. Second, the general dispatch method only regulates and controls relatively small water level fluctuations. This approach can result in an excessive frequency of throttle gate adjustments, greatly increasing the workload of the dispatcher. However, by adopting the adjusted interval method proposed in this paper, one can make full use of channel ponds to provide flexible storage, allowing for certain interval fluctuations in the water levels, reducing the frequency of throttling gate adjustments and reducing the workload of scheduling personnel. Third, in actual project calculations, adopting a quick computation formula for adjusting the interval can improve the response efficiency of the diversion canal pool, ensure its operational safety, and provide theoretical guidance and technical support for project construction, management, and operation.

2. Research Methods

During the operation stage of the water diversion project, to meet the goals of downstream water supply, ships passing through the locks, and ecological water replenishment, dispatchers need to adjust the rotation angle of the pumping station, the number of units to be turned on to meet the flow demand. Operation plays a vital role. To prevent the water level at the boundary of the canal and pond from exceeding the upper and lower limits due to the untimely adjustment of the pumping station, the earliest and latest regulation times of the pumping station based on the mechanism model are proposed. Aimed at preventing engineering safety accidents, which may be caused by large changes in the water level at the boundary of the canal and pond due to the increase in upstream water, by constructing a one-dimensional non-constant flow hydrodynamic numerical simulation model of the coupled pump station, the prediction of the water regime of each section can be realized, and it is feasible to combine the safety regulation time. The domain can provide technical guidance for the actual operation of the project.

2.1. Governing Equations and Solutions

Based on the Saint-Venant equations of one-dimensional unsteady flow, a one-dimensional hydrodynamic simulation model is constructed [26], including the continuity equation and the momentum equation [27]. The equations are expressed as follows:

$$\begin{cases} B\frac{\partial H}{\partial t} + \frac{\partial Q}{\partial x} = q \\ \frac{\partial Q}{\partial t} + \frac{\partial}{\partial x}\left(\frac{\alpha Q^2}{A}\right) + gA\frac{\partial H}{\partial x} + gAS_f = 0 \end{cases}, \tag{1}$$

where B is the width of the water-passing section, m; H is the water level, m; Q is the flow rate, m^3/s; g is the acceleration of gravity, m/s^2; x is the space coordinate of the section, m; t is the time coordinate, s; q is the side inflow, m^3/s; α is the momentum correction coefficient; A is the water passage area, m^2; and S_f is the friction ratio drop, which can be expressed by the following equation:

$$S_f = \frac{{n_c}^2 Q|Q|}{A^2 R^{4/3}}, \tag{2}$$

where n_c is the Manning roughness coefficient of the water conveyance channel and R is the hydraulic radius, m.

The above-mentioned Saint-Venant equations belong to the first-order quasi-linear hyperbolic partial differential equations. Generally, the analytical solution cannot be directly obtained, and the approximate solution can only be obtained by the numerical discrete method. After comprehensive comparison, this model uses the Preissmann four-point weighted implicit difference scheme with good stability, high calculation accuracy, and fast convergence speed to discretize the Saint-Venant equations [28] with the chase method [29]. The discrete grid form [30] is shown in Figure 1.

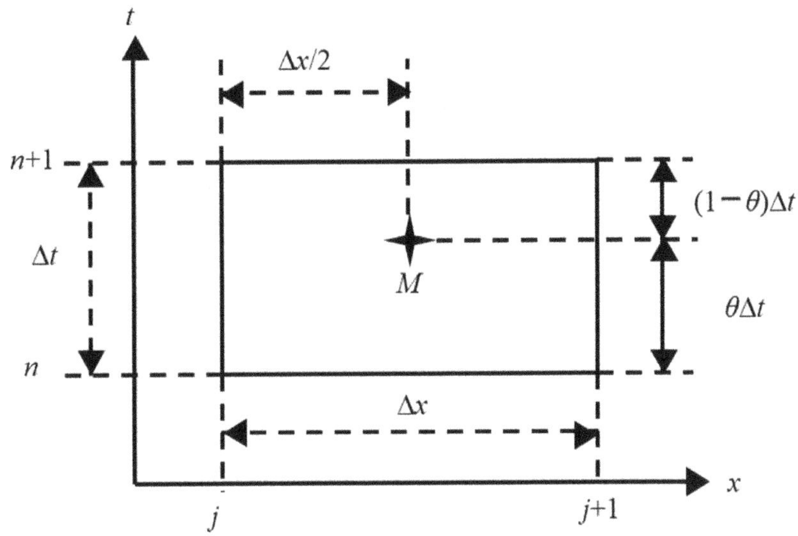

Figure 1. Preissmann weighted implicit four-point scheme.

2.2. Model Internal Boundary Conditions' Generalization

The solution to one-dimensional unsteady flow in an open channel involves the setting of external and internal boundary conditions. The internal boundary refers to the nodes with obvious changes in hydraulic properties or geometric shapes, but the Saint-Venant equations do not apply to such nodes, so the internal boundary needs to be generalized [31]. As one of the important control buildings in the open-channel water diversion project, the pumping station can raise the water resources from the lower ground to a higher place and effectively improve the distribution of water resources [32]. In the pumping station, the combined operation of multiple units is usually used to lift water. The dispatcher adjusts the operating conditions of the pumping station by changing the number of units, the type of units, the rotation angle, and the speed of the unit.

When the pump station is used as the inner boundary, the continuity equation and the flow–head–rotation angle characteristic curve of each pumping device can be used for processing, and the losses in the station can be ignored at the same time. From a macro point of view, multiple units in the same pumping station work in parallel, and they should be treated as a whole in the model. The horizontal addition method can be used to horizontally add the flows under the same head on each performance curve to obtain the lower head. There is information to explain the total flow of the pumping station [33]. When using the model, it is calculated according to different flow–head–rotation angle equations.

The equation of continuity is:

$$Q_k = Q_{k+1}, \tag{3}$$

where Q_k and Q_{k+1} represent the inlet and outlet flow of the pump (m^3/s), respectively.

The relationship between the pump flow and the water level of the canal pool satisfies Equation (4).

$$H_{k+1} - H_k = a_k Q_k{}^2 + b_k Q_k + c_k, \tag{4}$$

where H_{k+1} is the water level of the outlet pool (m); H_k is the water level of the inlet pool (m); and a_k, b_k, and c_k are the relevant parameters of the unit characteristic curve.

To solve the Saint-Venant equations simultaneously, it is necessary to linearize Equation (4), and the equation can be discretized according to the Preissmann four-point equation mentioned above:

$$-Q_k^{n+1} + 0 + Q_{k+1}^{n+1} + 0 = 0, \tag{5}$$

Equation (4) accords with the same method; the continuity equation is expanded by the Taylor series, and only the second-order terms are retained. Assuming that the inlet and outlet flows of the pumping station are equal, the governing equation of the pumping station is as follows:

$$0 - H_k^{n+1} - (2a_k Q_{k+1}^n + b_k)Q_{k+1}^{n+1} + H_{k+1}^{n+1} = -a_k Q_{k+1}^n{}^2 + c_k, \tag{6}$$

Equations (5) and (6) following the four-point Preissmann format on the continuity equation and momentum equations are dispersed into discrete form and Saint–Venant equations. The last parts of Equations (5) and (6) coupled to the channel model, a simultaneous solution, finally realize the solution of the pumping station. The Saint-Venant equations of the discrete process due to the derivation process are relatively complicated. Due to the limited space available, please refer to [34] for details.

2.3. Safety Regulation Time Frame

In this study, the channel hydrological simulation was carried out through the hydrodynamic model. The minimum and maximum operating water level of the single channel pond boundary composed of upstream control buildings and inner boundary control buildings in the study area were taken as the water level overrun threshold, and the water level overrun time was calculated by the model. When the upstream flow mutation occurs, to maintain the inflow and outflow balance of the canal pool, the downstream buildings must be regulated. Because the drainage pool flow wave transmission has a time lag, when the time is less than the flow downstream structure adjustment wave transmission time, it will inevitably produce buildings near the water level decrease, and the drainage pool is the lowest water level operation, so for the structure of the control time, there must be a minimum regulation time to ensure the water level after regulation or just close to the lowest water level operation. This is the earliest regulation time. When the regulation time of the downstream building is greater than the flow propagation time, the water level near the building must rise, and then, there is the highest operating water level in the channel pool. Therefore, there must be a maximum regulation time for the building to make the regulated water level just at or close to the highest operating water level, which is the latest regulation time. The time difference between the latest and the earliest regulation time is the safe regulation domain value. By calculating the earliest and latest regulation time, the safety regulation time range of the pump station can be obtained, which provides technical support for the actual operation of the project. The earliest and latest regulation times are shown in Figure 2.

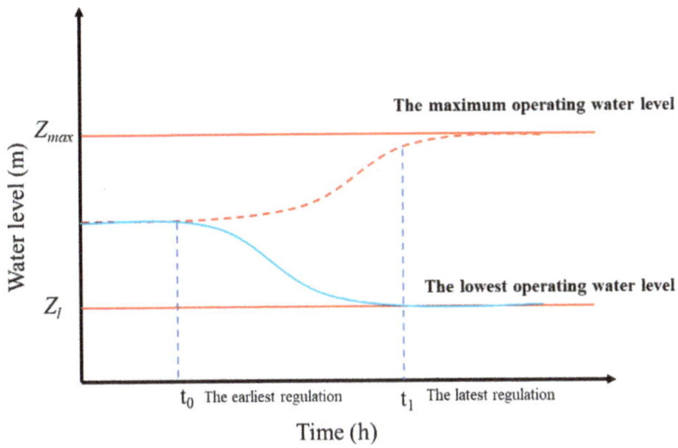

Figure 2. The earliest and latest regulation times.

2.4. Sobol Global Sensitivity Analysis

The Sobol sensitivity analysis was proposed by Sobol in 1993. This method is centered on decomposing the total variance of the model inputs into the sum of the variance of each parameter and the variance of the parameter interactions. Then, the sensitivity of each parameter is identified by calculating its contribution to the variance of the output parameter.

Suppose the model is $Y = f(x) = f(x_1, x_2,..., x_n)$, $X = (x_1, x_2,..., x_n)$ is the parameter vector, and n is the number of parameters; the variance decomposition equation is expressed as:

$$V = \sum_{i=1}^{n} V_i + \sum_{i<j}^{n} V_{i,j} + V_{1,2,\cdots,n}, \quad (7)$$

$$S_T = V_i/V, \quad (8)$$

where V represents the total variance of the model; V_i represents the variance for the effect of the i-th parameter; $V_{i,j}$ represent the variance of the i-th and j-th parameters acting together; $V_{1,2,...,n}$ represents the variance of all parameters acting together; and S_T represents the full sensitivity coefficient

2.5. Overview of the Region

The Yangtze River and Huaihe River Diversion Project connects the Yangtze and Huaihe River systems and is one of the 172 major water-saving and water-supply projects designated by the State Council. From south to north, the Yangtze River and Huaihe River Diversion Project (Anhui section) is divided into three major sections: the Yangtze River and the Chaohu Lake Diversion Project, the Chaohu Lake and Huaihe River Diversion Project, and the Huaihe River and Henan Province Diversion Project. A total of 12 pumping stations are being set up during the first phase of the project, including Fenghuangjing pumping station, Zongyang pumping station, Paihekou pumping station, Shushan pumping station, etc. The project covers Anhui and Henan provinces, including Anqing, Wuhu, Ma'anshan, and Hefei in Anhui Province and Zhoukou and Shangqiu in Henan Province, covering a total of 15 cities and 55 counties. After the completion of the project, it will be able to supply water from the Yangtze River to the water-scarce areas in the north, promote the development of navigation in Jianghuai, and improve the water ecological environment of Caizi Lake, Chaohu Lake, Wabu Lake, and Huaihe River. In addition, it will inject new vitality into the social and economic development of Anhui and Henan Provinces.

The water diversion project from the Yangtze River to the Huaihe River has a large water flow, long lines, and a wide range. The Xiaohe branch line controls the intake gate to divert the water from the Yangtze River and Chaohu Lake from the Baishitian River mouth to the Paihekou pumping station. It is the key interval for the whole project to link the past and the future. This section is a newly excavated open channel. Affected by the topography, the hydraulic response of the canal pool is more obvious than in other canals, and this section involves the joint regulation and single-stage regulation of the gate pump group. The normal operation of the project is of great significance. Therefore, this section was selected as the research area in this study, and the control inlet gate of the Xiaohe branch line is used as the upstream control node, while the Shushan pumping station is the downstream control node. An overview of the canal section is shown in Figure 3.

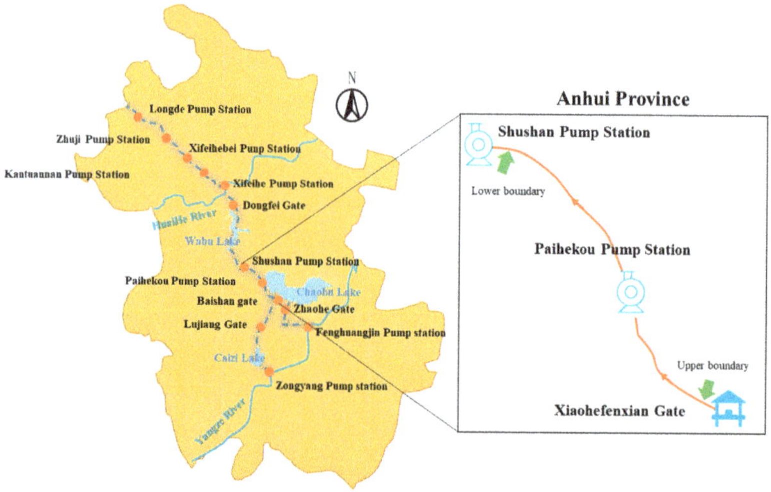

Figure 3. Study area graph.

At the same time, the Paihekou pumping station also undertakes the task of controlling the water level of the Paihekou ship lock. When a ship passes through the lock, the pumping station needs to be turned on to raise the water level of the lock chamber to the Paihe water level, which can effectively connect with the Jianghuai shipping industry. The designed water supply flow of the ship lock is 6.5 m^3/s.

The design of the flood control level in front of the Xiaohe dividing line control inlet gate is 11.46 m; the designed water diversion level behind the gate is 6.0 m, and the bottom elevation is 0.6 m. The designed water level under the Shushan pumping station is 7.6 m; the lowest operating water level is 5.8 m, and the highest operating water level is 10.7 m. Paihekou pumping station has a single design flow rate of 37.69 m^3/s and operates in 8 use modes and 1 standby mode, and the designed water level below the station is 4.1 m. The specific information is shown in Table 1 and Figure 4.

Table 1. Engineering foundation information of Paihekou pumping station and Shushan pumping station.

The Pump Station	Type	Design Flow (m^3/s)	Number	Design Flow (m^3/s)	Water Level in Front of Station		
					Design (m)	Max (m)	Min (m)
Paihekou	Axial flow	37.69	9	301.5	4.10	5.80	3.60
Shushan	Axial flow	36.25	8	290	7.60	10.70	5.80

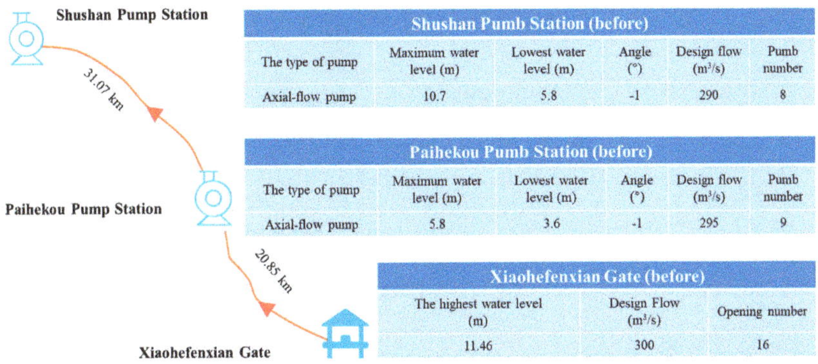

Figure 4. Specific parameters of the study area.

2.6. Hydrodynamic and Parameter Verification

The core of the verification of the hydrodynamic model is to verify the parameters of the hydrodynamic model. The parameters of the hydrodynamic model are mainly divided into fixed parameters and variable parameters, and the fixed parameters are mainly the parameters of the hydraulic components related to the line. The variable parameters mainly include hydrodynamic initial conditions, boundary conditions, calculation time, calculation step, pump station angle, river roughness, etc. This paper mainly verifies the variable parameters of the hydrodynamic model.

In order to test the simulation effect of the hydrodynamic model, a hydraulic model of the Shushan pumping station controlled by small joint and branch lines was constructed. The upstream flow through the sluice controlled by small joint and branch lines was used as the flow boundary; the downstream water level in front of the Shushan pumping station was selected as the water level boundary, and the measured water level of the Paihekou pumping station was used as the comparison to verify the accuracy of the model. Therefore, the output step size of the model was selected as 2 h/time; the simulation time was 60 h, and the roughness of the canal was the designed value of 0.015. The simulation results are shown in Figure 5.

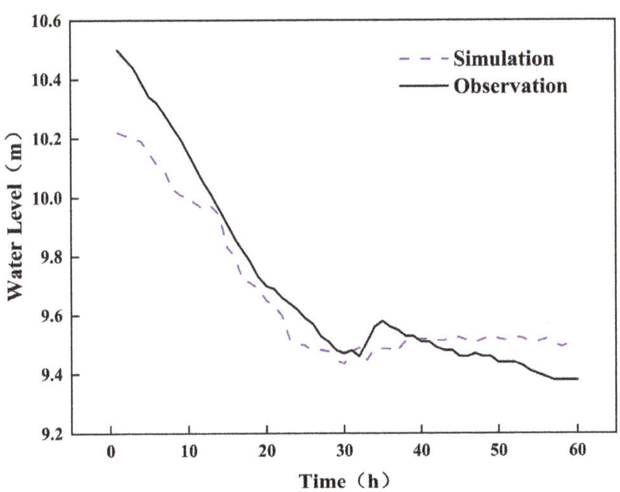

Figure 5. Comparison of the actual observation and simulation results.

As can be seen from the above figure, the simulation accuracy of the hydrodynamic model tends to be consistent with the measured value. The maximum deviation of the simulation was 0.26 m, and the minimum deviation was only 0.01 m; the NSE = 0.86. The simulation accuracy was expected and met the requirements of subsequent hydraulic response analysis.

3. Project Case

3.1. Operation Setting

This study took the upstream and downstream canals of the Paihekou pumping station as an example, using the Xiaohe branch line to control the flow of the intake gate as the upper boundary, the Paihekou pumping station as the inner boundary, and the water level in front of the Shushan pumping station as the lower boundary. The influence of the pumping station unit adjustment on the hydraulic response characteristics of the channel was studied. During the actual operation of the project, affected by the time-of-use electricity price, the pumping stations along the line usually chose asynchronous regulation to reduce the costs. At the same time, to make full use of the self-adjusting ability of the canal and pond, the dispatcher will also adjust the time according to the actual choice. However, when the water intake of the channel increases and the downstream pumping station is not adjusted in time, the water level in front of the station will gradually rise until it exceeds the limit. According to the actual needs of maintaining the safe operation of the project, the earliest and latest regulation time of the pump station were studied, and the adjustable time range was provided for the dispatchers. It was assumed that the current flow in and out of the channel is balanced, and the overflow of the inlet sluice is controlled by adjusting the small junction line according to the single flow value of the Paihekou pumping station. The water level of the downstream Shushan pumping station was fixed, including six working conditions of 5.8, 6.6, 7.6, 8.6, 9.6, and 10.7 m. Table 2 shows all the conditions of the flow changes at the upstream boundary for the regulation time.

Table 2. Flow changes at the upstream boundary.

Initial Upstream Flow (m^3/s)	0	37.69	75.38	113.07	150.76	188.45	226.14	263.83
Flow change (m^3/s)	37.69	–	–	–	–	–	–	–
	75.38	75.38	–	–	–	–	–	–
	113.07	113.07	113.07	–	–	–	–	–
	150.76	150.76	150.76	150.76	–	–	–	–
	188.45	188.45	188.45	188.45	188.45	–	–	–
	226.14	226.14	226.14	226.14	226.14	226.14	–	–
	263.83	263.83	263.83	263.83	263.83	263.83	263.83	–
	301.5	301.5	301.5	301.5	301.5	301.5	301.5	301.5

3.2. Feasible Domain of Safety Regulation Time

Through the hydrodynamic model calculations, the downstream overrun time is significantly shorter than the upstream overrun time, so the water level in front of the Paihekou pumping station is not overrun as the regulation target. The earliest regulation time is 0 h, and the latest regulation time is calculated following the order of the working conditions set in Table 2, as shown in Figure 6.

As can be seen from Figure 6a,b, for a drainage basin, as long as the initial flow rate, downstream water level, and the amount of flow variation are determined, there will be the latest regulation time, and for the same initial flow rate and downstream water level, the regulation time will decrease as the amount of flow variation increases; at the same initial flow rate and amount of flow variation, the latest regulation time will decrease as the water level increases. For the same flow change, the relationship between the downstream water level and the initial flow and the latest regulation time, as can be seen from Figure 6c, when the flow change is the same, with the same downstream water level, the latest regulation time with the initial flow increases and then decreases the trend.

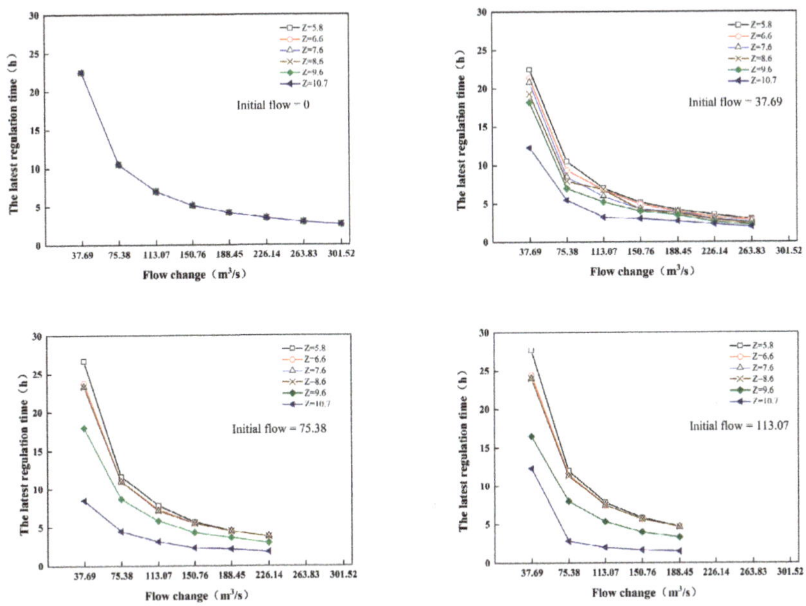

(a)

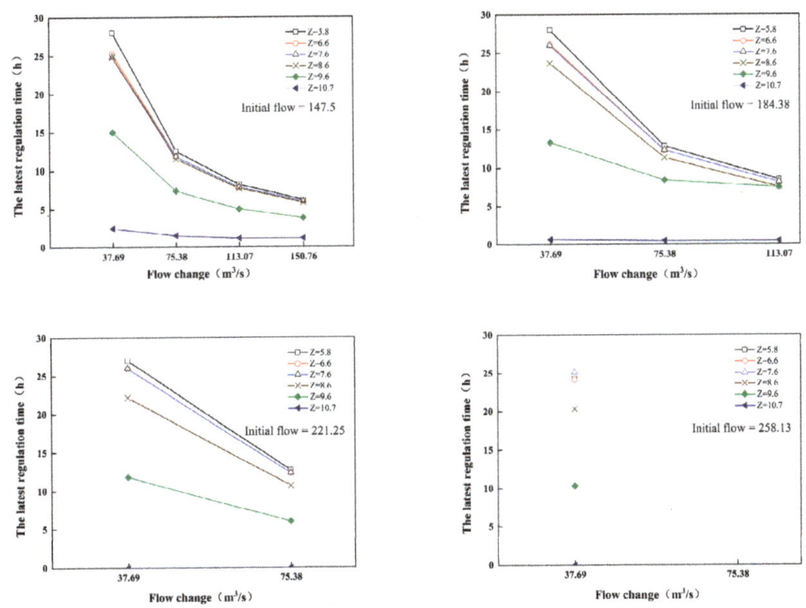

(b)

Figure 6. *Cont.*

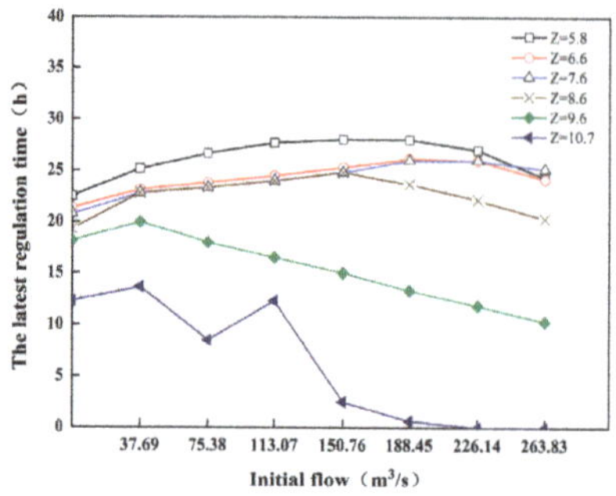

(c)

Figure 6. The result chart of the latest regulation time. (**a**) Initial flow of 0–113.07 m^3/s. (**b**) Initial flow of 145.5–258.13 m^3/s. (**c**) Relationship between downstream water level, initial flow, and latest regulation time for the same amount of flow change.

As can be seen from Figure 6, the hydraulic response characteristics can be expressed as a function of some kind, so a numerical method can be used to fit the hydraulic response characteristics to the relational equation, thus providing theoretical support for the rapid regulation of the channel.

3.3. *Parametric Sensitivity Analysis*

To quickly calculate the regulation time, the Sobol global sensitivity analysis method was used to analyze the most-sensitive factor for the regulation time, which results in a fast prediction equation for calculating the regulation time that is applicable in different situations.

3.3.1. Sobol Global Sensitivity Analysis Method

Parameter sensitivity analysis refers to the degree of contribution of the independent variables of the function to the dependent variable of the function. The more-sensitive independent variables are also more important for the dependent variable, and the selection of the more-sensitive independent variables is particularly important for generating a fast calculation of the regulation time equation.

In the process of calculating the regulation time for the hydrodynamic model, the amount of flow variation, the initial flow, and the downstream boundary level of the model all affect the regulation time, so the Sobol method was used to analyze the sensitivity of the three variables above and select the best-fitting variable. The process used the Sobol sequence sampling method to obtain 1200 parameter samples, which were substituted into the hydrodynamic model for cyclic calculations. The Nash–Sutcliffe efficiency coefficient was chosen to characterize the degree of agreement between the calculated and measured values of the model. As the focus of the pumping station is to consider the effect of different parameters on the water level changes, the pre-station water level at the Paihekou pumping station was the objective function.

3.3.2. Analysis of the Results of the Sobol Method

By comparing the total sensitivity index of the Sobol method, the multi-factor problem can be transformed into a fewer-factor problem or a single-factor problem, which greatly simplifies the work. The results showed that the total sensitivity of the amount of flow variation is higher than that of the initial flow and downstream water level with the water level in front of the Paihekou pumping station as the objective function. Therefore, the amount of flow variation was determined to be the most-sensitive factor affecting the regulation time. The sensitivity indicators for each parameter of the Sobol method are shown in Figure 7.

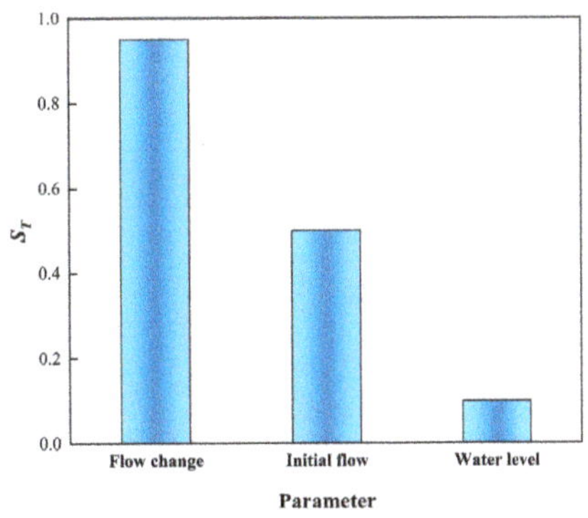

Figure 7. Parameter-sensitive indicator graph.

3.4. Fast Computational Equation Fitting

In the actual project, to protect the safety of the channel, the Paihekou pumping station often runs three units and stabilizes the water level to operate at the designed level, thus saving energy and achieving safe water transfer. For this reason, the fitting equation was selected for the working conditions of a selected initial flow rate of 113.03 m^3/s and a downstream water level of 7.6 m.

According to the results of the parameter sensitivity analysis in Section 3.1, it is known that the amount of change in flow is the most-sensitive factor for the regulation time, so the amount of change in flow was used as the independent variable, and the latest regulation time was used as the dependent variable to fit the fast equation for the regulation time. The univariate fitting equation due to the form of the fitting function has a variety of methods to select the best-fitting equation, so there is a need to determine the specific form of the fitting equation. For this purpose, the linear function, exponential function, power function, and logarithmic function were selected for the initial flow rate of 113.07 m^3/s. The downstream water level of 7.6 m was chosen for the conditions to fit using Matlab's toolbox. The results of the fitting are shown in Table 3.

Table 3. Equation fitting form.

Number	Variable	Fitting Equation Form	R^2
1	Flow change	Linear function fitting	0.77
2		Exponential function fitting	0.92
3		Logarithmic function fitting	0.90
4		Power function fitting	0.99

As shown in Table 3, in the form of the fitting equation, the single variable should be fit in the form of a power function.

A univariate equation was used for the fit, and the equation for the fit is:

$$H_t = 961.68x^{-1.022} \tag{9}$$

where H_t is the latest time for regulation, h, and x is the flow change, m^3/s.

3.5. Test of the Fitting Equation

To test the accuracy of the fitting equation, different amounts of flow variation were selected to test the general applicability of the fitting equation, and the selected working conditions are shown in Table 4, while the results of the test are shown in Figure 8.

Table 4. Flow switching conditions tested.

Number	Flow Change (m^3/s)	Number	Flow Change (m^3/s)
1	10	6	60
2	20	7	70
3	30	8	80
4	40	9	90
5	50	10	100

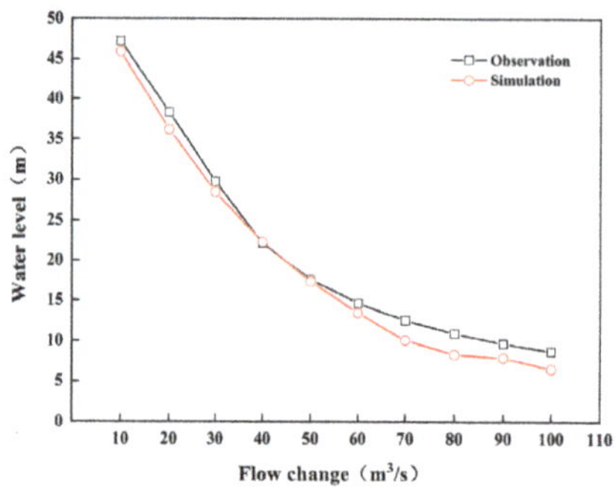

Figure 8. The model validation results.

As demonstrated in Figure 8, the simulation performed better with the NSE = 0.98. The latest regulation time derived by fitting the equation tended to be spatially consistent with the latest regulation time predicted by the model. Thus, it can be demonstrated that the engineering dispatching clock can be perfectly adapted to the most-recent regulation time determined by interpolation through the method of fitting the equation, replacing the conventional hydraulics model, and offering safe dispatching decisions for the dispatchers. This can lessen the workload of the schedulers.

4. Conclusions

This paper presented the idea of the earliest and latest regulation time of the regulation building under the flow switching conditions of the open-channel water transfer project with engineering safety as the objective and used the upstream and downstream of the Paihekou pumping station of the river diversion project as the study area for the example analysis. This concept is based on the one-dimensional non-constant flow dynamics numerical simulation model. The key findings are listed below:

(1) The latest regulation time at the Paihekou pumping station upstream and downstream was calculated in this study to determine the effects of the flow change, downstream water level, and upstream initial flow. It was found that, for the same initial flow and downstream water level, the regulation time decreased with the increase in the flow variation volume; for the same initial flow and flow variation volume, the latest regulation time decreased with the increase in the water level. The latest regulation time with the starting flow increased initially and subsequently reduced in a trending format for the same flow variation and downstream water level conditions.

(2) The overall sensitivity index of the flow variation was found to be at the maximum when using the Sobol global sensitivity analysis method at the latest regulation time, which is connected to the starting upstream flow, the flow variation, and the downstream water level. At the same time, it was suggested for the method of the fitting equation to be used for the calculation of the most-recent regulation time in order to calculate it rapidly. The hydrodynamic model was used for testing, and the amount of flow change was chosen as a one variate to suit the computation of the

most-recent regulation time. The results showed that the effect of the fit was better with the computed NSE = 0.98, and the fitting equation can be a perfect replacement.

Author Contributions: N.Z. and H.R.: conceptualization, data curation, formal analysis, investigation, methodology, software, validation, visualization, writing—original draft, and writing—review and editing. F.L.: resources, supervision, funding acquisition, and project administration. All authors have read and agreed to the published version of the manuscript.

Funding: This research was funded by the National Key Research and Development Program of China, Grant Number 2017YFC0404504, and the National Water Pollution Control and Treatment Science and Technology Major Project, Grant Number 2017ZX07602-002.

Data Availability Statement: Not applicable.

Conflicts of Interest: The authors declare no conflict of interest.

References

1. Zhang, Z.; Tian, Y.; Lei, X.H. Study and application of general unsteady flow simulation model for complex open channel water transfer Project. In Proceedings of the 2018 CHES Annual Conference, Amsterdam, The Netherlands, 9–12 September 2018.
2. Litrico, X.; Fromion, V. Simplified Modeling of Irrigation Canals for Controller Design. *J. Irrig. Drain. Eng.* **2004**, *130*, 373–383. [CrossRef]
3. Qi, J.S.; Cui, Y.R.; Wang, S.Y. Analysis of Hydraulic Response in Emergency Gate Control of the Water Supply Project in North Xinjiang. *J. Water Resour. Archit. Eng.* **2021**, *19*, 250–256.
4. Li, K.B.; Shen, B.; Li, Z.L. Open channel hydraulic response characteristics in irrigation area based on unsteady flow simula-tion analysis. *Trans. Chin. Soc. Agric. Eng.* **2015**, *31*, 107–114.
5. Schuurmans, J.; Clemmens, A.J.; Dijkstra, S.; Hof, A.; Brouwer, R. Modeling of Irrigation and Drainage Canals for Controller Design. *J. Irrig. Drain. Eng.* **1999**, *125*, 338–344. [CrossRef]
6. Bautista, E.; Clemmens, A.J. Volume Compensation Method for Routing Irrigation Canal Demand Changes. *J. Irrig. Drain. Eng.* **2005**, *131*, 494–503. [CrossRef]
7. Parrish, J. *A Methodology for Automated Control of Sloping Canals*; University of Iowa: Lowa City, IA, USA, 1997.
8. Zheng, H.Z.; Tian, Y.; Wang, C. Study on the Predictive Control Method of Cascade Pumping Stations in Water Transfer System. *Yellow River* **2018**, *40*, 140–143.
9. Wahlin, B.T.; Clemmens, A.J. Automatic downstream water-level feedback control of branching canal networks: Simula-tion results. *J. Irrig. Drain. Eng.* **2006**, *132*, 208–219. [CrossRef]
10. Lu, L.B.; Lei, X.H.; Tian, Y. Research on the Characteristic of Hydraulic Response in Open Channel of the Cascade Pumping Station Project in an Emergent State. *J. Basic Sci. Eng.* **2018**, *26*, 263–275.
11. Li, M.X.; Tian, Y.; Lei, X.H. Analysis onemergency response time for accidental pump—Stop at multi—Stage pumping station of Water Diversion Project from Yangtze River to Huaihe River. *Yangtze River* **2021**, *52*, 212–217.
12. Liu, Z.Y.; Liu, M.Q. Analysis and Control of Pump-stopping Hydraulic Transient in Cascade Water Supply Pumping Station. *China Rural Water Hydropower* **2006**, *7*, 32–34.
13. Zhang, Z.; Lei, X.; Tian, Y.; Wang, L.; Wang, H.; Su, K. Optimized Scheduling of Cascade Pumping Stations in Open-Channel Water Transfer Systems Based on Station Skipping. *J. Water Resour. Plan. Manag.* **2019**, *145*, 05019011. [CrossRef]
14. Yan, P.; Zhang, Z.; Lei, X.; Zheng, Y.; Zhu, J.; Wang, H.; Tan, Q. A Simple Method for the Control Time of a Pumping Station to Ensure a Stable Water Level Immediately Upstream of the Pumping Station under a Change of the Discharge in an Open Channel. *Water* **2021**, *13*, 355. [CrossRef]
15. Zhu, J.; Zhang, Z.; Lei, X.; Yue, X.; Xiang, X.; Wang, H.; Ye, M. Optimal Regulation of the Cascade Gates Group Water Diversion Project in a Flow Adjustment Period. *Water* **2021**, *13*, 2825. [CrossRef]
16. Belaud, G.; Litrico, X.; Clemmens, A.J. Response Time of a Canal Pool for Scheduled Water Delivery. *J. Irrig. Drain. Eng.* **2013**, *139*, 300–308. [CrossRef]
17. Burt, C.M.; Feist, K.E.; Piao, X.S. Accelerated Irrigation Canal Flow Change Routing. *J. Irrig. Drain. Eng.* **2018**, *144*, 6. [CrossRef]
18. Liao, W.; Guan, G.; Tian, X. Exploring Explicit Delay Time for Volume Compensation in Feedforward Control of Canal Systems. *Water* **2019**, *11*, 1080. [CrossRef]
19. Kong, L.; Li, Y.; Tang, H.; Yuan, S.; Yang, Q.; Ji, Q.; Li, Z.; Chen, R. Predictive control for the operation of cascade pumping stations in water supply canal systems considering energy consumption and costs. *Appl. Energy* **2023**, *341*, 121103. [CrossRef]
20. Khatavkar, P.; Mays, L.W. Real-Time Operation of Water-Supply Canal Systems under Limited Electrical Power and/or Water Availability. *J. Water Resour. Plan. Manag.* **2020**, *146*, 04020012. [CrossRef]
21. Kucuk, H.; Turan, M.; Yarali, K.; Al-Sanabani, H.; Iskefiyeli, M. A new algorithm for load shifting operation of water pumping stations. *J. Fac. Eng. Archit. Gaz.* **2021**, *36*, 2087–2093.
22. Shahdany, S.M.H.; Maestre, J.M.; van Overloop, P.J. Equitable Water Distribution in Main Irrigation Canals with Constrained Water Supply. *Water Resour. Manag.* **2015**, *29*, 3315–3328. [CrossRef]

23. Wang, J.P.; Lv, W.B.; Fan, Q.F. New Empirical Equations of Equilibrium Flow Scour Depth of Open Channel Outlet. *Water Resour. Power* **2013**, *31*, 139–140.
24. Xia, W.Y.; Zhao, X.D.; Zhang, X.Z. Application of Empirical Equation of Sediment Transport Capacity in Forecast of Riv-erbed Erosion and Deposition of Jiaojiang River Estuary. *Water Resour. Power* **2017**, *35*, 101–103.
25. Li, G.H.; Zhang, Z.; Kong, L.Z. Research on the Hydraulic Control Method of Shaping II Hydropower Station. *China Rural Water Hydropower* **2022**, *1*, 196–199.
26. Malaterre, P.-O.; Rogers, D.C.; Schuurmans, J. Classification of Canal Control Algorithms. *J. Irrig. Drain. Eng.* **1998**, *124*, 3–10. [CrossRef]
27. Mu, J.-B.; Zhang, X.-F. Real-Time Flood Forecasting Method With 1-D Unsteady Flow Model. *J. Hydrodyn.* **2007**, *19*, 150–154. [CrossRef]
28. Wang, C.H.; Li, G.Z.; Xiang, X.H. *Practical River Network Flow Calculation*; Hohai University Press: Nanjing, China, 2015.
29. Litrico, X.; Fromion, V.; Baume, J.-P.; Arranja, C.; Rijo, M. Experimental validation of a methodology to control irrigation canals based on Saint-Venant equations. *Control Eng. Pract.* **2005**, *13*, 1425–1437. [CrossRef]
30. Mehmed, Y. *Steady Non-Uniform Flow in Open Channel*; Water Resources and Electric Power Press: Beijing, China, 1987.
31. Wang, C.; Yang, J.; Nilsson, H. Simulation of Water Level Fluctuations in a Hydraulic System Using a Coupled Liquid-Gas Model. *Water* **2015**, *7*, 4446–4476. [CrossRef]
32. Yan, P.; Zhang, Z.; Lei, X.; Hou, Q.; Wang, H. A multi-objective optimal control model of cascade pumping stations considering both cost and safety. *J. Clean. Prod.* **2022**, *345*, 131171. [CrossRef]
33. Zhang, Z. Research on Perception and Regulation Model of Water Delivery Process in Open Channel Water Transfer Project. Ph.D. Thesis, Hohai University, Nanjing, China, 2020.
34. Olszewski, P. Genetic optimization and experimental verification of complex parallel pumping station with centrifugal pumps. *Appl. Energy* **2016**, *178*, 527–539. [CrossRef]

water

MDPI

Article

Flow Field and Gas Field Distribution of Non-Submerged Cavitation Water Jet Based on Dual-Nozzle with Concentric Configuration

Yun Luo [1,2,*], Jingyu Zang [1] and Hongxiang Zheng [1]

1 College of New Energy, China University of Petroleum (East China), Qingdao 266580, China; zhenghongxiang1@163.com (H.Z.)
2 State Key Laboratory of Heavy Oil Processing, China University of Petroleum (East China), Qingdao 266580, China
* Correspondence: luoyun1003@163.com or luoyun@upc.edu.cn; Tel./Fax: +86-532-86983481

Abstract: Cavitation water jet peening is an efficient and green surface treatment technology. The dual-nozzle can realize a cavitation water jet in air (non-submerged condition), which can be used for the surface treatment of large structures. The flow field characteristics of the dual-nozzle determine the cavitation effect. In this paper, the simulation of a cavitation water jet in air is carried out using Fluent software. The flow field characteristics containing velocity distribution, impact pressure, and gas phase volume fraction distribution are studied in detail. Furthermore, the effects of the nozzle structure parameters and incidence pressure on flow field characteristics are discussed. It was found that the structure parameters of the inner nozzle have a great influence on the flow field characteristics. Setting a contraction segment and expansion segment can improve the impact pressure and increase the intensity of the cavitation jet. Increasing the throat diameter and incidence pressure of the internal nozzle is also beneficial to improve the impact pressure and cavitation intensity. In order to assure a good cavitation effect, nozzle optimization should be performed. This study has guiding significance for the design of the dual-nozzle for a non-submerged cavitation water jet.

Keywords: composite nozzle; cavitation water jet; flow field characteristics; gas distribution; simulation

1. Introduction

Water jet technology is a rapidly developing, environmentally friendly machining process [1]. At present, branch technologies such as the abrasive water jet, pulsed water jet, and cavitation water jet have been gradually developed [2–4]. Water jet technology is used in a wide range of fields such as cutting, cleaning, rock breaking, and well drilling [5–7]. Water jet peening is a new surface strengthening technology, which originated in the 1980s. Water jet peening can introduce compressive residual stresses on surfaces and subsurfaces, and improve the fatigue life of components [8]. Cavitation peening, which is a new and more efficient surface treatment technology, was developed on the basis of water jet peening [9]. Compared with traditional water jet peening, cavitation water jet peening is a more effective surface strengthening technique with the characteristics of large process range and high treatment efficiency.

Cavitation shot peening uses the huge impact force generated by the collapse of cavitation bubbles to strengthen the surface of the workpiece [10]. Therefore, the number of cavitation bubbles determines the impact pressure and range of action of the cavitation jet. In order to improve the efficiency of cavitation peening, it is necessary to enhance the cavitation intensity in the flow field. Wenqiang Dong established a model of a submerged angular cavitation nozzle, and the flow field of the submerged cavitation nozzle was studied by numerical simulation [11]. It was proved that optimizing the nozzle structure can improve the cavitation effect (the size of the cavitation cloud) in the flow field. Mingming Ge investigated the effect of temperature on the intensity and the dynamics of cavitation [12]. It is found that when the temperature reaches the range of 55–60, the cavitation extent is larger and the vapor cloud shedding is more vigorous [13].

Many attempts have been made to investigate the capabilities of cavitation water jet peening for treating different materials by experimental method. During the initial stage of cavitation water jet peening development, cavitation bubbles are generated by injecting a high-speed water jet into water. Enomoto et al. [14] first reported the injection of a high-speed water jet into water to form a submerged jet, which was used to mitigate the stress corrosion cracking of pressure vessels. Odhiambo and Soyama [15] revealed that the fatigue strength of carbonized chrome–molybdenum alloy steel was improved by submerged cavitation water jet peening. Han et al. [16] demonstrated experimentally that cavitation jets improved the fatigue strength and life of SAE1045 steel. Ju and Han [17] performed microstructure analysis of titanium and concluded that the strengthening mechanism of cavitation jets was caused by the observed twin-crystal and twin-crystal interactions. Hutli et al. [18] found that the high-speed submerged cavitating jets could be used for the surface modification of mechanical and electrostatic properties of aluminum alloy. Balamurugan et al. [19] found that the introduction of compressive residual stresses in gear teeth by cavitation peening can increase their life. To sum up, the abovementioned cavitation peening is a conventional submerged jet, i.e., a cavitation jet in water.

However, it is difficult to treat big pressure vessels and pipes using submerged cavitation jet due to the limitation of the water tank. In response to this problem, Soyama [20] used a low-speed water jet to replace the static water in the water tank, thus realizing the non-submerged cavitation water jet. The high-speed water jet and low-speed water jet were transported by an inner nozzle and external nozzle, respectively. This dual-nozzle realized cavitation jet in air without the submerged environment and treated any-size components in the site. Afterwards, Soyama [21,22] analyzed the beneficial effects of the non-submerged cavitation water jet on the residual stresses and fatigue strength of SUS316L stainless steel and demonstrated the effectiveness of the cavitation jet in air. An important parameter for the cavitation flow is cavitation number, which is a ratio of dynamic pressure and static pressure. To enhance the effect of cavitation peening, a pressurized chamber was used to optimize the cavitation condition considering the cavitation number [23]. After that, the geometry of the nozzle and the outlet bore of the nozzle throat were optimized, thereby enhancing the aggressive intensity of the jet by about 20 times [24,25]. Soyama et al. [26,27] made a detailed comparison between the compressive and the fatigue strength of stainless steel by cavitation peening, water jet peening, shot peening, and laser peening. They concluded that the fatigue strength of the cavitation-peened specimen is the highest. However, most of the current studies have not reported the details of the characteristics in the convection field.

Besides, as an impact treatment method, cavitation peening can also cause erosion damage to equipment surfaces if the working parameters are not appropriate. Yamauchi et al. [28] found that the useful function of cavitation water jet peening depends on the standoff distance. Smaller working distances are suitable for water jet cutting because of the concentration of jet energy and the small treatment area. Longer working distances are suitable for cavitation peening, which guarantees a certain treatment area without damaging the integrity of the surface. Therefore, it is important to accurately clarify the flow field distribution of cavitation water jet peening to determine the optimum conditions without inducing detrimental material erosion. Marcon et al. [29] investigated the effect of concentric jet velocity on cavitation intensity and peening performance for a jet formed by a coaxial double nozzle. It was also pointed out that the jet velocity and the working distance have a great influence on the cavitation intensity and impact performance. The higher the jet velocity, the greater the residual compressive stress that can be generated and the thicker the compressive stress layer. Subsequently, they studied the effects of nozzle dimensions and shape on cavitation intensity and peening performance in co-flow cavitation water jet [30]. However, how the nozzle size and jet pressure affect the flow field distributions are still unrevealed. No studies were performed on the flow field analysis of non-submerged cavitation water jet peening.

Therefore, the aim of this study is to clarify the flow field characteristics of the non-submerged cavitation water jet peening with dual-nozzle based on numerical simulation. The influences of the nozzle dimension and the internal and external pressure on the flow field characteristics were also discussed. This study is of great significance for its application in surface strengthening of actual welded structures.

2. Numerical Simulation

2.1. Geometric Model

Figure 1 shows the schematic of the setup for cavitation jets in air and the geometric dimensions of the concentrically configured dual nozzles. As shown in Figure 1a, the whole device consists of a high-pressure water pump, a low-pressure water pump, a nozzle, two pressure gauges, a sample holder, and water tanks A and B. When the device is started, the high-pressure water pump injects a high-speed water jet with a pressure of p_h (20 MPa) into a low-speed water jet with a pressure of p_w (0.08 MPa) to form a cavitation water jet.

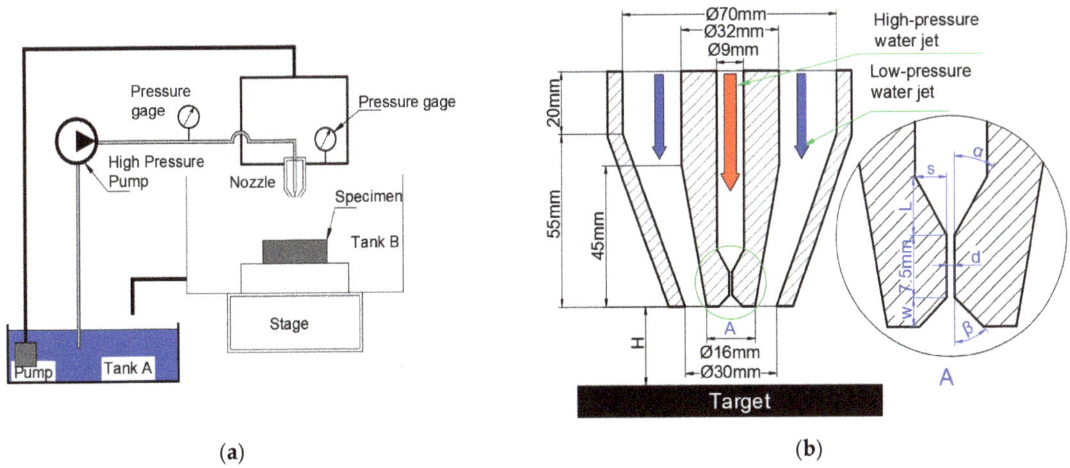

Figure 1. (**a**) Schematic diagram of cavitation water jet device. (**b**) Schematic diagram of the geometric size of the nozzle structure.

The nozzle is the key component to generate the cavitation jet, and the nozzle in the device is composed of two parts: the inner nozzle and the external nozzle. The diameter of the inner and external nozzle are d_H = 9 mm and d_L = 70 mm, respectively. The standoff distance is H = 25 mm, which is the same for the inner and external nozzle. The outer diameter of the inner nozzle is 32 mm, the linear transition segment of the outer wall is 30 mm, and the length of the outer wall constriction segment is 45 mm; the straight segment of the outer nozzle is 20 mm and the inner wall constriction segment is 55 mm. For the inner nozzle, the length of the constriction segment is L = 7 mm, the height is s = 4 mm, and the angle of the contraction segment is α = 30°. The length of the expansion segment is w = 3.5 mm and the expansion angle is β = 45°. The diameter of the throat is d = 1 mm and the length is 7.5 mm.

2.2. Meshing and Boundary Condition

The calculation domain of the flow field is shown in Figure 2a. AB is the high-pressure water inlet. CD and EF are low-pressure water inlets. IJ, KL, IG and KH are pressure outlets. GH is the target surface.

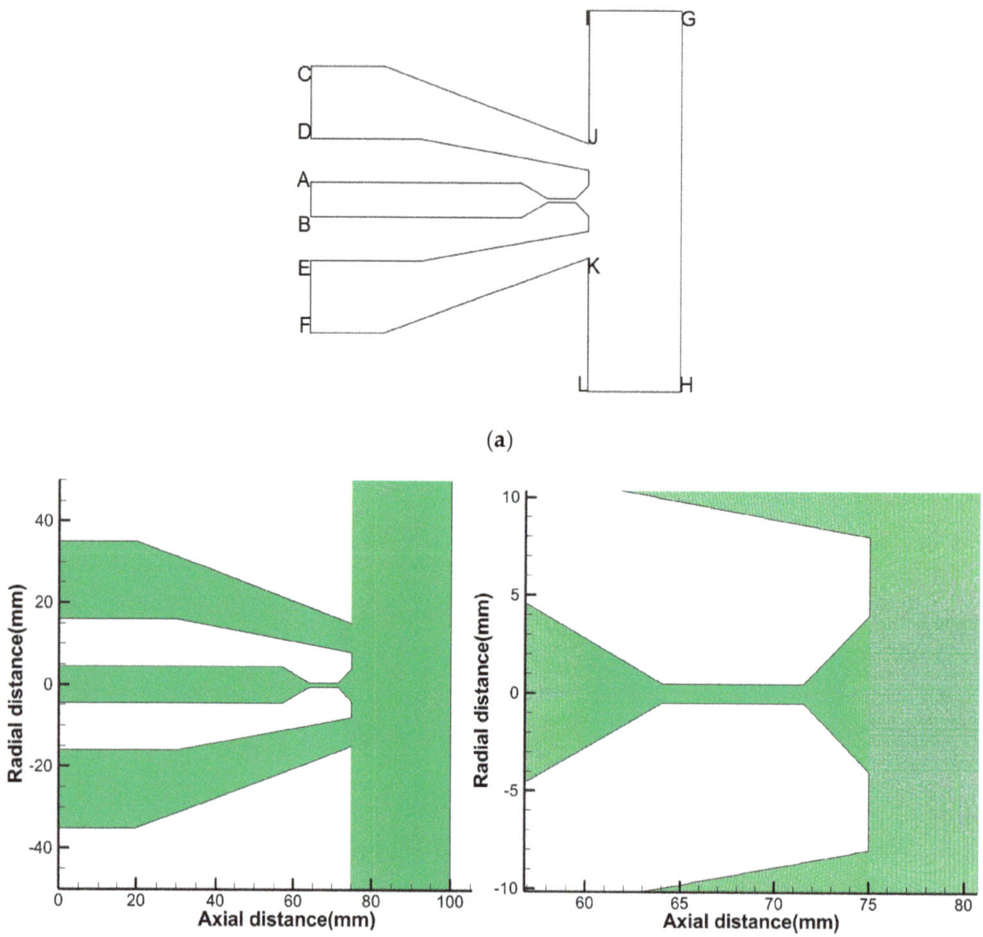

(a)

(b)

Figure 2. Flow field calculation area (**a**) and mesh division (**b**).

The calculation domain is divided into structured grids by GAMBIT 2.4.6 software, and the core area of the jet and the shear layer near the nozzle outlet is encrypted. The grid of the calculation domain is shown in Figure 2b. In order to reduce the amount of calculation as much as possible while ensuring the calculation effect, the distance between the grid nodes is gradually increased from the center of the nozzle outlet to the downstream and the boundary. Keeping the Courant number within a reasonable range, the grid size at the nozzle outlet shear layer is controlled at about 0.01 mm and the overall grid number in the final calculation domain is 122,840. The values of Equisize Skew and EquiAngle Skew are below 0.54, indicating good grid quality.

The high-pressure inlet pressure is set to 20 MPa, and the low-pressure inlet pressure is set to 0.08 MPa. The initial gas phase fraction of the inlet is 0, the pressure outlet pressure is 101,325 Pa at atmospheric pressure, and the gauge pressure is 0. The solver uses a pressure-based solution. The explicit relaxation factors for momentum and pressure are both 0.75. The second-order windward format is used for fitting, where the time step is 1×10^{-6} s. When the residual error is controlled within 1×10^{-6}, the calculation is considered to be basically convergent. The CPU used for the calculation is Intel(R) Xeon(R) Gold 6242R. The simulation calculation in this paper used a total of 2000 core hours.

2.3. Control Equation

2.3.1. Multiphase Flow Model

The common multiphase flow model contains the mixture model, VOF model, and Euler model. The mixture model or Euler model should be used to solve the calculation problems of the fluid containing bubbles, droplets, and particles with mass transfer between the fluids, as the VOF model is not suitable. Generally, the calculation results of the Euler model are more accurate than those of the mixture model but the latter requires less computer resources and has relatively high stability of the calculation process. In this study, the mixture model [31] is used:

$$\frac{\partial}{\partial t}\left(\rho_m \vec{v}_m\right) + \nabla \cdot \left(\rho_m \vec{v}_m\right) = -\nabla P + \nabla\left[\mu_m\left(\nabla \vec{v}_m + \nabla \vec{v}_m^{\,T}\right)\right] + \rho_m \vec{g}_m + F + \nabla \cdot \left(\sum_{k=1}^{n} \alpha_k \rho_k \vec{v}_{dr,k} \vec{v}_{dr,k}\right) \tag{1}$$

where $\vec{F}$ is the volume force, and $\vec{v}_{dr,k}$ is drift velocity at k phase, which is described as follows:

$$\vec{v}_{dr,v} = \vec{v}_{lv} - \sum_{k=1}^{n} \frac{\alpha_k \rho_k}{\rho_m} \vec{v}_{vk} \tag{2}$$

Then, the gas phase volume fraction equation can be obtained as follows:

$$\frac{\partial}{\partial t}(\alpha_v \rho_v) + \nabla \cdot (\alpha_v \rho_v v_m) = -\nabla \cdot (\alpha_v \rho_v v_{dr,v}) \tag{3}$$

2.3.2. Turbulence Model

The basic turbulence models include the k-ε model and k-ω model. The k-ε model has better performance and accuracy in high-speed calculation. The SST k-ω model has a good calculation for vortex structures. The velocity gradient of the cavitation jet is quite large at the nozzle and there will be cavitation vortex distribution in the flow field. The capture of such cavitation is very important for the presentation of simulation results. Therefore, SST k-ω turbulence model is selected for simulation calculation.

The equations are as follows [32]:

$$\frac{\partial}{\partial t}(\rho k) + \frac{\partial}{\partial x_i}(\rho k u_i) = \frac{\partial}{\partial x_j}\left(\Gamma_k \frac{\partial k}{\partial x_j}\right) + G_k - Y_k + S_k \tag{4}$$

$$\frac{\partial}{\partial t}(\rho \omega) + \frac{\partial}{\partial x_i}(\rho \omega u_i) = \frac{\partial}{\partial x_j}\left(\Gamma_\omega \frac{\partial \omega}{\partial x_j}\right) + G_\omega - Y_\omega + D_\omega + S_\omega \tag{5}$$

$$\Gamma_k = \mu + \frac{\mu_t}{\sigma_k} \tag{6}$$

$$\Gamma_\omega = \mu + \frac{\mu_t}{\sigma_\omega} \tag{7}$$

where G_k is the generating term of turbulent kinetic energy; G_ω is the generating term of dissipation rate ω; Y_k and Y_ω represent the dissipation of k and ω caused by turbulence, respectively; Γ_k and Γ_ω are the effective diffusivity of k and ω, respectively; and σ_k and σ_ω are the turbulent Prandtl numbers of k and ω, respectively. μ_t represents the turbulent viscosity, which is calculated by the following equation.

$$\mu_t = \frac{\rho k}{\omega} \frac{1}{\max\left[\frac{1}{\alpha^*}, \frac{SF_2}{a_1 \omega}\right]} \tag{8}$$

$$F_2 = \tanh\left(\Phi_2^2\right) \tag{9}$$

$$\Phi_2 = \max\left[2\frac{\sqrt{k}}{0.09\omega y}, \frac{500\mu}{\rho y^2 \omega}\right] \tag{10}$$

2.3.3. Cavitation Model

Cavitation models include Singhal et al.'s model, the Zwart–Gerber–Belamri model, Schnerr and Sauer model, Kunz model, etc. The Zwart–Gerber–Belamri model can be used to simulate the cavitation effect in the mixture model and Euler model. It is easier to converge and saves computational resources. In this paper, the Zwart–Gerber–Belamri model is used as the cavitation model. The convergence result can be obtained quickly.

In cavitation, the mass transfer between the liquid and gas phases is controlled by the vapor transport equation:

$$\frac{\partial}{\partial t}(\alpha \rho_v) + \nabla \cdot \left(\alpha \rho_v \vec{V}_v\right) = R_e - R_c \tag{11}$$

where

v = vapor phase;
α = vapor volume fraction;
ρ_v = vapor density;
$\vec{V}_v$ = vapor phase velocity;
R_e, R_c = mass transfer source terms connected to the growth and collapse of the vapor bubbles, respectively.

Assuming that all the bubbles in a system have the same size, Zwart–Gerber–Belamri [33] proposed that the total interphase mass transfer rate per unit volume (R) is calculated using the bubble number densities and the mass change rate of a single bubble:

$$R = \frac{3\alpha \rho_v}{R_b}\sqrt{\frac{2}{3}\frac{P_b - P}{\rho_l}} \tag{12}$$

where

α = vapor volume fraction;
ρ_v = vapor density;
R_b = bubble radius;
P_b = bubble surface pressure;
P = local far-field pressure;
ρ_l = liquid density.

Equation (12) is derived assuming bubble growth. In order to apply it to the bubble collapse process, Zwart–Gerber–Belamri proposed to replace it with Equation (13). Finally, they obtained the cavitation model as follows:

The Zwart–Gerber–Belamri model is described as follows:

If $P \le P_v$

$$R_e = F_{vap}\frac{3\alpha_{\text{nuc}}(1-\alpha_v)\rho_v}{R_b}\sqrt{\frac{2(P_v - P)}{3\rho_l}} \tag{13}$$

If $P \ge P_v$

$$R_c = F_{cond}\frac{3\alpha_v \rho_v}{R_b}\sqrt{\frac{2(P - P_v)}{3\rho_l}} \tag{14}$$

where

P_v = saturation vapor pressure;
F_{vap} (evaporation coefficient) = 50;
α_{nuc} (volume fraction of gas nuclei) = 5×10^{-4};
ρ_v = vapor density;
R_b (bubble radius) = 1×10^{-6} m;
P = local far-field pressure;
ρ_l = liquid density;
F_{cond} (condensation coefficient) = 0.01.

2.4. *Mesh Refinement Analysis*

For the simulation of fluid dynamics, grid-independence verification is essential. In this paper, the minimum grid size was controlled in order to multiply the number of grids. Finally, six different numbers of grids were divided, as shown in Table 1.

Table 1. Number of grids at different grid sizes.

	Mesh-1	Mesh-2	Mesh-3	Mesh-4	Mesh-5
Number of grids	14,860	22,200	79,200	122,840	245,400

In this paper, the velocity and the gas phase distribution in the flow field were the main focus. Therefore, the velocity along the axial direction and the distribution of the gas phase in the flow field were used as a measure of whether the number of meshes is sufficient.

As shown in Figure 3, as the number of grids multiplied, the maximum velocity increased and reached 200 m/s. It can be found that the velocity profiles overlapped when the number of grids reached 79,200. At this grid number, the velocity results in the flow field are already unaffected by the grid number.

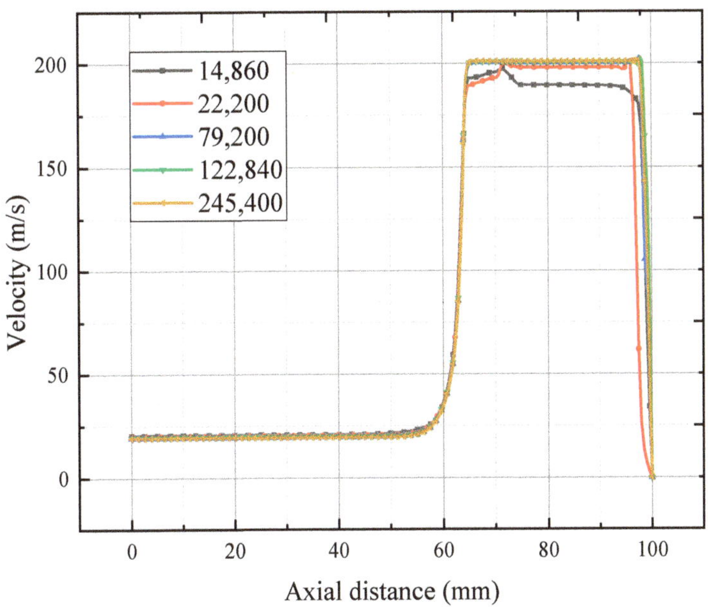

Figure 3. Variation in axial velocity and axial velocity in the flow field.

As shown in Figure 4, as the number of grids increased, the morphology of the gas phase within the flow field also changed. When the mesh number is 79,200, the separation of cavitation bubbles can already be simulated. When the grid number is encrypted from 122,840 to 245,400, the difference in calculation results is not much. Therefore, the simulation of the gas phase distribution can be achieved at the grid number of 122,840.

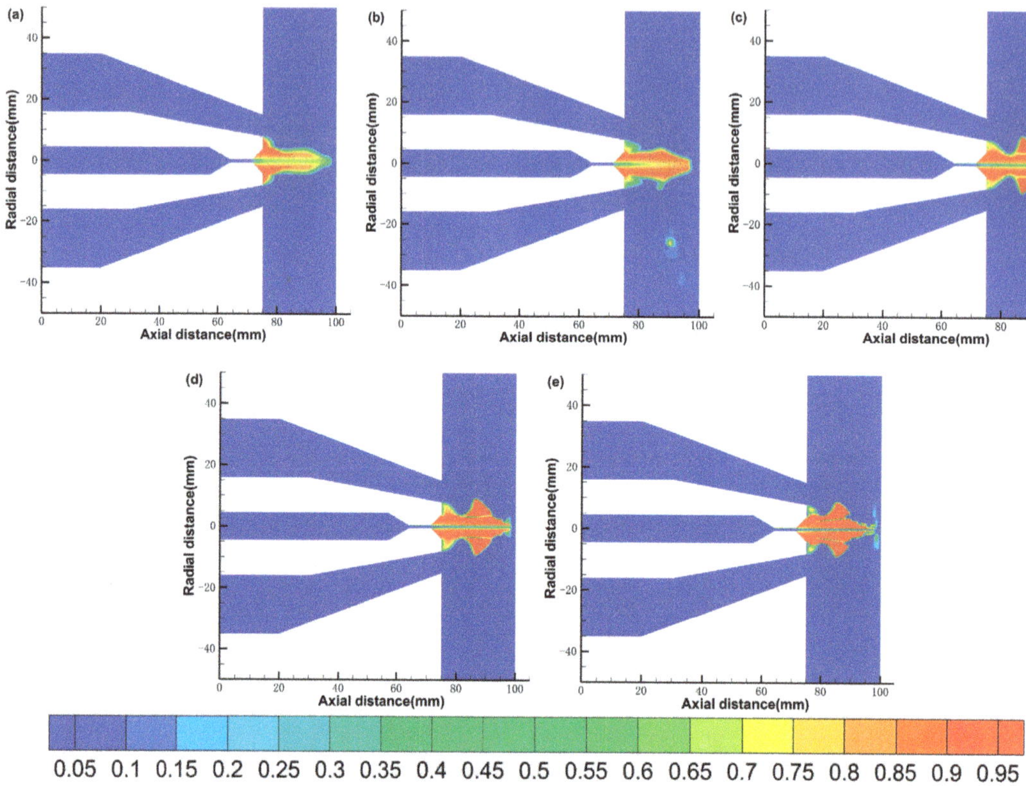

Figure 4. Cloud map of gas phase distribution in the flow field with different number of grids: (**a**) 14,860, (**b**) 22,200, (**c**) 79,200, (**d**) 122,840, (**e**) 245,400.

Therefore, combining the velocity results with the gas-phase cloud map results, the simulations in this paper all use a grid number of no less than 122,840 to exclude the effect of grid number on the simulation results.

2.5. Model Experimental Validation

In order to validate the reasonableness of simulations using the model above, some experiments results were compared with simulation results.

Figure 5a shows the morphology of the cavitation jet captured by Yang et al. [34] using a high-speed camera. The nozzle used in the experiment was modeled, and the parameters in the experimental environment were simulated as the initial boundary conditions. The simulation results are shown in Figure 5b.

Comparing the two figures, it can be found that the cavitation bubbles show a gradual growth trend in the range of x = 0~60 mm; after x = 70 mm, the vacuoles start to shed significantly. The general morphology of the cavitation jet and the process of vacuole growth, development, and shedding shown in the two figures are basically consistent. The morphological verification of the cavitation jets by simulation and experiment verifies that the model used in this paper is applicable to simulate cavitation jets.

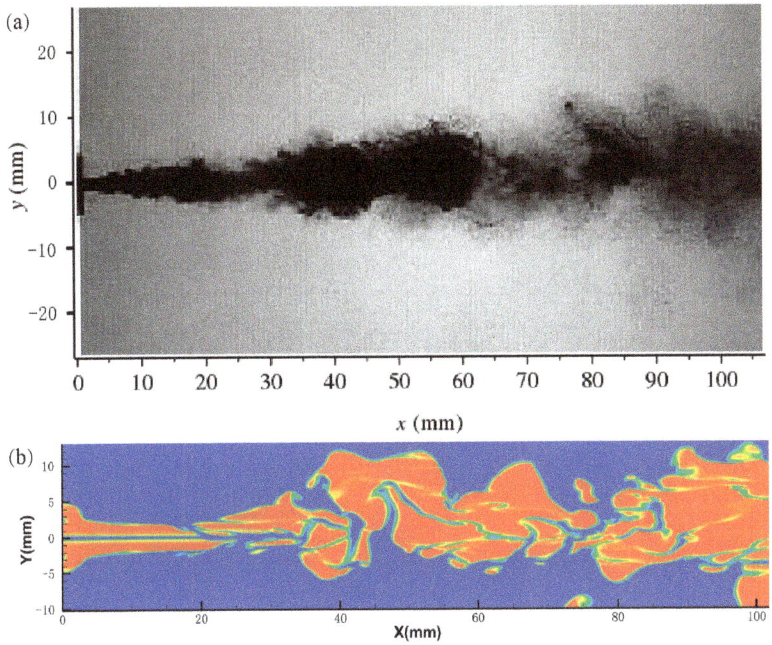

Figure 5. Comparison of cavitation jet morphology from high-speed photography (**a**) and numerical simulation (**b**).

3. Results and Discussions

3.1. Analysis of Flow Field Characteristics

3.1.1. The Gas Phase Distribution

The calculated results were processed using post-processing software to obtain a cloud diagram of the gas phase distribution within the flow field, as shown in Figure 6. In the figure, the darker the degree of red, the larger the gas content in the region, and closer to blue indicates lower gas content. From the figure, we can see that the maximum gas content reached 0.9 and there is a large range of gas phase area, which indicates that the nozzle can effectively produce the cavitation effect.

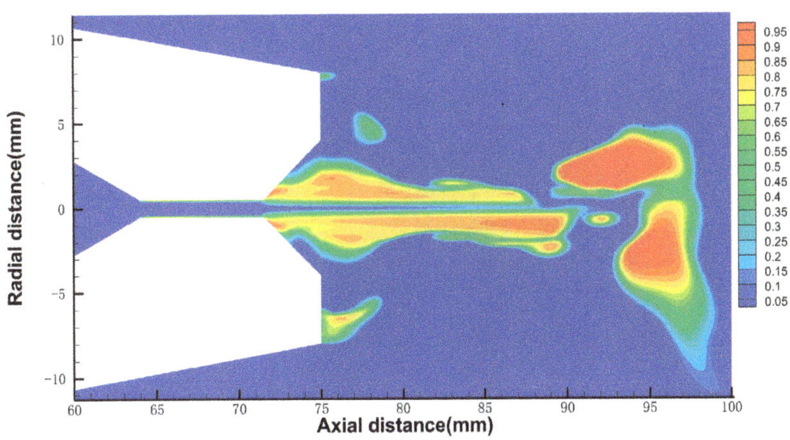

Figure 6. Cloud map of gas phase distribution in the flow field.

3.1.2. The Velocity Distribution

Figure 7 shows the contour distribution of axial velocity within the flow field. It can be seen that the jet velocity increases rapidly when flowing through the throat and the velocity reaches its maximum at the throat exit. From the throat exit to the outside of the flow field, the jet velocity presents obvious stratification due to the occurrence of shear. The middle part has the highest velocity, while the velocity in the surrounding part decreases rapidly. When the jet impinges on the surface of the specimen, the axial velocity decreases to 0 and the kinetic energy is transformed into pressure potential energy.

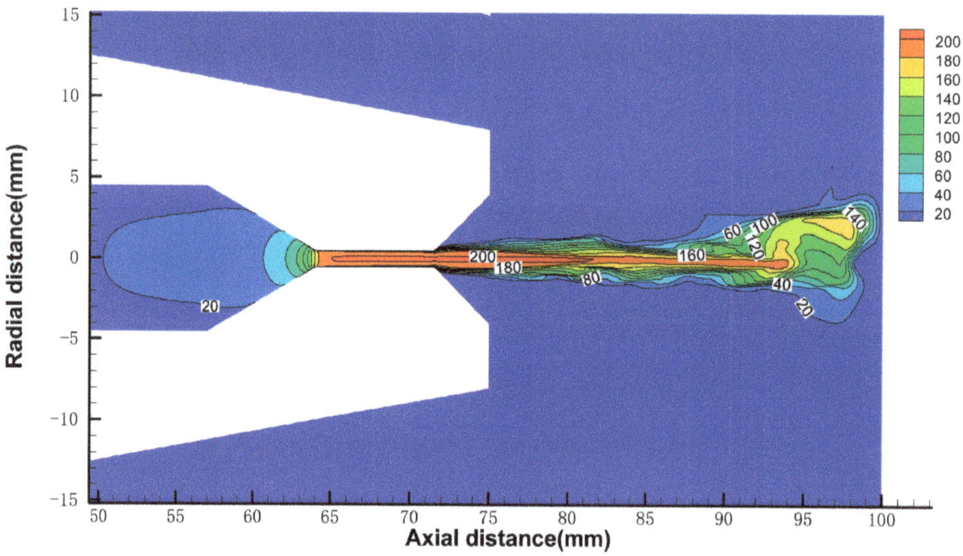

Figure 7. Velocity distribution cloud in the flow field.

3.1.3. The Pressure Distribution

Figure 8 shows the impact pressure on the surface of the specimen with the radial distance from the center of the surface. The maximum impact pressure appears in the center of the water-jet-impacted position (the center of the surface of the specimen) and its value is about 9 MPa. Then, it rapidly declines with the increase in radial distance. However, the process of bubbles collapse was ignored in this simulation. The actual impact pressure on the surface of the specimen induced by the bubble collapses is not included in the simulation results. So, the simulation results may have deviations from the actual results.

3.2. Effects of Geometric Parameters of Inner Nozzle

On the basis of the conical composite nozzle shown in Section 2 (hereinafter referred to as nozzle A), a series of conical composite nozzles were designed by changing the structural parameters of the inner nozzle. The structure of each nozzle is shown in Table 2. Compared with nozzle A, the throat diameter of nozzle B was increased to 2 mm. On the basis of nozzle A, the contraction angle of nozzle C was decreased to 18°. Compared with nozzle A, nozzle D did not have the expansion segment. Based on nozzle D, nozzle E omitted the contraction segment. The structural mesh was divided using Gambit 2.4.6 software, as shown in Figure 9. The built model is imported into Fluent software for simulation calculations. The boundary conditions, model selection, and iteration time step for the simulation of nozzles B, C, D, and E are identical to those used for nozzle A.

3.2.1. The Gas Phase Distribution

Figure 10 shows the gas phase distribution contour in the flow field of the five nozzles. The colors of the five gas phase distribution clouds correspond to the same gas content values, with red indicating a high gas content and blue indicating a low gas content. In general, all five nozzles are effective for generating cavitation. However, the flow field characteristics show some differences due to the different structures of the inner nozzles. The gas content rate and the area of gas phase region were used to measure the severity of the cavitation. The higher the gas content, the larger the area of the gas phase region, representing a more intense cavitation phenomenon.

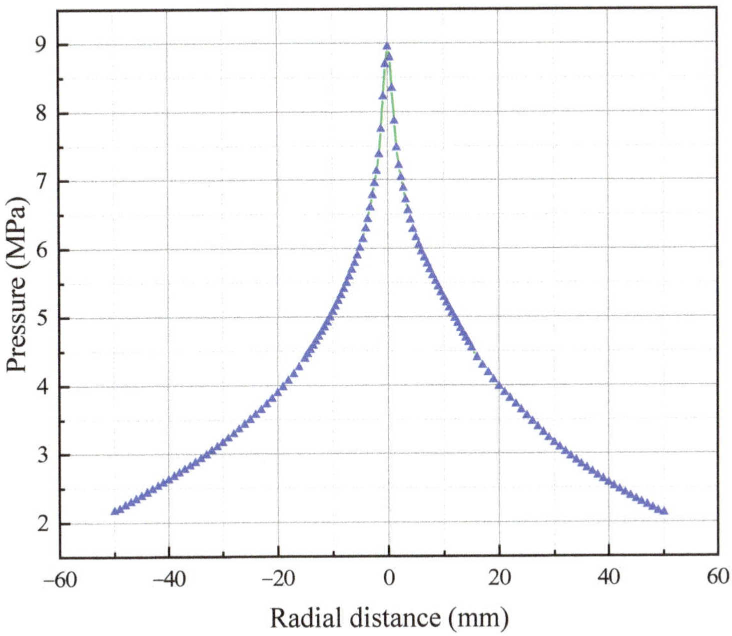

Figure 8. Impact pressure distribution on the surface of the specimen.

Table 2. Nozzle structure within the conical composite nozzle.

	Contraction Segment	Expansion Segment	Throat Diameter *d*	Target Distance *H*
Nozzle A	$L = 7$ mm, $s = 4$ mm ($\alpha = 30°$)	$w = 3.5$ mm, $\beta = 45°$	1 mm	25 mm
Nozzle B	$L = 6.125$ mm, $s = 3.5$ mm ($\alpha = 30°$)	$w = 3.5$ mm, $\beta = 45°$	2 mm	25 mm
Nozzle C	$L = 12$ mm, $s = 4$ mm ($\alpha = 18°$)	$w = 3.5$ mm, $\beta = 45°$	1 mm	25 mm
Nozzle D	$L = 7$ mm, $s = 4$ mm ($\alpha = 30°$)	No expansion segment	1 mm	25 mm
Nozzle E	No contraction segment	No expansion segment	1 mm	25 mm

Comparing the gas phase distribution of nozzle B, the gas phase region of nozzle A with a smaller throat diameter is larger and the distribution of cavitation bubble is more uniform. Comparing the gas phase distribution in the flow fields of nozzle A and nozzle C, it can be found that the shape and area of the gas phase region in the flow fields of the two nozzles are similar. The gas content rate in the flow field of nozzle C is higher than that in nozzle A. The nozzle with a smaller contraction angle has the higher gas content rate of the flow field. Comparing the gas distribution of nozzle A and nozzle D, the area of the gas phase region within the flow field of the nozzle with an expansion segment (nozzle A) is larger. The gas phase content rate of nozzle D is higher than that of nozzle A. The nozzle with an expansion segment has a larger gas phase region. Comparing the gas distribution in the flow fields of nozzle D and nozzle E, it is found that the gas phase region within the flow field of the nozzle with a contraction segment (nozzle D) is larger; however, the improving effect of gas phase distribution is not as big as that of other parameters.

From the gas phase distribution, the cavitation phenomenon is more intense in the nozzle flow field with expansion and contraction segments. The cavitation effect of the contraction segment with a smaller contraction angle is better. The expansion section and throat diameter have a great influence on the gas phase distribution in the flow field. The gas phase area in the flow field can be significantly increased by setting the expansion section. Increasing the throat diameter will somewhat reduce the gas phase area but will increase the gas phase content rate. Setting the contraction segment can also increase the gas phase region area within the flow field. However, the influence on the flow field is not as significant as that of other parameters.

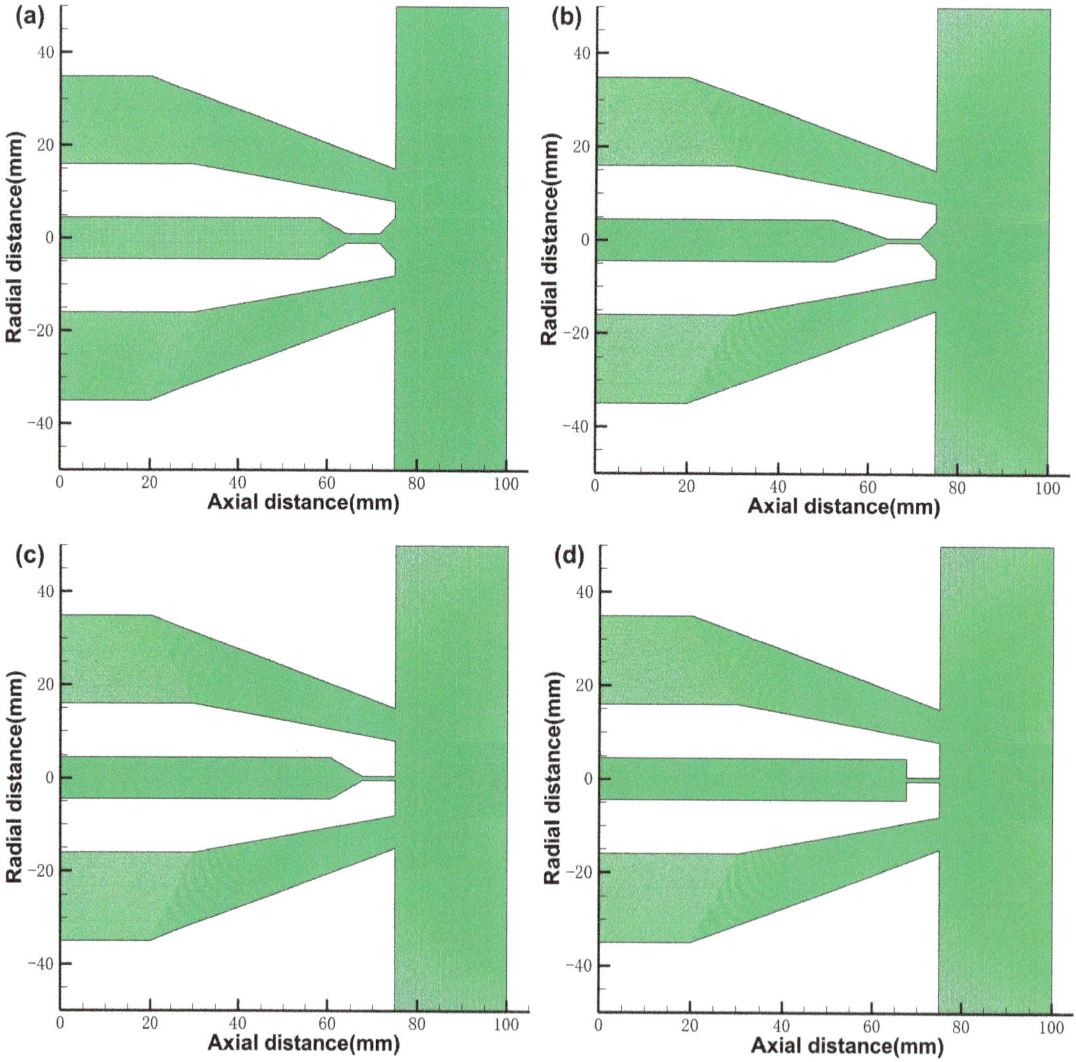

Figure 9. Conical composite nozzle structure and grid schematic: (**a**) nozzle B, (**b**) nozzle C, (**c**) nozzle D, (**d**) nozzle E.

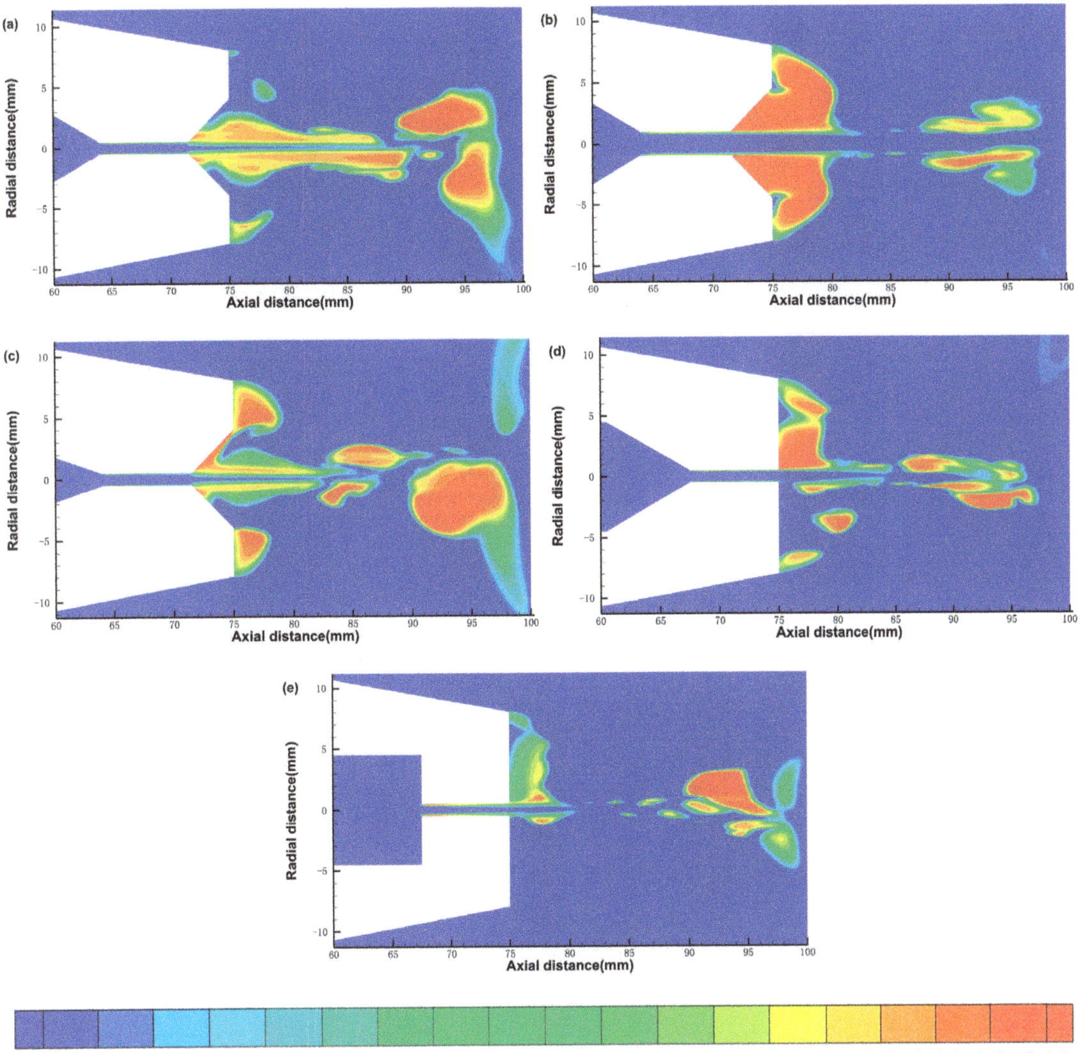

Figure 10. Cloud diagram of gas phase distribution in the flow fields of five nozzles: (**a**) nozzle A, (**b**) nozzle B, (**c**) nozzle C, (**d**) nozzle D, (**e**) nozzle E.

3.2.2. The Velocity Distribution

Figure 11 shows the variation in the axial velocity of the jet at the central axis of the five nozzles with the axial distance. The solid blue line in the figure is the horizontal line where the nozzle outlet is located (at 75 mm), and the nozzle outlet to the surface of the specimen is the external flow field. The water jet is distributed with the central axis as the axis of symmetry. The axial velocity decreases from the center to all around; so, the velocity on the central axis is maximum. In the external flow field, as the axial distance continues to increase, the water jet velocity is first maintained at the maximum velocity for a distance and then decreases, reaching the specimen when the velocity drops to 0. Regardless of the structure of the nozzle, the maximum velocity in the flow field is almost the same, which is consistent with the conclusion calculated by the Bernoulli equation.

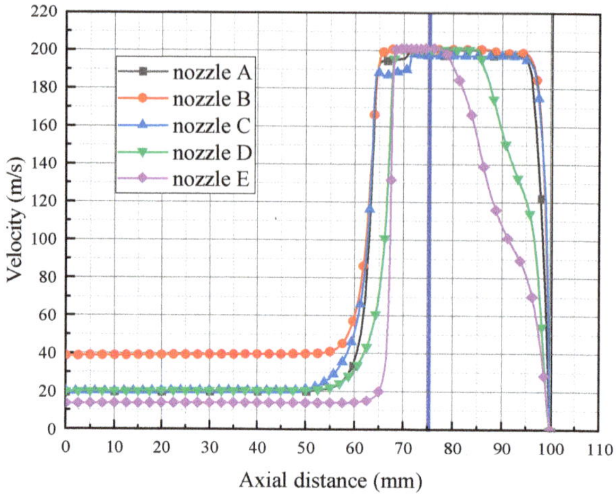

Figure 11. Axial velocity distribution of the central axis of the five nozzles.

The velocity in the flow field of nozzle E without the constriction segment rises very sharply, while the velocity in the flow field of the nozzle D with the constriction segment rises relatively gently. Comparing the velocity curves in the flow fields of nozzle A and nozzle C, the velocity of nozzle C with a smaller constriction angle rises more gently. It can be found that the constriction segment is equivalent to a transition segment, which can reduce the loss caused by the rapid change of speed; the smaller the constriction angle of the nozzle, the better the effect. In the external flow field, the jet velocity in the flow field of nozzle A with the expansion segment maintains a distance of 20 mm before it begins to drop rapidly. This distance is much larger than that of nozzle D without the expansion segment, which shows that the expansion segment can be effective to increase the cavitation jet's working distance.

Comparing the axial velocity change curves in the flow fields of nozzle A and nozzle B, in the internal flow fields of the nozzles, the initial velocities of the inner straight sections of the two nozzles are different. The initial velocity of nozzle A with the smaller throat diameter is lower. However, after passing through the constriction segment, the velocity of nozzle B in the external flow field is only slightly higher than that of nozzle A, and the two curves are not very different. Therefore, changing the throat diameter has a limited effect on the velocity in the flow field and the range of action. From the velocity distribution in the flow fields of different nozzles, the nozzle with a smaller contraction segment angle and expansion segment can reach the best velocity distribution. The expansion segment has a more positive influence on the velocity distribution in the flow field than the contraction segment. In addition, changing the diameter has a minimal influence on the velocity in the external flow field.

3.2.3. The Pressure Distribution

Figure 12 shows the impact pressure distribution of the five nozzles on the sample surface. It can be seen from the figure that, in the flow field of all nozzles, the impact pressure near the axis is the largest and the impact pressure decreases with the increase in the radial distance. However, it is worth noting that nozzle C and nozzle D do not strictly follow this law and the impact pressure is the largest at about 2 mm from the center axis of the nozzle. In addition, the speed at which the impingement pressure decreases with the increase in the radial distance is also different for the nozzles of different structures.

Comparing the impact pressure distributions of nozzle A and nozzle B shown in the figure, the impact pressure of a nozzle with a throat diameter of 2 mm (nozzle B) is approximately twice that of a nozzle with a throat diameter of 1 mm (nozzle A). Therefore, increasing the throat diameter will increase the impingement pressure of the jet. The impact pressure of nozzle C with a smaller constriction angle is much greater than that of nozzle A, and the two curves intersect at 28 mm. It can be found that the jet from nozzles with smaller constriction angles have stronger concentration and can achieve higher impact pressures.

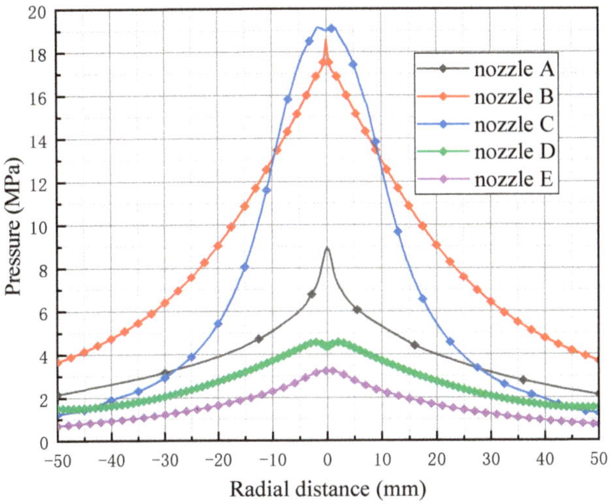

Figure 12. Impact pressure distribution of five nozzles on the surface of the specimen.

The impact pressure of the nozzle with the expansion segment (nozzle A) is always greater than that of the nozzle without the expansion segment (nozzle D). The impingement pressure of nozzle D is greater than the pressure of nozzle E (without constriction). For nozzles with no constriction segment and no expansion segment, the increase in the impact pressure caused by the constriction segment is relatively limited and the impact of the expansion segment on the impact pressure is more significant. From the impact pressure distribution, the nozzle with a smaller contraction angle, contraction segment, and expansion segment has the highest impact pressure. Increasing the throat diameter can significantly increase impact pressure. For a nozzle without a contraction segment and expansion segment, setting the expansion segment increases the impact pressure more obviously.

3.3. Effects of Pressure Ratio of Inner Nozzle to External Nozzle

Nozzle A was used to study the effect of the pressure ratio of the inner nozzle to the external nozzle on flow field characteristics. The rest of the conditions were kept constant. The ratios of the internal incidence pressure to external incidence pressure were selected as 20 MPa:0 Pa, 20 MPa:0.04 MPa, 20 MPa:0.08 MPa, 30 MPa:0 Pa, and 40 MPa:0 Pa, respectively. The simulation results were compared to study the effect of the incidence pressure on the flow field characteristics of the water jet.

3.3.1. The Gas Phase Distribution

Figure 13 shows the gas phase distribution clouds in the flow field at different incidence pressures. Comparing panels a, b, and c, it can be found that the area of the gas phase region in the flow field is largest when the incidence pressure of the outer nozzle is 0. Comparing panels a, d, and e, we can find that when the incidence pressure of the outer nozzle is 0, with the increase in the incidence pressure of the inner nozzle, the maximum gas content rate in the flow field basically remains the same but the area of the gas phase region increases.

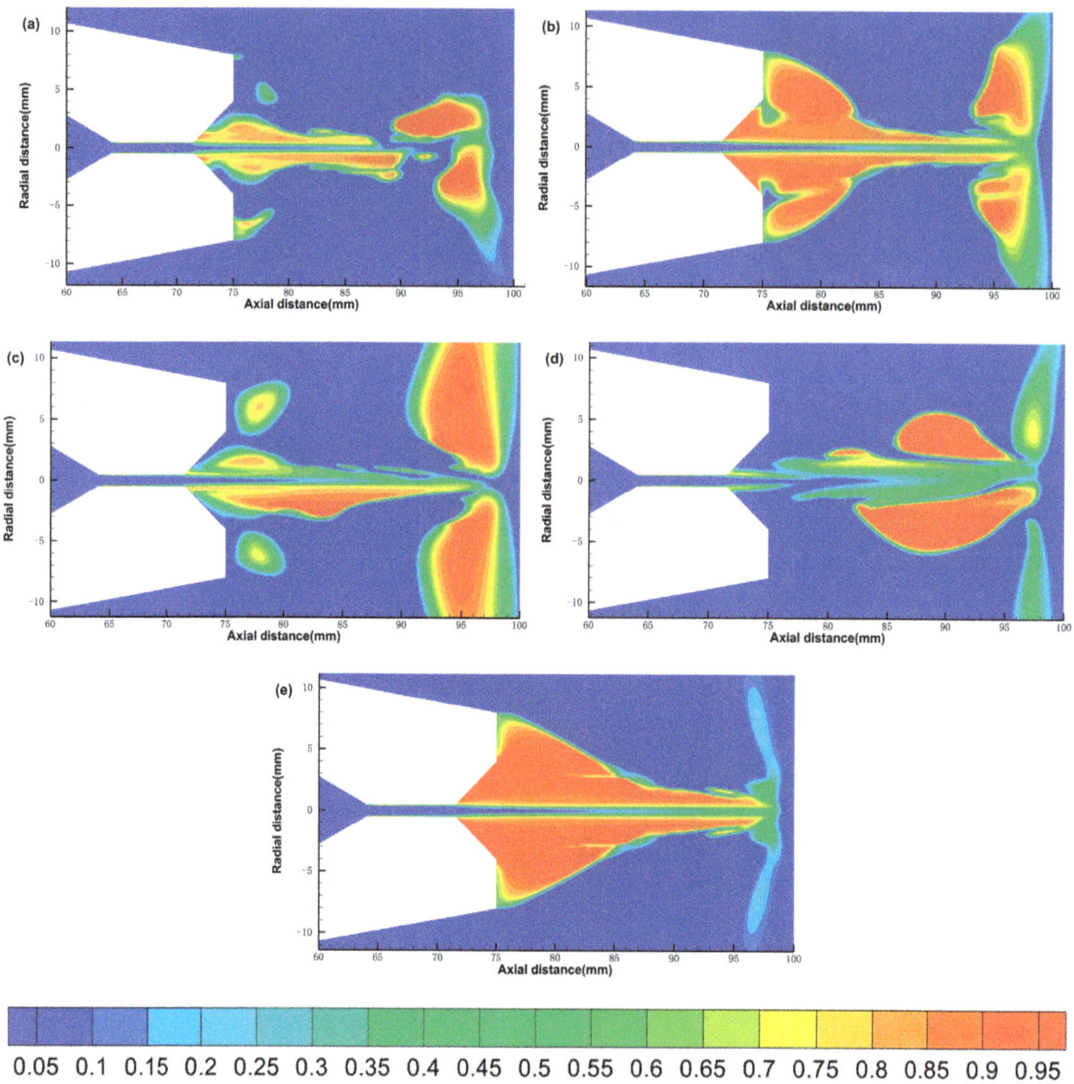

Figure 13. Flow field gas phase distribution clouds at different incident pressures: (**a**) 20 MPa:0.08 MPa, (**b**) 20 MPa:0.04 MPa, (**c**) 20 MPa:0 Pa, (**d**) 30 MPa:0 Pa, (**e**) 40 MPa:0 Pa.

Changing the internal and external pressure ratio actually changes the cavitation number σ, which is of great importance in cavitation jets. The smaller the cavitation number, the more intense the cavitation phenomenon, and when the cavitation number is larger, the cavitation phenomenon is less likely to occur. The cavitation number can be determined by Equation (15).

$$\sigma = \frac{p - p_v}{\frac{1}{2}\rho_L v^2} = \frac{p_2 - p_v}{p_1 - p_2} \approx \frac{p_2}{p_1} \tag{15}$$

where p_1 is the upstream pressure of the jet, p_2 is the downstream pressure, and p_v is the saturated vapor pressure of the environment in which it is located.

As can be seen from Equation (15), when increasing the inner nozzle incidence pressure, the cavitation number decreases, the degree of cavitation is enhanced, and the vacuole cloud in the flow field increases (the area of the gas phase region increases). While increasing the incidence pressure of the outer nozzle, the cavitation number decreases and the degree of cavitation within the flow field decreases accordingly.

3.3.2. The Velocity Distribution

Figure 14 shows the axial velocity distribution along the central axis for different incident pressures. The effect of incident pressure of the outer nozzle on the axial velocity is very small when the incident pressure of the inner nozzle is the same. Keeping the incident pressure of the outer nozzle the same, the axial velocity distribution curves have the same trend for different incident pressures of the inner nozzle. The difference lies only in the maximum value of the velocity. The greater the incident pressure of the inner nozzle, the greater the maximum velocity in the flow field.

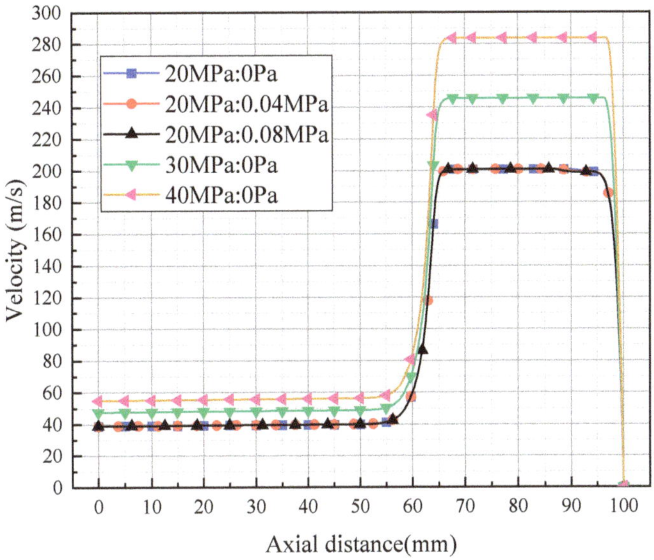

Figure 14. Axial velocity distribution at different incident pressures.

3.3.3. The Pressure Distribution

Figure 15 shows the impact pressure distribution at different incidence pressures. It can be seen that the three curves almost overlap when the incidence pressure is 20 MPa:0 Pa, 20 MPa:0.04 MPa, and 20 MPa:0.08 MPa. The impact pressure on the specimen surface is less affected by the incidence pressure of the outer nozzle when the incidence pressure of the outer nozzle is lower. Comparing the impact pressure distribution when the incidence pressure is 20 MPa:0 Pa, 30 MPa:0 Pa, and 40 MPa:0 Pa, the impact pressure on the specimen surface increases as the incidence pressure increases. The maximum impact pressure on the specimen surface is about 2 Mpa lower than the incidence pressure.

3.4. Discussions

In this study, finite element simulations of the flow field of nozzles with different nozzle structures and incidence pressures are performed to obtain the gas phase distribution, velocity distribution, and impact pressure within the flow field. Based on the above analysis, we can conclude that the inner nozzle with a smaller contraction segment, smaller expansion segment, and bigger throat diameter has the better cavitation effect (the cavitation cloud area in the flow field is larger) and impact pressure. Increasing the inner nozzle incidence pressure and reducing the outer nozzle pressure can also improve the cavitation effect and enhance the impact pressure. The contraction segment and expansion segment promote the occurrence of cavitation phenomena. Zhang et al. [35] used pressure-sensitive paper to measure the impact pressure distribution of cavitation water jets along the jet injection direction (axial direction). The study showed that different nozzle shapes can have a significant effect on the cavitation phenomenon. Pressure drop occurs in the throat of nozzles with constricted and expanded segments but not in nozzles without expanded segments. This verifies that our simulation results are right. In addition, Soyama et al. [27] also proved that the cavitation jet of the nozzle with an expanding segment had the highest intensity of erosion on the specimen compared with other conventional water jet nozzles. These results are also consistent with the conclusions of this paper.

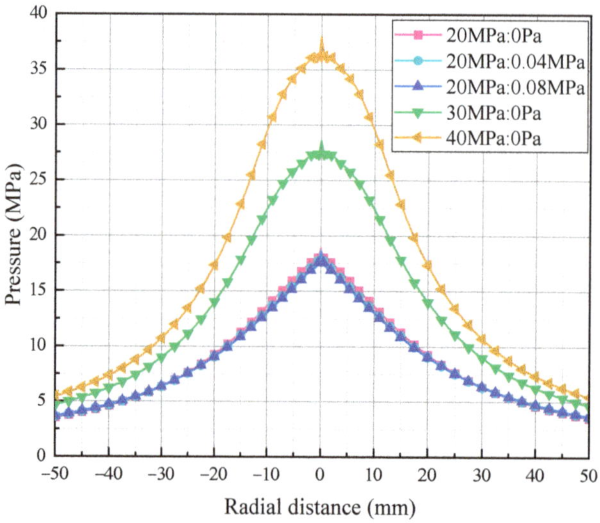

Figure 15. Impact pressure distribution at different incidence pressures.

It should be noted that the erosion intensity and impact effect of the cavitation water jet are judged mainly by gas phase distribution, velocity distribution, and impact pressure. However, the impact pressure is not accurate enough because the cavitation bubbles' collapse process is not simulated. So, the impact pressure obtained may not be the actual impact pressure. The modeling of the generation of cavitation bubbles is challenging because of the complexity of the cavitation phenomenon [29]. In addition, only simulation analysis was performed on the cavitation phenomenon and no experiments were carried out. Further numerical simulations and experimental verification of cavitation bubble collapse are required. However, this study still has a great guideline for the design of the conical composite nozzle and the selection of incident pressure.

4. Conclusions

In previous studies, scholars mainly focused on the cavitation phenomenon occurring in venturi tubes and submerged cavitation jets. The more widely used non-submerged cavitation jets are understudied. In this paper, the flow field characteristics containing gas phase distribution, velocity distribution, and impact pressure of the cavitation water jet in air based on a dual-nozzle design with concentric configuration was investigated in detail by numerical simulation. The simulation study of the non-submerged cavitation jet flow field has reference significance and guiding significance for the design of concentric double nozzles. The effects of inner nozzle structure parameters and the incidence pressures on the flow field characteristics were also compared. The following conclusions can be drawn:

The structure of the inner nozzle has a significant impact on the flow field characteristics, and the selection of a suitable inner nozzle structure can improve the action distance of the jet as well as the impact pressure.

The contraction segment of the inner nozzle has the effect of increasing the area of gas phase in the flow field and increasing the impact pressure; the smaller the contraction angle of the contraction segment, the stronger the effect.

The expansion segment of the inner nozzle has a great influence on the characteristics of the flow field. Setting the expansion segment can significantly improve the gas content rate in the flow field and the impact pressure of the jet.

The internal and external incident pressures also affect the flow field characteristics. When increasing the incident pressure of the inner nozzle, the cavitation number decreases and the cavitation phenomenon in the flow field is enhanced. The gas phase area and gas holdup in the flow field will increase, and the impact pressure will increase. In contrast, increasing the incident pressure of the outer nozzle will increase the cavitation number, which in turn will weaken the cavitation in the flow field.

Author Contributions: Conceptualization, Y.L. and J.Z.; methodology, Y.L.; software, J.Z.; validation, J.Z. and H.Z.; formal analysis, J.Z. and H.Z.; investigation, J.Z.; resources, Y.L.; data curation, H.Z.; writing—original draft preparation, J.Z.; writing—review and editing, Y.L.; visualization, Y.L.; supervision, Y.L.; project administration, H.Z.; funding acquisition, Y.L. All authors have read and agreed to the published version of the manuscript.

Funding: This research was funded by [National Key R&D Program of China] grant number [2021YFB4001503], [National Natural Science Foundation of China] grant number [51905545] and [Fundamental Research Funds for the Central Universities] grant number [20CX02219A].

Data Availability Statement: Data available on request due to restrictions eg privacy or ethical. The data presented in this study are available on request from the corresponding author. The data are not publicly available due to the temporary need for secrecy of the project.

Acknowledgments: The authors gratefully acknowledge the support provided by National Key R&D Program of China (2021YFB4001503), National Natural Science Foundation of China (51905545), Fundamental Research Funds for the Central Universities (20CX02219A).

Conflicts of Interest: The authors declare no conflict of interest.

References

1. Zhang, Q.; Shi, Z.; Shi, W.; Xie, Z.; Tan, L.; Yang, Y. Research on Flow Field Characteristics in Water Jet Nozzle and Surface Damage Caused by Target Impact. *Sustainability* **2022**, *14*, 9074. [CrossRef]
2. Chlupová, A.; Hloch, S.; Nag, A.; Šulák, I.; Kruml, T. Effect of pulsating water jet processing on erosion grooves and microstructure in the subsurface layer of 25CrMo4 (EA4T) steel. *Wear* **2023**, *524*, 204774. [CrossRef]
3. Natarajan, Y.; Murugesan, P.K.; Mohan, M.; Khan, S.A.L.A. Abrasive Water Jet Machining process: A state of art of review. *J. Manuf. Process.* **2020**, *49*, 271–322. [CrossRef]
4. Zhao, H.; Jiang, H.; Warisawa, S.; Li, H. Numerical study of abrasive water jet rotational slits in hard rock using a coupled SPH-FEM method. *Powder Technol.* **2023**, *426*, 118622. [CrossRef]
5. Liu, Z.; Ma, Z.; Liu, K.; Zhao, S.; Wang, Y. Coupled CEL-FDEM modeling of rock failure induced by high-pressure water jet. *Eng. Fract. Mech.* **2023**, *277*, 108958. [CrossRef]
6. Luo, X.; Zhang, J.; Yang, F.; He, F.; Xia, Y. Research on the hard rock cutting characteristics of disc cutter under front-mounted water jet precutting kerf conditions. *Eng. Fract. Mech.* **2023**, *287*, 109330. [CrossRef]
7. Nadolny, K.; Romanowski, M.; Sutowski, P. Assessing the technological quality of abrasive water jet and laser cutting processes by geometrical errors and a multiplicative indicator. *Measurement* **2023**, *217*, 113060. [CrossRef]
8. Winter, S.; Smith, A.; Lappin, D.; McDonagh, G.; Kirk, B. Failure of non-vacuum steam sterilization processes for dental handpieces. *J. Hosp. Infect.* **2017**, *97*, 343–347. [CrossRef]
9. Yang, Y.F.; Shi, W.D.; Li, W.; Chen, S.P.; Zhang, W.Q.; Pan, B. Experimental study on the surface property changes of aluminum alloy and stainless steel after impingement with submerged cavitation jet. *Strength Mater.* **2021**, *53*, 353–363. [CrossRef]
10. Murugesan, P.; Jung, S.; Lee, H. Inner surface peening for long and narrow helical pipe using acoustic cavitation. *Mater. Des.* **2023**, *229*, 111906. [CrossRef]
11. Dong, W.; Yao, L.; Luo, W. Numerical Simulation of Flow Field of Submerged Angular Cavitation Nozzle. *Appl. Sci.* **2023**, *13*, 613. [CrossRef]

12. Ge, M.; Manikkam, P.; Ghossein, J.; Subramanian, R.K.; Coutier-Delgosha, O.; Zhang, G. Dynamic mode decomposition to classify cavitating flow regimes induced by thermodynamic effects. *Energy* **2022**, *254*, 124426. [CrossRef]
13. Ge, M.; Petkovšek, M.; Zhang, G.; Jacobs, D.; Coutier-Delgosha, O. Cavitation dynamics and thermodynamic effects at elevated temperatures in a small Venturi channel. *Int. J. Heat Mass Transf.* **2021**, *170*, 120970. [CrossRef]
14. Enomoto, K.; Hirano, K.; Mochizuki, M.; Kurosawa, K.; Saito, H.; Hayashi, E. Improvement of residual stress on material surface by water jet peening; Water jet peening ni yoru zairyo hyomen no zanryu oryoku kaizen koka no kento. *Zair. J. Soc. Mater. Sci. Jpn.* **1996**, *45*, 734–739. [CrossRef]
15. Giorgetti, V.; Santos, E.A.D.; Marcomini, J.B.; Sordi, V.L. Stress corrosion cracking and fatigue crack growth of an API 5L X70 welded joint in an ethanol environment. *Int. J. Press. Vessel. Pip.* **2019**, *169*, 223–229. [CrossRef]
16. Ryakhovskikh, I.V.; Bogdanov, R.I. Model of stress corrosion cracking and practical guidelines for pipe-lines operation. *Eng. Fail. Anal.* **2021**, *121*, 105134. [CrossRef]
17. Shakeri, I.; Shahani, A.R.; Rans, C.D. Fatigue crack growth of butt welded joints subjected to mixed mode loading and overloading. *Eng. Fract. Mech.* **2021**, *241*, 107376. [CrossRef]
18. Xu, S.; Wang, C.; Wang, W. Failure analysis of stress corrosion cracking in heat exchanger tubes during start-up operation. *Eng. Fail. Anal.* **2015**, *51*, 1–8. [CrossRef]
19. Lv, F.; Hu, X.; Ma, C.; Yang, B.; Luo, Y. Failure analysis on cracking of backing plate of lifting lug for air preheater. *Eng. Fail. Anal.* **2020**, *109*, 104395. [CrossRef]
20. Soyama, H. Introduction of compressive residual stress using a cavitating jet in air. *J. Eng. Mater. Technol.* **2004**, *126*, 123–128. [CrossRef]
21. Soyama, H. Improvement of fatigue strength by using cavitating jets in air and water. *J. Mater. Sci.* **2007**, *42*, 6638–6641. [CrossRef]
22. Soyama, H.; Chighizola, C.R.; Hill, M.R. Effect of compressive residual stress introduced by cavitation peening and shot peening on the improvement of fatigue strength of stainless steel. *J. Mater. Process. Technol.* **2021**, *288*, 116877. [CrossRef]
23. Soyama, H.; Hoshino, J. Enhancing the aggressive intensity of hydrodynamic cavitation through a Venturi tube by increasing the pressure in the region where the bubbles collapse. *AIP Adv.* **2016**, *6*, 045113. [CrossRef]
24. Nakashima, K.; Ebi, Y.; Shibasaki-Kitakawa, N.; Soyama, H.; Yonemoto, T. Hydrodynamic cavitation reac-tor for efficient pretreatment of lignocellulosic biomass. *Ind. Eng. Chem. Res.* **2016**, *55*, 1866–1871. [CrossRef]
25. Soyama, H. Enhancing the aggressive intensity of a cavitating jet by means of the nozzle outlet geometry. *J. Fluids Eng.* **2011**, *133*, 101301. [CrossRef]
26. Soyama, H. Comparison between the improvements made to the fatigue strength of stainless steel by cavitation peening, water jet peening, shot peening and laser peening. *J. Mater. Process. Technol.* **2019**, *269*, 65–78. [CrossRef]
27. Soyama, H. Cavitation peening: A review. *Metals* **2020**, *10*, 270. [CrossRef]
28. Yamauchi, Y.; Soyama, H.; Adachi, Y.; Sato, K.; Shindo, T.; Oba, R.; Oshima, R.; Yamabe, M. Suitable region of high-speed submerged water jets for cutting and peening. *JSME Int. J. Ser. B Fluids Therm. Eng.* **1995**, *38*, 31–38. [CrossRef]
29. Marcon, A.; Melkote, S.N.; Castle, J.; Sanders, D.G.; Yoda, M. Effect of jet velocity in co-flow water cavita-tion jet peening. *Wear* **2016**, *360*, 38–50. [CrossRef]
30. Marcon, A.; Melkote, S.N.; Yoda, M. Effect of nozzle size scaling in co-flow water cavitation jet peening. *J. Manuf. Process.* **2018**, *31*, 372–381. [CrossRef]
31. Zhu, R.; Zhu, H.; Zhang, X. Numerical investigation about the unsteady behavior of a free submerged cavitation jet using the SBES approach. *Ocean. Eng.* **2023**, *281*, 115010. [CrossRef]
32. Liu, M.; Li, W.; Li, H.; Xu, X.; Tian, F.; Agarwal, R.K.; Ji, L. Numerical simulation of cryogenic cavitating flow by an extended transport-based cavitation model with thermal effects. *Cryogenics* **2023**, *133*, 103697. [CrossRef]
33. Zwart, P.J.; Gerber, A.G.; Belamri, T. A two-phase flow model for predicting cavitation dynamics. In Proceedings of the Fifth International Conference on Multiphase Flow, Yokohama, Japan, 30 May–3 June 2004.
34. Yang, Y.F. Research on the Mechanism of Submerged High-Pressure Water Cavitation Jet and Hollow Bubble Shock Wave Strengthening Metal Properties. Ph.D. Thesis, Jiangsu University, Zhenjiang, China, 2020. Available online: https://kns.cnki.net/KCMS/detail/detail.aspx?dbname=CDFDLAST2021&filename=1020758544.nh (accessed on 20 June 2020).
35. Zhang, H.; Han, B.; Yu, X.G.; Ju, D.Y. Numerical and experimental studies of cavitation behavior in water-jet cavitation peening processing. *Shock Vib.* **2013**, *20*, 895–905. [CrossRef]

water

MDPI

Article

Rivers under Ice: Evaluating Simulated Morphodynamics through a Riffle-Pool Sequence

Karine Smith [1,2], Jaclyn M. H. Cockburn [1,*] and Paul V. Villard [2]

1 Department of Geography, Environment & Geomatics, University of Guelph, Guelph, ON N1G 2W1, Canada
2 GEO Morphix Ltd., Cambellville, ON L0P 1B0, Canada; paulv@geomorphix.com
* Correspondence: jaclyn.cockburn@uoguelph.ca

Abstract: Modeling in ice-covered rivers is limited due to added computational complexity, specifically challenges with the collection of field calibration data. Using River2D, a 2-dimensional hydrodynamic modeling software, this study simulates depth-averaged velocity and shear stress distributions under ice cover and in open-water conditions during varying flow conditions in a small, shallow riffle-pool sequence. The results demonstrated differences in velocity distribution throughout the channel and increases in discharge were found to impact the velocity magnitude under ice cover, while the spatial distribution remained consistent. A recirculating eddy found along the pool's left bank was exacerbated under ice cover, with potential implications for silver shiner habitat suitability. Bed shear stress magnitude did not vary significantly between ice and open water, although the spatial distribution differed notably. Model validation demonstrated success in simulating water depth and velocities, and the shear stress was estimated within a reasonable margin. Using hydrodynamic models provides valuable insight into seasonal changes in velocities and shear stress when ice is present.

Keywords: hydrodynamics; channel resilience; riffle-pool; River2D; erosion; deposition; velocity distribution; recirculation

1. Introduction

Fluvial processes are studied through field research using direct measurements to increase the understanding of channel hydrodynamics under various conditions with different morphological characteristics [1–3]. However, field measurements are time consuming, and it is challenging to cover larger areas in short periods of field time. Ice cover on rivers exacerbates the challenges in field data collection and increases the complexity associated with channel fluvial processes [3–8]. Increasingly, one-, two- and three-dimensional models are used alongside field studies to understand, simulate, and predict the fluvial processes in rivers to support channel design projects, management, and hazard-mitigation efforts [3,6,9–14]. Specifically, modeling is relied on to simulate, predict, and evaluate ice processes that occur on a river. CRISSP2D, RIVICE, DynaRICE, and MESH-RBM are modeling frameworks that simulate the ice formation on large rivers and predict the location of ice jams in an effort to reduce flooding and infrastructure damage [14–17]. Although efficient in simulating ice formation, it is a challenge for models to incorporate hydrodynamics or morphodynamics within a channel, limiting their use in investigating the impacts of ice on fluvial processes (e.g., velocity and shear stress). Additionally, despite the numerous models available for flow modeling under open-channel conditions, there are few models that consider both ice processes and fluvial dynamics, especially models that handle the parameters in small channels (e.g., less than 1 m deep).

Fluvial processes occurring under ice cover require more investigation, especially in the context of flow distribution, thalweg concentration, and shear stress at the bed [2,5,6,10,12,18]. Enhancements in methods to analyze these processes under ice cover using instrumentation such as acoustic Doppler velocimeters (ADVs) and acoustic Doppler current profilers (ADCPs) provide inputs for flow modeling under ice cover [2,5,6,11]. Numerical models are powerful tools for simulating river processes using computational methods to solve non-linear equations that describe the hydro-morphodynamics of rivers [19]. Numerical modeling incorporates empirical relationships for roughness, velocity, sediment transport capacity, and shear stress to predict flow characteristics under ice cover [20]. Hydrodynamic modeling continues to develop to address the limitations associated with uncertainties related to fluvial dynamics under ice [7]. Lotsari et al. [6] aimed to improve our understanding of the spatial variation of ice-covered mid-winter flow and the erosion and sedimentation capacity of ice-covered flow as compared to open-water using the two-dimensional (2D), hydrodynamic model River2D. River2D has the capacity to simulate direction and magnitude for time- and depth-averaged velocities and to identify recirculating flow structures [21]. The 2D hydrodynamic model produces results that efficiently compare the differences in the near-bed velocities and the spatial distribution of flow under ice-covered conditions and open-channel conditions for meandering rivers [6]. However, changes in the velocity distribution and shear stress under different flow conditions and the influence of a variable bed morphology were not investigated.

The objectives of this study are to simulate ice-covered velocities and shear stress (magnitudes and distributions) under various flow conditions, and to assess the model's performance in a small, ice-covered riffle-pool sequence (Supplementary Materials Figure S1). For this purpose, River2D was used as it includes ice as an upper boundary layer and models depth-averaged velocity and shear velocity distributions [22]. Two scenarios were employed to conduct direct comparisons and investigate the impacts of ice on flow and shear stress: simulation 1 used field data collected during ice cover in lower flow conditions to calibrate an ice-free simulation, and simulation 2 used field data collected during open water in higher flow conditions to calibrate an ice-covered simulation. Hydrodynamic modeling, such as what is presented here, helps users to confidently evaluate processes that are difficult to measure directly (e.g., under ice-covered conditions and dangerous high flow conditions).

2. Study Site and Field Measurements

Sixteen Mile Creek in southern Ontario, Canada, has a total watershed area of 372 km^2, and it empties into Lake Ontario (Figure 1). It flows southeast through the city of Milton, and experiences impacts from a variety of land-use types including urban, agriculture (cropping and grazing), and forests. The study reach drains an area of 108 km^2 and is a 75 m segment, which is ~20 m at its widest (Figure 1b) and ~1 m at its deepest. Based on the Water Survey of Canada gauging station (02HB005, period of record: 1957–present), located approximately 4.5 km upstream of the study reach, the average winter discharge is 1.35 m^3/s, the average spring discharge is 2.3 m^3/s, and the average summer discharge is 0.8 m^3/s [23]. The 2021 average winter discharge was 1.2 m^3/s, the average spring discharge was 2.3 m^3/s, and the average summer discharge was 0.8 m^3/s [23]). Bed roughness was quantified using roughness lengths (y_0), which is defined as the height above the bed where the mean velocity is theoretically 0 m/s [24]. The roughness length estimates along seven cross-sections throughout the study site indicated that in the upstream and downstream, most sections of the study reach exhibit bed roughness features that are typical of riffles, while the lower section along cross-sections 4, 5, and 6 exhibit lower levels of roughness, which is typical of pools (Table 1). Wolman pebble counts determined that all cross-sections had a large range in substrate size but were typical for riffles and pools (Table 1). During site visits, sediment transport was limited to suspended sediment transport, and the bedload was negligible [5].

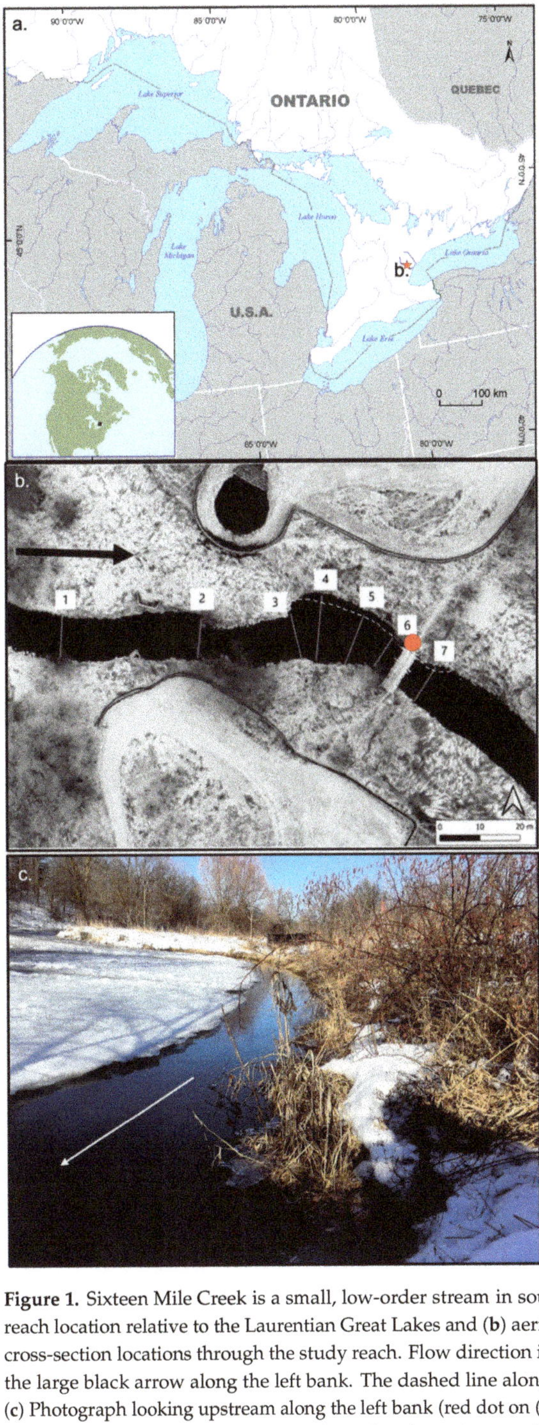

Figure 1. Sixteen Mile Creek is a small, low-order stream in southern Ontario, Canada. (**a**) Study reach location relative to the Laurentian Great Lakes and (**b**) aerial photograph showing numbered cross-section locations through the study reach. Flow direction is from left to right, as indicated by the large black arrow along the left bank. The dashed line along the left bank indicates ice extent. (**c**) Photograph looking upstream along the left bank (red dot on (**b**)) showing the thin (~1 m) strip of open water on 25 February 2021 (two days after the major sampling campaign on 23 February 2021). The white arrow in the photograph indicates flow direction.

Table 1. Channel geometry and bed substrate characteristics.

Cross-Section	Width	Average Depth	d/D	Median Grain Size (D50)	Roughness Lengths	Notes
1—Riffle	12 m	0.35 m	0.85	412 mm	2 mm	Large boulders near left bank
2—Riffle	10 m	0.45 m	3	150 mm	2–2.5 mm	Embedded, poorly sorted fine material found between cobles
3—Riffle	10 m	0.28 m	0.95	295 mm	2.4–2.8 mm	Large, flat boulder 3 m from the right bank, ~25% embedded
4—Pool	18 m	0.48 m	10	48 mm	0.5–0.7 mm	Fine material (sand sized and fine) along the left bank, embedded gravel, cobbles, and boulders toward the right bank
5—Pool	14 m	0.52 m	100	5.2 mm	0.5 mm	Dominated by fine-grained bed material (sand-sized and smaller)
6—Pool	10 m	0.6 m	125	4.8 mm	0.4 mm	Dominated by fine-grained bed material (sand-sized and smaller)
7—Riffle	9 m	0.5 m	1.14	439 mm	2.5–2.7 mm	Cobbles and boulders with embedded pebbles and coarse sand

Topographic site surveying using a real-time kinematic global position system (RTK GPS) at 3–5 m intervals along the top and bottom of the channel banks, the top of the slope, the edge of the water, and along the channel bed was conducted across the site. Measurements were differentiated based on location using codes: edge of water (EOW), bottom of bank (BOB), top of bank (TOB), channel bed (CB), and central line (CL). Additionally, the ice extent was mapped at 3–5 m intervals throughout the study site on field collection visits [5,25]. Boulders with a b-axis exceeding 0.25 m were surveyed and fed into the elevation model to describe bed heterogeneity throughout [26]. The bed slope was less than 1% through the entire reach (~0.9%), 1.06% through the riffle bed slope, and 1.09% through the pool bed slope. For model calibration and validation purposes, the water levels were continuously monitored from January to June, and the velocity measurements were collected throughout February and March along seven delineated cross-sections (Figure 1) [5,22,25].

The velocity data collected on 23 February 2021 (under ice cover and lower flow conditions) used a Sontek Flowtracker2 Handheld ADV and the Sontek S5 ADCP (where depth was sufficient), and on 2 March 2021 (open water, higher flow conditions), a Sontek S5 ADCP was used. The procedures outlined in Lotsari et al. [7] and Demers et al. [8] were followed on both days. The Flowtracker2 ADV used for this project has an acoustic frequency of 10.0 MHz, a minimum depth requirement of 0.02 m, and depth and velocity resolutions of 0.001 m and 0.0001 m/s [27]. Despite the minimum depth requirement of 0.02 m for the ADV, the measurements could only be completed for depths of 0.05 m or more during this study. The Flowtracker2 ADV also recorded the Signal Noise Ratio (SNR) to aid in detecting measurement errors and/or sensor blockages [27]. The average sampling time was kept relatively short in order to collect data over a wide spatial extent [5]. The S5 ADCP has a minimum depth requirement of 0.2 m and thus is limited in shallow waters. The Sontek S5 has a vertical resolution of ±0.001 m and a velocity resolution of ±0.0001 m/s, and the velocities are averaged between all five transducers [28]. While collecting samples in open water, the ADCP sensor automatically adjusted the velocity cell sizes based on instrument movement speed, water depth, and flow velocity [28]. Sixteen Mile Creek is a shallow channel with a maximum depth of less than a meter, resulting in an average cell size of 0.02 m. The S5 had the blanking distance set at 0.05 m. The average measurement height above the bed was 0.082 m and the average measurement depth below the surface was 0.06 m in open water.

Velocity measurements were collected along all seven cross-sections using a combination of the ADV and ADCP (Figure 1). ADV measurements were collected in 0.05 or 0.10 m-depth increments at 1 m lateral intervals when possible, and at 2 m lateral intervals when time constraints applied. These measurements were taken along each cross-section under both open-water and ice-covered conditions, where holes were drilled through the ice (when present) using augers [7,8]. The ADV was used in combination with the ADCP to collect data in both shallow and deeper locations throughout the channel. Under ice cover, stationary ADCP measurements were taken through the hole in the ice, from which the flow direction, water depth, and maximum velocity were calculated, and the mean velocity profiles were determined. In ice-free conditions, the ADCP was mounted to a float and manually towed along each cross-section. Where ice was present, the ice thickness and visual descriptions of roughness were also manually measured and recorded.

Since the ADCP records velocities in the East, North, and Up (ENU) directions [29], the E and N vectors were rotated around their axis at an angle corresponding to the streamwise direction to yield u and w velocity components. This was determined using the Velocity Mapping Toolbox (VMT) developed by USGS [29]. The vertical velocity vector was unchanged by the rotation of the E and N velocity rotations. The ADV collects u, v, and w (streamwise, vertical, and lateral) velocities and it did not require additional data processing prior to analysis.

3. Pre-Processing, Data Analysis, and Model Validation

The 2-dimensional model River2D was used for hydrodynamic simulations under ice and open water across the study site. River2D requires channel bed and ice surface topography converted into a discrete mesh, roughness estimates, initial flow, and boundary conditions [22]. The boundary conditions were defined as discharge along the inflow boundary, and fixed water surface elevations at the outflow boundary [22]. Depth-averaged velocities were calculated based on the St. Vernant equations: conservation of mass, conservation of x-direction momentum, and conservation of y-direction momentum [22]. A singular velocity value was assigned at each node, and the depth specific velocities were not computed. The model runs on the following three basic assumptions: (1) a hydrostatic pressure distribution across the water column depth, (2) constant distribution of horizontal velocities over depth, and (3) Coriolis and wind forces are negligible [22]. When ice cover was present, River2D used an ice and bed resistance model, where the bed and ice roughness values were combined to create composite roughness used in the hydraulic calculations [22].

The topography data obtained for the ice extent and the high-resolution channel topography survey were converted into Digital Elevation Models (DEMs) using the Triangular Irregular Network (TIN) interpolation tool in River2D_Bed [30]. All topography points were identified as fixed nodes, with elevation and roughness values assigned from the underlying bed topography layer [30]. Hard breaklines were delineated along the deepest location of the channel, as well as the edge of the water and the top of the banks, to prevent 'leakage' from the model [6]. Breaklines ensured the appropriate interpolation of values along linear features in a finalized channel bed elevation TIN (Figure 2) [30]. Throughout winter 2021, full ice cover was present up- and downstream of the pool, with partial cover over the pool, leaving a strip of open water along the left bank (Figures 1c and 2c). The open-water section along the pool is considered within the model by assigning ice thickness values to ice nodes and assigning 0 to no-ice nodes [22]. Breaklines were then established along the ice edge to ensure interpolation along the ice–water boundary. The fixed nodes and breaklines were subsequently smoothed into a mesh element (10 cm resolution) with all the necessary components, allowing the topographical data to be fed into the River2D software [22]. Velocity bins measured using the ACDP under the ice cover ranged from 2 cm–10 cm. The mesh resolution was set to provide sufficient details on velocity and shear stress distribution while still allowing for validation with field data.

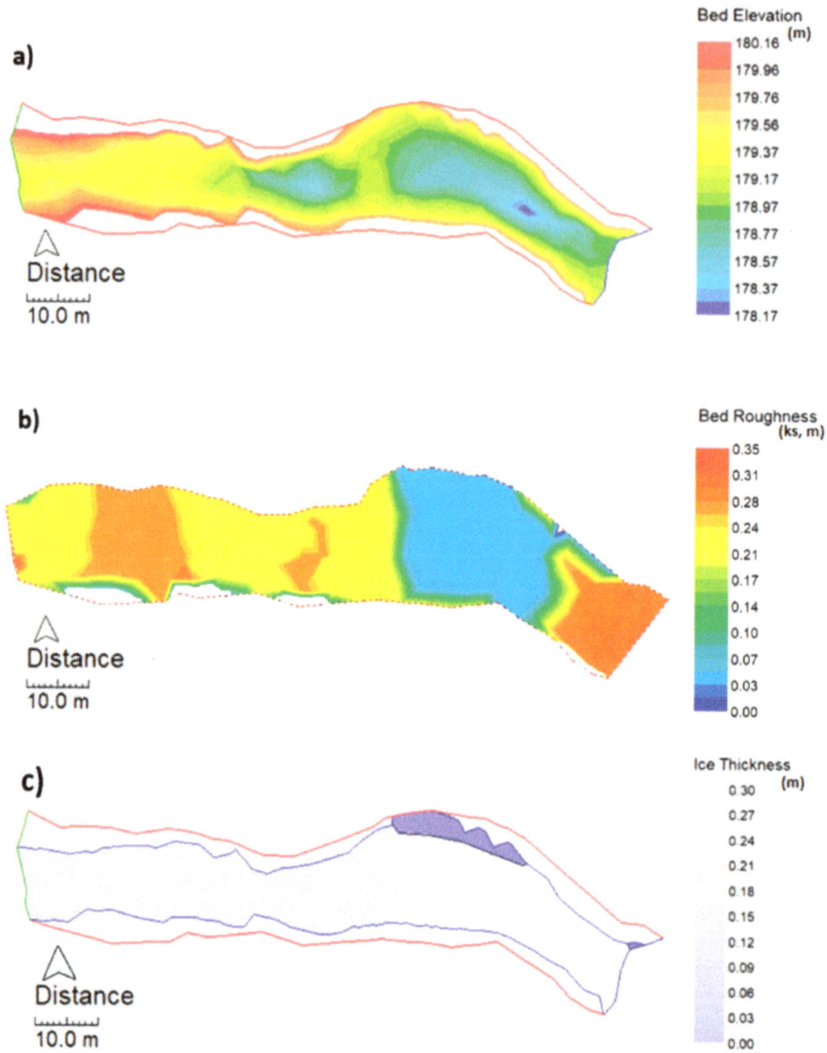

Figure 2. Interpolated model surfaces. (**a**) Channel bed elevation, (**b**) channel bed roughness, and (**c**) ice thickness (dark colors represent minimal (or no) ice thickness). Bed roughness was manually prepared in the River2D_bed extension and extends past water edge to ensure roughness values are assigned along the water edge breaklines.

All triangulation points in the mesh element layer require equivalent sand roughness height (k_s) estimates, which serve as representations of hydraulic friction across the surface [31]. Roughness heights (k_s) were calculated for all measured cross-sections based on estimated roughness lengths from logarithmic velocity profiles seen in Equations (1)–(3):

$$k_s = 30.1 y_0, \tag{1}$$

$$y_0 = e^{-a/b}, \tag{2}$$

$$u_y = a + b \ln y, \quad (3)$$

where u_y is the mean velocity at a given depth, a is the intercept, b is the slope and velocity gradient, and y_0 is the calculated roughness length [32]. Individual channel bed nodes throughout the mesh element were assigned roughness height values corresponding to the nearest cross-sections to yield a finalized channel bed roughness TIN. Visual observations and roughness length estimates of the ice surface demonstrated little change in the roughness across the ice surface, allowing for a singular roughness length assigned for the upper boundary ice surface [5,25]. The River2D software requires discharge estimates at the upper limit of the study site (cross-section 1, Figure 1). These estimates were calculated using the transect velocity–area method, based on velocity measurements collected throughout the February and March site visits [5,25].

Simulations were run for ice-covered and open-water conditions under two scenarios (lower water levels derived from ice-covered field data on 23 February 2021, and higher water levels derived from open-water field data on 2 March 2021) [5]. Each simulation yielded direct estimates of the shear velocity for each node along the lower boundary, which were used to calculate the shear stress along the channel bed using Equation (4):

$$\tau_b = \rho u_*^2, \quad (4)$$

where τ_b is the bed shear stress, ρ is the water density, and u_* is the shear velocity. When ice cover was present, the calculated shear velocity only applied to the channel bed and did not allow for the calculation of shear stress along the ice surface [22]. Model validation was conducted by calculating mean absolute error (MAE) values between simulated and measured velocities, ice surface elevations and water levels at calibration point locations [6]. MAE was calculated using Equation (5):

$$MAE = \frac{1}{N} \sum_J^N abs(M_j - P_j), \quad (5)$$

where N is the number of observations, M_j is the modeled value, and P_j is the measured value [33].

4. Results

4.1. Cross-Section Velocity Distribution

Cross-sectional geometry analyses and changes in velocity through each cross-section were achieved by creating a mesh grid with cross-section widths corresponding to x values, depths corresponding to y values, and streamwise velocities corresponding to z values. This grid was used to create cross-sectional maps of velocity to visualize the lateral distributions for each cross-section. The lateral and vertical velocities were represented using a vector plot overlain on the cross-sectional contour, demonstrating the direction and magnitude of secondary velocities [29]. For ice-covered velocities, the intervals between the measurement locations were omitted to avoid errors caused by interpolation. The largest changes in streamwise velocities were observed within the pool, especially along the left bank of cross-section 4, where upstream velocities exhibited a notable increase under ice and a shift in the secondary flow toward the right bank rather than the left (Figure 3). The flow structures observed in open water, such as the circulation seen in cross-section 5 (Figure 3) and the helix-shaped flow cell in cross-section 6 (Figure 3), were not readily observed under ice cover on 23 February, which was likely due to the sparse data availability under ice. The cross-sectional velocity analysis hints at potential changes in velocity distribution during ice cover in the pool (Figure 3), but because of the nature of data collection in the winter, it was difficult to generate meaningful results for riffle cross-sections (Supplementary Materials Figure S2).

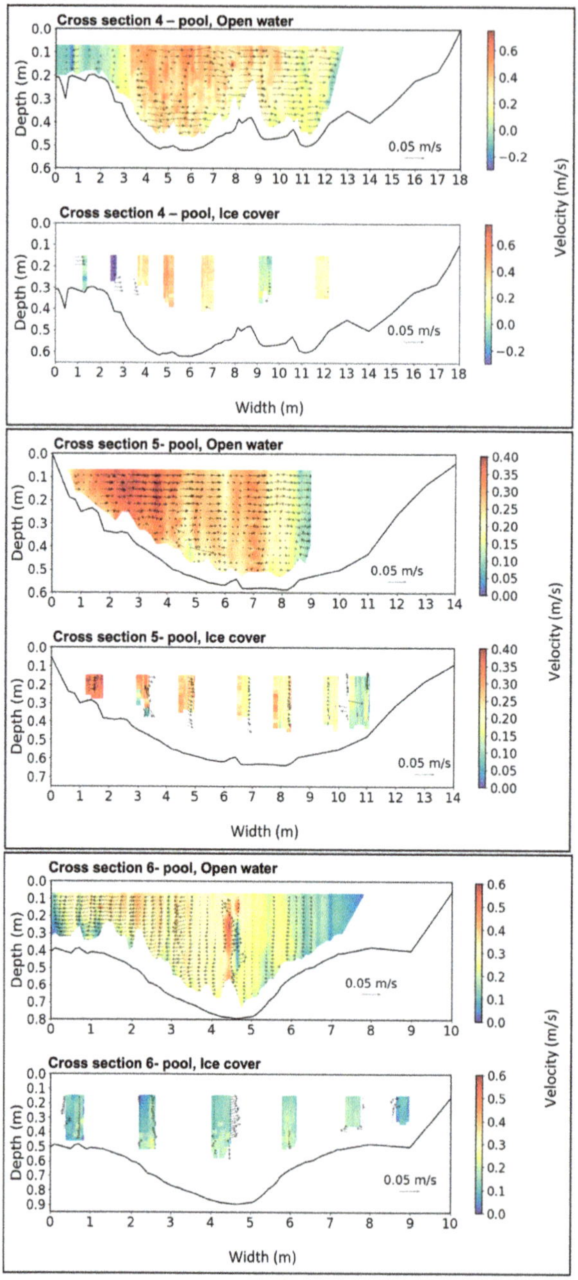

Figure 3. Velocity distribution maps along cross-section 4 (**upper panel**), 5 (**middle panel**) and 6 (**lower panel**) within the pool of the study site (Figure 1), from left bank to right bank under open-water conditions (top panel) and ice-covered conditions (bottom panel). Streamwise velocities are represented on a color scale and secondary velocities are represented by arrows showing direction and magnitude of the transverse and vertical velocity component. Data were collected using a Sontek S5 ADCP under ice-covered conditions on 23 February 2021, and open-water conditions on 2 March 2021.

4.2. Water Depth Distribution and Depth-Averaged Velocity

River2D modeled water depth under various conditions (Figure 4). In general, when ice cover was present, the water depth increased, with its maximum under higher flow, ice-covered conditions (Figure 4b), and at its minimum under lower flow, open-water conditions (Figure 4c). River2D does not allow for a fixed ice cover; thus, the water depths increased when ice was present to maintain flow continuity under a higher flow resistance. Outflow water depths were constant between the two lower flow simulations and the two higher flow simulations due to fixed water surface elevations as measured in the field (Figure 4).

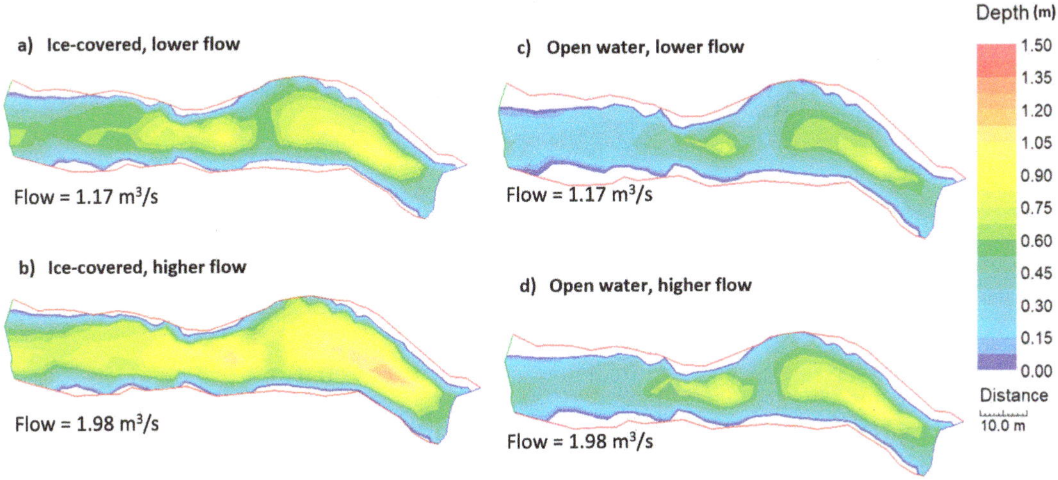

Figure 4. Modeled water depth under (**a**) lower flow, ice-covered conditions; (**c**) lower flow, open-water conditions; (**b**) higher flow, ice-covered conditions; and (**d**) higher flow, open-water conditions. Prepared in River2D.

Field data collection is challenging under ice cover; thus, models, such as River 2D, are useful tools to help characterize velocity flow patterns when ice is present. Simulations were run for ice-covered and open-water conditions based on lower flow and higher flow field data [5]. The lower flow, ice-covered and higher flow, open-water simulations were used for model calibration, validation purposes, and direct comparisons of depth-averaged velocities under different flow characteristics. No significant difference in the depth-averaged velocity magnitude was observed across the reach during lower flow conditions (Figure 5). However, depth-averaged velocities observed downstream of the pool directly above the outflow of the model under ice cover abruptly decreased and then increased (~0.4 m/s) in the River2D output (Figure 5a). There were slight changes to the wetted perimeter and thalweg width. Under lower flow, ice-covered conditions, the wetted perimeter increased (Figure 5a) when compared to the open-water, lower flow conditions (Figure 5c). The upstream thalweg width was largest under open-water conditions (Figure 5a vs. Figure 5c), indicating a wider spatial distribution of faster velocities. Along the left bank of the pool, a recirculating eddy was more obvious under ice-covered conditions compared to open-water conditions (e.g., Figure 5a vs. Figure 5c). Flow direction vectors were a direct output of River2D, and the depth-averaged flow directions were predominantly streamwise throughout the length of the channel, except for the recirculating eddy (Figure 5).

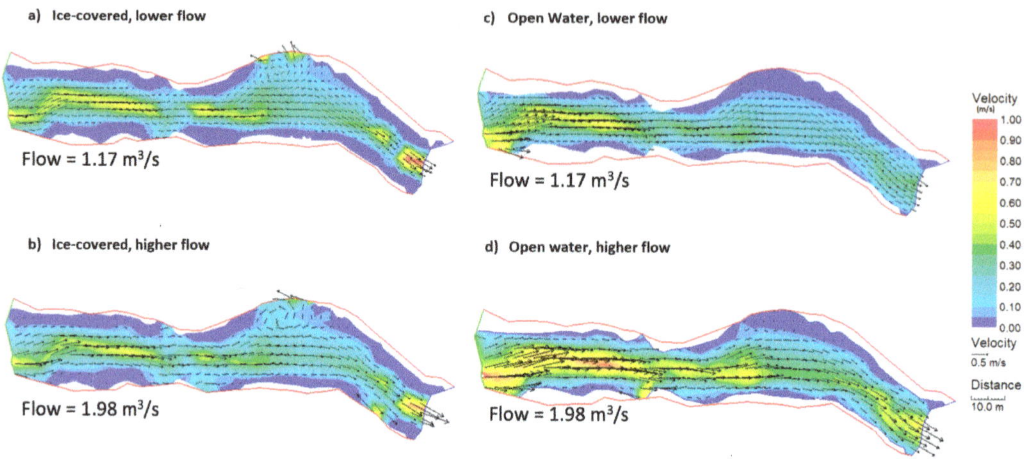

Figure 5. Modeled depth-averaged velocities during (**a**) ice cover, lower flow conditions (1.17 m^3/s); (**b**) ice cover, higher flow conditions (1.98 m^3/s); (**c**) open-water, lower flow conditions (1.17m^3/s); and (**d**) open water, higher flow (1.98 m^3/s). Flow directions are indicated at 1 m intervals using vector arrows (if they were plotted at 0.1 m intervals it would be difficult to visualize), and no-flow boundaries are outlined in red; the inflow boundary is on the left and outflow boundary on the right. Prepared in River2D.

Similar results were generated for higher flow conditions in open water and ice-covered flow (Figure 5b,d). The wetted perimeter increased under ice as demonstrated by an increase in the wetted area from bank to bank (Figure 5b vs. Figure 5d). The thalweg widened under open water, and a notable increase was observed in velocities throughout the upstream section and within the pool in open-water higher flow (Figure 5d). The depth-averaged velocity magnitude under higher flow open-water conditions ranged from 0.6 m/s to 1 m/s, while ice-covered higher flow conditions exhibited a slower thalweg velocity range between 0.4 m/s and 0.6 m/s (Figure 5b,d). As demonstrated under lower flow, ice-covered conditions (Figure 5a), an abrupt decrease followed by an increase in velocities was observed under higher flow, ice-covered conditions within the downstream, bottleneck section of the channel. This change exceeded velocities in the same location under open-water, higher flow conditions (Figure 5d). The eddy along the left bank of the pool was also present under higher flow conditions, and was exacerbated by the presence of ice cover.

The recirculating eddy along the left bank of the pool exhibited notable differences in velocities between open-water and ice-covered conditions for both lower and higher flow simulations (Figure 6). While still present under open-water conditions, the recirculating eddy along the left bank of the pool only exhibited velocities ranging between 0 and 0.1 m/s (Figure 6). Conversely, under ice-covered conditions, velocities within the eddy ranged from 0.1 to 0.6 m/s. Although higher flow open-water conditions did result in increased velocities when compared to ice-covered conditions, this increase in discharge under open-water conditions was concentrated within the thalweg across the pool, rather than in the eddy (Figure 6). Additionally, the eddy demonstrated slightly faster depth-averaged velocities under lower flow ice-covered conditions than under higher flow ice-covered conditions (Figure 6). However, velocities within the pool thalweg during higher flow, ice-covered conditions exceeded those seen in the same location but with a lower flow (Figure 6). The largest velocities within the pool thalweg occurred under higher flow, open-water conditions, ranging between 0.35 and 0.65 m/s (Figure 6).

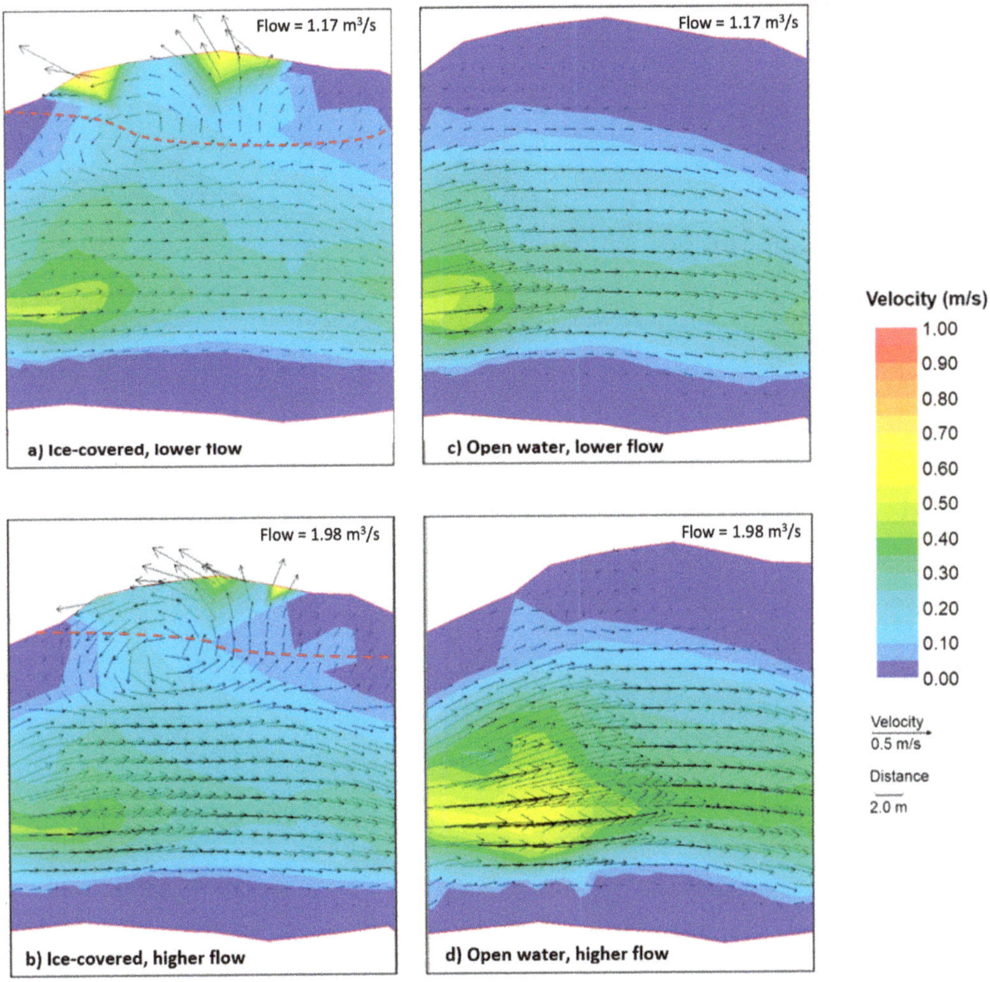

Figure 6. Depth-averaged velocity magnitude and directions within the pool under (**a**) lower flow, ice-covered conditions; (**c**) lower flow, open-water conditions; (**b**) higher flow, ice-covered conditions; and (**d**) higher flow, open-water conditions. Flow directions are indicated at 1 m intervals using vector arrows (if they were plotted at the mesh resolution of 0.1 m intervals it would be difficult to visualize). Ice extent along the left bank is delineated by dashed line in (**a**,**b**). Prepared in River2D.

The number of data points collected on 2 March 2021 during higher flow conditions without ice cover were more numerous than the data collected in February (with ice cover). Specifically, the ADCP collected data continuously throughout the site, as it was mounted to a float and towed throughout the reach. There were minimal differences in average velocities when the 2 March 2021 data were 'resampled' to data densities similar to the ice-covered collection. However, the velocity direction within the previously identified recirculation zone along the left bank of the pool was less obvious (Figure 7). These differences highlight the potential biases due to sample density during the open-water conditions.

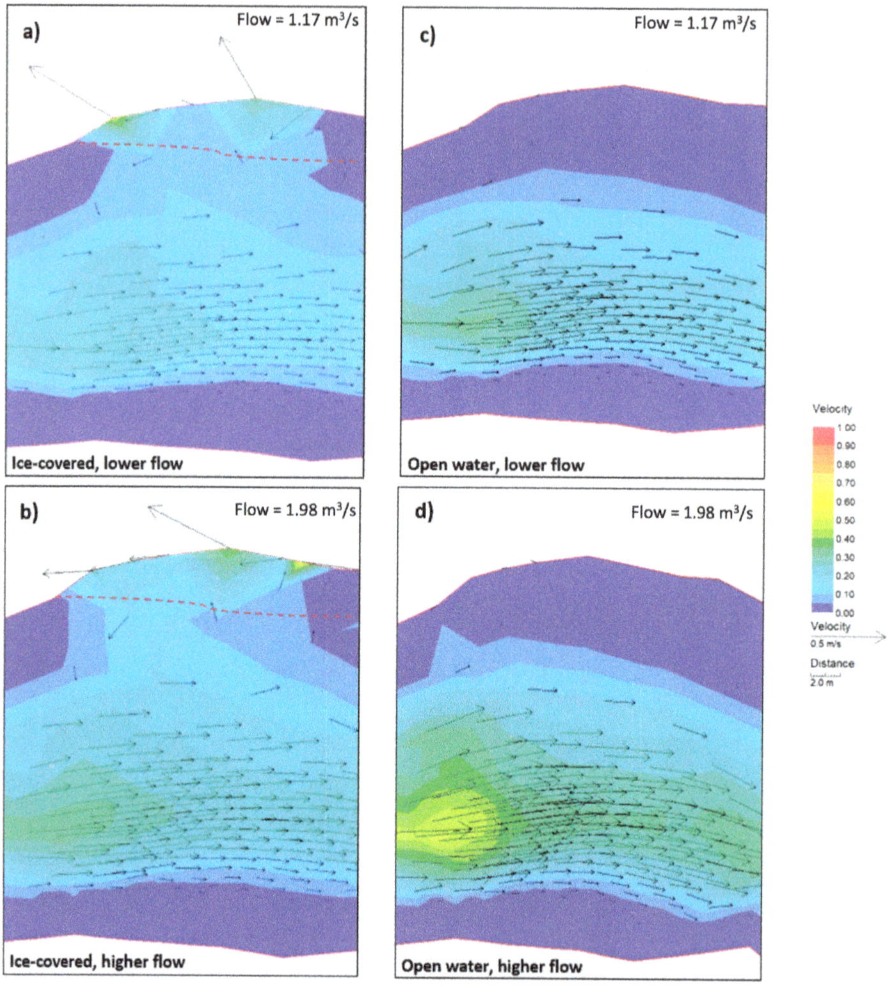

Figure 7. Resampled depth-averaged velocity magnitude and directions within the pool under (**a**) lower flow, ice-covered conditions; (**c**) lower flow, open-water conditions; (**b**) higher flow, ice-covered conditions; and (**d**) higher flow, open-water conditions. Flow directions are indicated using vector arrows at each mesh point, set at 1 m intervals. Ice extent along the left bank is delineated by dashed line in (**a**,**b**). Prepared in River2D.

4.3. Shear Stress Magnitude and Distribution

The shear stress values were calculated for all mesh nodes based on the River2D shear velocity output. Under all simulations, the upstream section of the river segment exhibited bed shear stress values exceeding 1 N/m^2 (Figure 8). Under open-water conditions, the shear stress hotspots spanned a wider area of the channel, while under ice-covered conditions, the hotspots narrowly follow the thalweg (Figure 8). The bed shear stress magnitudes were larger under higher flow conditions, ranging from 2 to 16 N/m^2 within the riffles, and from 0 to 1.5 N/m^2 within the pool (Figure 8b,d). Under ice-covered lower flow conditions, the bed shear stress values remained low throughout the pool, only increasing above 1 N/m^2 within the recirculating eddy along the left bank. The maximum bed shear stress values increased under lower flow ice-covered conditions in the upstream and downstream sections, while they remained smaller within the pool when compared to lower flow open-water conditions (Figure 8a,b). Overall, the widest distribution of high bed shear stress values occurred under open-water, higher flow conditions, exceeding those seen in all other simulations (Figure 8d). Lower flow, open-water conditions experienced a similar distribution but at a lower magnitude (Figure 8c). When ice cover was present, regions exhibiting a high bed shear stress shifted, with hotspots observed along the left bank of the pool and upstream and downstream of the pool (Figure 8a,b). The most notable differences between the ice-covered and open-water shear stress distributions were noted along the recirculating eddies within the pool and the downstream section located above the outflow boundary (Figure 8). Overall, the range in the bed shear stress magnitude did not vary much between open-water and ice-covered conditions, but larger bed shear stress values were observed under higher flow conditions.

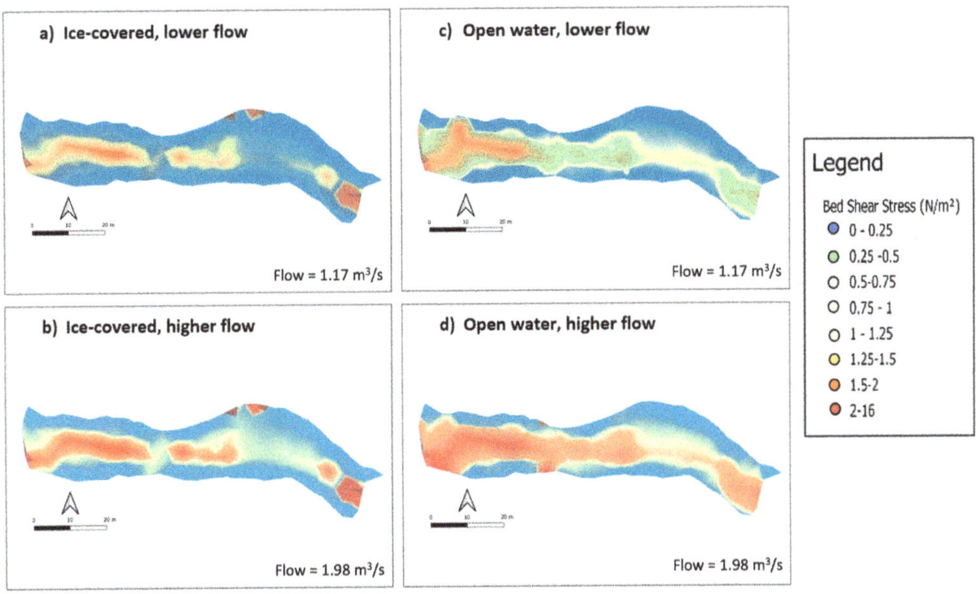

Figure 8. Bed shear stress distribution under lower flow, ice-covered conditions (**a**); lower flow, open-water conditions (**c**); higher flow, ice-covered conditions (**b**); and higher flow, open-water conditions (**d**). Prepared in QGIS.

Based on field bed and channel geometry surveys (Table 1), nodes within the modeled mesh were classified using the cross-sectional depth to median grain size ratio (d/D) to evaluate the shear stress magnitude variability as a function of the geomorphic unit (e.g., riffle vs. pool) (Figure 9) [34]. The shear stress magnitudes for pool cross-sections (d/D > 10) were lower than the shear stress for riffle cross-sections (d/D < 10) (Figure 9). The shear stress along pool cross-sections was less than 1 N/m^2, indicating low shear stress values, while the shear stress at riffle cross-sections ranged above 1 N/m^2 (Figure 9). *p*-values from a two-tailed t-test indicate significant differences in the bed shear stress values within a 95% confidence interval for the pool under higher flow conditions and the riffle under lower flow conditions (Table 2, Figure 9). However, the *p*-values for the riffles under higher flow conditions and the pool under lower flow conditions did not indicate significant differences between the shear stress magnitude under open-water and ice-covered conditions (Table 2, Figure 9).

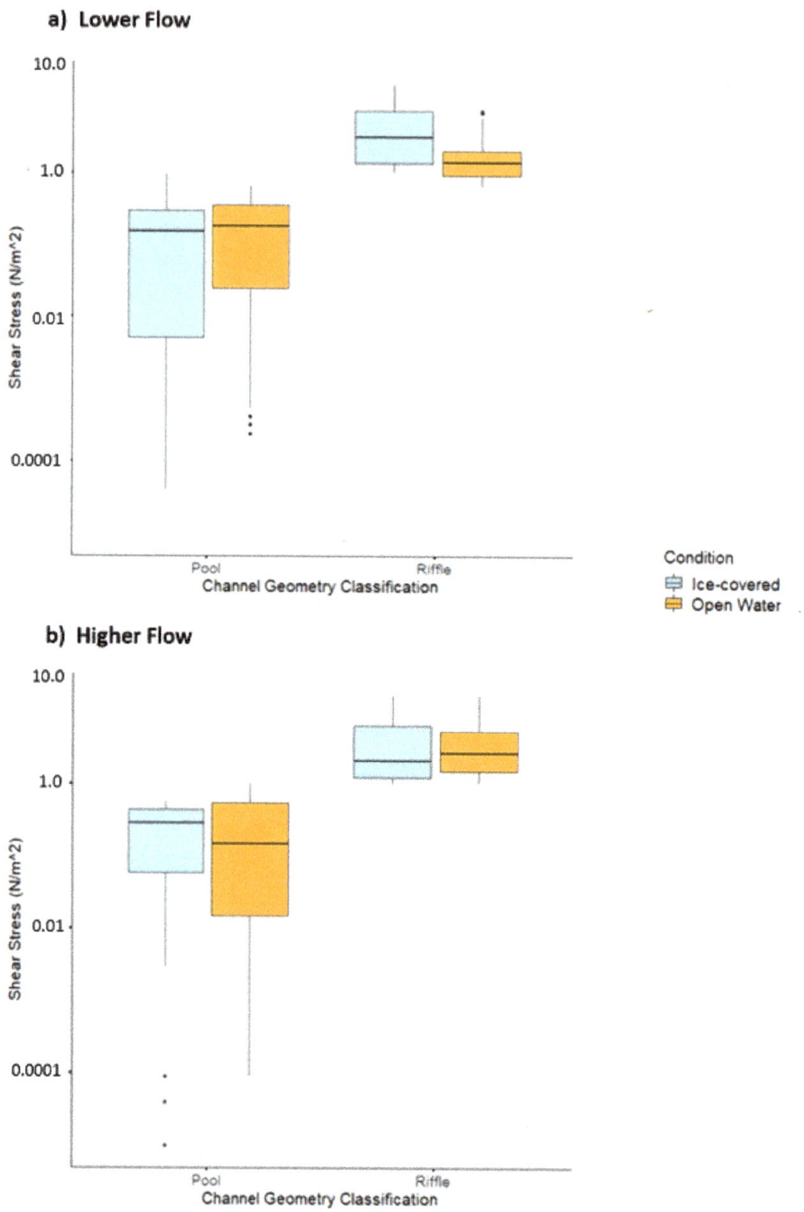

Figure 9. Bed shear stress distribution under (**a**) lower flow and (**b**) higher flow conditions based on the channel geometry classification of depth over median grain size (Table 1).

Table 2. *p*-values from difference of means test conducted between ice-covered and open-water conditions in the pool (d/D > 10) and riffle (d/D < 10) for lower and higher flow conditions.

	Lower Flow	**Higher Flow**
Pool (d/D > 10)	0.26	0.041
Riffle (d/D < 10)	<0.0001	0.053

4.4. Model Validation Results

A model validation demonstrated that the lower flow simulation results were moderately stronger than those from the higher flow (Table 3). No real-time observations were conducted for lower flow, open-water conditions or higher flow, ice-covered conditions; therefore, model validation was not calculated for those simulations. The lower flow model was most successful in simulating depth-averaged velocities, within 0.046 m/s, with water depths being simulated within 0.098 m and shear stress within 0.147 N/m^2 (Table 3). Under higher flow, open-water conditions, the depth-averaged velocities exhibited a marginally lower MAE, while the MAE for water depth was halved, and the MAE for shear stress was slightly larger (Table 3). Overall, the shear stress demonstrated the lowest success in modeled values for both lower flow and higher flow simulations, and depth-averaged velocities demonstrated the highest success (Table 3). Under the higher flow simulation, all average MAE values fell under 0.2, and the mean absolute percentage errors fell below 20% (Table 3). Additionally, the MAE values on the resampled datasets were conducted and follow the same patterns as the denser dataset analysis. The resampled MAE values also indicated a lower accuracy when compared to field values.

Table 3. Calculated MAE values for the modeled water depth, depth-averaged velocities, and shear stress values under lower flow, ice-covered conditions and measured data on 23 February 2021, and higher flow, open-water conditions and measured data on 2 March 2021. MAE values for the downsampled simulations models are shown in parentheses.

Cross-Section	Ice-Covered Conditions, Feb 23 Field Data (Lower Flow)			Open-Water Conditions, Mar 2 Field Data (Higher Flow)		
	Water Depth (m)	Depth-Averaged Velocity (m/s)	Shear Stress (N/m^2)	Water Depth (m)	Depth-Averaged Velocity (m/s^2)	Shear Stress (N/m^2)
1	0.038 (0.12)	0.018 (0.19)	0.22 (0.36)	0.035 (0.15)	0.018 (0.13)	0.25 (0.17)
2	0.162 (0.22)	0.065 (0.18)	0.25 (0.27)	0.098 (0.11)	0.033 (0.16)	0.13 (0.20)
3	0.177 (0.24)	0.079 (0.21)	0.39 (0.19)	0.063 (0.18)	0.14 (0.15)	0.36 (0.39)
4	0.043 (0.17)	0.036 (0.13)	0.11 (0.23)	0.031 (0.12)	0.026 (0.09)	0.073 (0.16)
5	0.134 (0.23)	0.058 (0.16)	0.15 (0.18)	0.012 (0.08)	0.011 (0.10)	0.11 (0.13)
6	0.157 (0.28)	0.077 (0.21)	0.098 (0.21)	0.014 (0.15)	0.052 (0.14)	0.093 (0.18)
7	0.066 (0.15)	0.016 (0.11)	0.083 (0.14)	0.091 (0.11)	0.012 (0.09)	0.047 (0.16)
Reach Average	0.11 (0.20)	0.050 (0.17)	0.19 (0.23)	0.047 (0.129)	0.041 (0.123)	0.152 (0.199)

Model validation demonstrates that the model output accuracy varied between the criteria and location throughout (Table 3). Pool cross-sections exhibited lower MAE values, while riffle cross-sections exhibited higher MAE values (Table 3). These results indicate that the River2D-simulated ice-covered flow in this reach of the Sixteen Mile Creek is sensitive to water depth, and that shallower flows are more problematic. River2D employs the St. Vernant equation to calculate depth-averaged velocities by considering the conservation of mass and momentum in the x and y direction [22,30]. Studies have demonstrated that in shallow, turbulent waters, with significant secondary flow components, the accuracy of this equation may be affected [35]. Furthermore, locations with shallower, slower-moving waters will be proportionally more sensitive to mean absolute error values.

5. Discussion

Higher and lower flow conditions largely influence velocities, flow directions, and shear stress in channels. There are limited data on the interaction between discharge and ice cover as it relates to fluvial processes in small channels [36,37]. The lower and higher flow scenarios tested in this study were representative of a range of conditions experienced in winter 2021. The nearest Water Survey of Canada (WSC) gauging station is 4.5 km upstream of the study site (Station ID 02HB005); from this record, the winter 2021 average discharge was 1.2 m^3/s (long-term winter season average was 1.4 m^3/s, 65 years) [23]. At the WSC gauging station, the discharge was 0.625 m^3/s on 23 February 2021 and 1.22 m^3/s on 2 March 2021 [23]. These values are slightly lower than the conditions measured at the study site and are reflective of the upstream position of the WSC gauge. However, given the relative differences between our study site and the WSC data, we can conclude that the February 23 field data coincided with lower flow conditions, likely below the seasonal winter average, and the March 2 field observations coincided with average winter flow conditions but represent the higher conditions of the two comparison datasets.

5.1. Spatial Distribution of Depth-Averaged Velocities and Bed Shear Stress under Ice Cover

Under both higher and lower flow conditions, the outflow water surface elevation was constant between ice-covered and open-water trials, therefore reasonably simulating a fixed ice cover (Figure 4; Supplementary Material Figure S3). Calibrated model-simulated depth-averaged velocities and water levels were within a 10% margin of error for both simulations (Table 3). Experimental flume studies demonstrated that a fixed ice cover leads to confined flow and increased depth-averaged velocities [12], which were similar to the observed values downstream of the pool in this study (Figure 5). However, water depth within the pool and in the upstream riffle increased under ice cover, indicating that incorporating ice cover in River2D does not explicitly consider bed- or bank-fast ice unless the water depth is maintained constant, as seen near the outflow (Figure 5) [38]. Throughout winter 2021, the ice cover did not fully extend across the pool, leaving a strip of open water along the left bank and preventing fixed ice and confined flow from occurring in the pool (Figure 2). The results from the velocity simulations demonstrated overall slower depth-averaged velocities under ice cover when compared to open water for both the lower flow and higher flow outputs (Figure 5). However, the recirculating eddies along the left bank of the pool had higher velocities when ice was present, and the segment downstream of the pool presented exceptions to these results (Figure 5). In pool-riffle sequences, increases in velocity downstream of pools occur to maintain flow continuity following channel narrowing [38]. Peters et al. [20] found that when ice cover was not fixed to the bed or banks, increases in the flow resistance from the upper boundary increased the water depths and reduced the bulk velocities. The changes in depth-averaged velocities in higher flow, open-water conditions as compared to higher flow, ice-covered conditions were more notable than under lower flow conditions, suggesting that in these simulations, the ice cover impact on velocities is exacerbated under higher flow conditions (Figure 5).

The thalweg location remained relatively consistent in all simulations, although the thalweg was distinctly wider under open-water conditions (Figure 5c,d). The 'sandwiching' effect that the ice, bank, and bed have on velocities prevents faster velocities from being located close to the banks, which effectively narrows the thalweg in the channel when ice cover is present [39,40]. Previous studies found changes in the thalweg location under ice cover conditions, noting that faster flows were concentrated in narrower areas [6,11]. However, in this study, there was limited evidence from the measured data (Figure 4) to suggest a concentration of the thalweg when ice was present. Overall, the simulated spatial distribution of velocities under ice remained constant under both flow conditions (Figure 5). However, recirculation and upstream flow were more prevalent under higher flow conditions, particularly when ice cover was present (Figures 5 and 6). This indicates that although changes in discharge may not significantly impact the distribution of slower and faster velocities under ice, they can enhance recirculation and cause faster bulk velocities. Turbulent flow in riffle-pool sequences typically increases in prevalence and magnitude under higher discharges, and this was observed between lower and higher flow conditions, which was further exacerbated by the presence of ice [38].

Bed roughness did not have a consistent impact on depth-averaged velocities under ice cover (Figure 5a,c). Field observations at the same location [5] suggest no reduction and, in some cases, slight increases in the maximum velocities under ice at cross-sections within riffles, while cross-sections exhibiting lower roughness (e.g., in the pool) noted a decrease in maximum velocity magnitudes (Figure 4). Conversely, model simulations only noted increases in depth-averaged velocities under ice along the riffle cross-section downstream of the pool (cross-section 7 (Figure 1), Figure 5a,b). These results demonstrate that while ice-covered depth-averaged velocities may be reduced in magnitude along riffle cross-sections, maximum velocities may not show the same trend [13,41]. Furthermore, as River2D is able to simulate depth-averaged velocities, the ability to interrogate changes in maximum velocities within the water column is limited.

Another notable change in the depth-averaged velocity magnitude under ice cover was observed along the left bank of the pool (Figure 6). In each model output, a recirculating eddy was present along the left bank edge of the pool (Figures 5 and 6). Vertical and horizontal eddies commonly occur in pools located directly downstream of fast-flowing riffles as a result of sudden changes in the channel geometry [42]. However, during both lower and higher flow simulations, the eddy exhibited faster depth-averaged velocities when ice cover was present (Figure 6). This is supported by the velocity mapping work for each of the data collection dates (Figure 4, upper panels in a–c) and from velocity magnitude and direction mapping across each cross-section in the field [5]. The location of the eddy coincided with the strip of open water located along the left bank during ice-covered conditions (Figures 1 and 2) [5]. Partial ice cover can cause a redistribution of depth-averaged velocities, resulting in slower velocities directly under the ice cover and displacing faster flow to uncovered (unconfined) locations within the channel [20]. The velocities within the pool were reduced under ice cover due to increases in resistance, and faster velocities were redistributed from the thalweg toward the ice-free left bank, therefore intensifying the eddy (Figure 6). An experimental study by Peters et al. [20] had similar findings, where faster velocities were located under open-water segments of a partially covered flume.

Bed shear stress is dependent on multiple variables, including shear velocity magnitude, bed roughness, and the relative depth of maximum velocity within the water column [43]. While the presence of ice has no immediate impact on bed roughness, the upper and lower boundaries' interacting influence on depth-averaged velocities can influence velocity gradients and, therefore, shear velocities [4]. This causes different impacts on the bed shear stress under the ice within riffles and pools [7]. The results from the simulations demonstrated that under both lower and higher flow, open-water conditions exhibited a wider distribution of bed shear stress values exceeding 0.5 N/m^2, following closely along the thalweg (Figure 8). These results mirror the narrowing in thalweg observed under ice-covered conditions, which is similar to other studies that confirmed that bed shear stress values are dependent on the thalweg location [44]. The bed shear stress did not exceed 16 N/m^2 in any simulation or scenario and is consistent with shear stress values observed in small channels exhibiting riffle-pool sequences [5,24,45]. The higher flow simulations demonstrated the similar maximum and minimum values as lower flow conditions, although bed shear stress values exceeding 1 N/m^2 covered a larger area under the higher flow conditions (Figure 8). This suggests a wider distribution of large bed shear stress values under higher flow conditions (Figure 8). Bed shear stress is known to increase with higher velocities, therefore increasing the shear stress under higher flow conditions [18]. For both higher and lower flow conditions, the results from the bed shear stress analysis indicated that when ice cover was present, the shear stress increased in the riffle directly upstream of the pool (cross-section 3, Figure 1), as well as downstream of the pool (Figure 8). Additionally, the bed shear stress was greater along the left bank of the partially covered pool in the same location as the recirculating eddy, which was caused by faster depth-average velocities (Figure 8).

Changes in depth-averaged velocities alone do not explain the changes in the shear stress observed through these simulations (Figure 8). Previous work has demonstrated that ice cover can push the maximum velocities closer to the bed, increasing the velocity gradient and resulting in faster shear velocities [5,18,25]. Field studies at Sixteen Mile Creek demonstrated that maximum velocities were located closer to the bed under ice-covered conditions and coupled with negligible changes in magnitudes of maximum velocities, the bed shear stress was found to increase both upstream and downstream of the pool cross-sections [5]. Conversely, in cross-sections with a lower level of roughness within the pool, maximum velocities were notably reduced and did not result in an increase in shear stress under ice [5]. Modeled shear stress values were not consistently different between ice cover and ice-free conditions (Figure 9; Table 2). Similar to the field results [5] and previously published studies including flume experiments [13,18], the bed shear stress was larger in channels with greater roughness (e.g., riffles) than in lower roughness channel segments (e.g., pools). River2D does not take into consideration the depth of maximum velocities when calculating the shear velocities [22]. Additionally, the shear velocity output is derived using the composite roughness of the upper and lower boundaries and it may not present an accurate picture of the shear velocities along the bed under the ice [6]. This could explain why the modeled shear stress values within rough and smooth cross-sections under ice and open-water conditions do not agree with the findings from the field [5]. However, while the shear stress magnitude is not as accurately calculated by the model, the simulations provide detailed insight into the distribution of shear stress. This enables the identification of high and low shear stress zones and demonstrates changes caused by ice cover. Ice-impacted river management can benefit from understanding where zones of high shear stress shift when ice cover is present, therefore informing seasonal erosion and deposition processes.

5.2. Implications for Ecosystem Services

There is a known population of silver shiners, an endangered minnow species, that utilizes the left bank of the pool during open-water conditions [46]. Their preferred habitat includes riffle-pool sequences, which are generally wider than 30 m, although they are occasionally found in smaller tributaries such as Sixteen Mile Creek [47]. Previous work on silver shiners reports that they are commonly found within the water column where velocities are ~0.12 m/s (±0.02 m/s; [47]). The velocities along the left bank at this site ranged between 0.01 and 0.18 m/s in open water but exceeded 0.40 m/s when ice partially covered the pool (Figure 6), potentially reducing the usable silver shiner habitat during the winter season. Furthermore, previous works (e.g., Huusko et al. [48]) have demonstrated the need for flow variability such as this, to offer refuge, but also, to offer a steady supply of materials from upstream. The recirculation zone along the left bank offers a point for fish to station hold while materials move downstream closer to the middle of the channel, acting as an almost food conveyor belt.

Increased velocities along the eddy in the partially covered pool may present insight into pool maintenance mechanisms during ice-covered conditions. Riffle-pool sequences are common geomorphic units in low- to moderate- gradient channels and are important influences on local hydraulics, sediment transport, and aquatic habitat [49,50]. The periodic removal of fine sediment deposited along pools is essential for maintaining pool morphological characteristics [51]. Geomorphologists have presented several potential mechanisms for riffle-pool maintenance, including velocity reversal, stormflow, flow convergence, and natural hydrogeomorphic maintenance [38,49,52,53]. Within the pool, the presence of partial ice cover increased the depth-averaged velocities along the left bank (Figure 6). This suggests a potential increase in sediment transport and the removal of fine-grained sediment from the pool under both lower and higher flow conditions, especially those with a partial ice cover. The grain size along the left bank of the pool was dominated by sand and silt-sized material (Table 1), and the critical velocity required for entrainment was ~0.19 m/s [54]. While open-water depth-averaged velocities may be insufficient for entrainment, velocities along the left bank when ice cover was present exceeded the critical velocity of 0.19 m/s under both lower and higher flow conditions (Figure 6). If consistent over a long period of time, the impact of ice cover on pool depth-averaged velocities could play a role in pool maintenance in ice-affected riffle-pool systems. Additionally, the recirculation zone found along the pool's left bank may provide an important ecosystem function for silver shiners by transporting nutrients and food sources into the channel [55].

5.3. Modeled Shear Stress Distribution and Magnitude: Advantages and Drawbacks

Simulated shear stress during ice presence in the channel reached up to 2 N/m^2, suggesting variability in the erosional potential between ice-covered and open-water conditions (Figure 8). A similar study by Lotsari et al. [6] investigated spatial variations in depth-averaged velocities under ice along a Pulmanki River meander bend. The results from the Pulmanki River study indicated that near-bed velocities during all seasons would be sufficient for incipient motion, but that the spatial variability in erosion–deposition locations increased under ice [6]. Lotsari et al. [6] stated that ice cover should be considered as an important factor in morphological changes to meander bends, and the findings from the Sixteen Mile Creek suggest that the same is true for riffle-pool sequences [5]. Moreover, the value in mapping shear stress across the channel using River2D highlights the opportunity to see changes in the shear stress distribution across the bed, which can be limited only when the field data are evaluated (e.g., Smith et al. [5]). High-resolution velocity data collection (e.g., Figure 3) is challenging with ice cover present, as the model facilitates an additional opportunity to interrogate the data.

The bed shear stress estimates exhibited higher error values than depth-averaged velocities and water levels (within a 30% margin of error) (Table 3). There are several factors that influence the accuracy of shear stress estimates. The field bed shear stress estimates were calculated using logarithmic velocity profiles to identify velocity gradients and calculate shear velocity and shear stress [4,5]. However, River2D calculates shear velocity using the Keulegan equation, which incorporates the mean velocity, water depth, and equivalent sand roughness (k_s) [22,56]. Methods that use different variables to calculate shear stress can lead to varying results and increase the margin of error between field-based and simulated estimations [24]. The logarithmic (law of the wall) method does not consider antecedents along the bed or acceleration and deceleration within riffle-pool transitions, while the Keulegan equation simplifies the influence of bed roughness using an equivalent roughness height coefficient [57,58]. Ice-covered simulated bed shear stress values may be further influenced by the composite roughness value used in River2D to calculate shear velocities, leading to a reduced accuracy [6]. Shear velocity model outputs can be used conservatively to estimate bed shear stress but should be coupled with field surveys to ensure accuracy in bed shear stress magnitude estimates.

6. Conclusions

The River2D hydrodynamic model successfully simulates flow characteristics in rivers of varying sizes and can be used for applications, such as habitat suitability models, river restoration plans, and sediment transport estimates [6,9]. This research simulated velocity distribution and shear stress in a small, shallow riffle-pool sequence under ice cover and through two different flow levels. The results indicated reduced depth-averaged velocities throughout the upstream riffle and pool, and increased depth-averaged velocities along the left bank and downstream of the pool under ice cover for both higher and lower flow simulations. The thalweg widened under ice cover, and the spatial distribution of the velocities shifted when ice cover was present, exhibiting a higher amount of recirculating flow along the partial ice cover in the pool. Increased discharge resulted in larger differences in water depth and velocities between open and ice-covered conditions, although the spatial distribution of flow remained consistent between the two ice-covered and open-water simulations. The recirculating eddy intensified with the presence of ice, increasing the depth-averaged velocities in known silver shiner habitats. Additionally, the findings from velocities within the pool suggest that ice cover may play a role in pool maintenance in ice-affected streams.

The shear stress analysis demonstrated no difference in the maximum and minimum shear stress values between open-water and ice-covered flow. However, there were clear differences in the distribution of bed shear stress under ice and open water for lower and higher flows. Overall, the model was successful in simulating depth-averaged velocities and water levels throughout the study site, while the shear stress values exhibited a higher and more variable margin of error. The shear stress values derived from River2D for small, ice-covered streams established a baseline approximation, but should be paired with field measurements for a more accurate picture. Ultimately, River2D was successfully calibrated to simulate depth-averaged velocities, flow directions, and water levels in a small and shallow riffle-pool sequence but was proven to be more effective in deeper sections.

Supplementary Materials: The following supporting information can be downloaded at: https://www.mdpi.com/article/10.3390/w15081604/s1. Figure S1: Problem schematic. Figure S2: Velocity distribution maps for each riffle cross-section (a is cross-section 1; b is cross-section 2; c is cross-section 3; d is cross-section 7; Figure 1) from left bank to right bank under open water conditions (top panel) and ice-covered conditions (bottom panel). Streamwise velocities are represented on a colour scale and secondary velocities are represented by arrows showing the direction and magnitude of the transverse and vertical velocity component. Data were collected using a Sontek S5 ADCP under ice-covered conditions on 23 February 2021, and open water conditions on 2 March 2021. Figure S3: Modelled water surface elevations for all four simulations. Water elevation decreases in the downstream direction according to the channel bed slope, and to simulated water depths.

Author Contributions: Conceptualization, K.S., J.M.H.C. and P.V.V.; methodology, K.S., J.M.H.C. and P.V.V.; data collection, K.S. and J.M.H.C.; formal analysis, K.S., P.V.V. and J.M.H.C.; writing—original draft preparation, K.S. and J.M.H.C.; writing—review and editing, J.M.H.C., K.S. and P.V.V.; supervision, J.M.H.C.; project administration, J.M.H.C.; funding acquisition, K.S., J.M.H.C. and P.V.V. All authors have read and agreed to the published version of the manuscript. **Funding:** This research was funded

by the Canadian Foundation for Innovation, grant number: 31341. K.S. received additional stipends from the Yukon Foundation: Dorreene and Herb Wahl Award and Yukon Order of Pioneers Yukon Grant. The APC was waived.

Data Availability Statement: This article is based on the MSc. Thesis research was conducted by K.S. and available https://hdl.handle.net/10214/26885. Some data are proprietary, please contact the corresponding author for details.

Acknowledgments: We express our sincere gratitude to Abby Smith for epic field support on the coldest and longest of days! Further, field and technical support by GEO Morphix staff, specifically, Patrick Padovan, Tye Rusnak, Lindsay Davis, Marie Hoekstra, and Bryce Molder is greatly appreciated. Marie Puddister and Skyler Barclay in the Department of Geography, Environment and Geomatics at University of Guelph, provided production support. Additionally, we offer our sincere thanks to Mattamy Homes for the site access to Sixteen Mile Creek.

Conflicts of Interest: The authors declare no conflict of interest. The funders had no role in the design of the study; in the collection, analyses, or interpretation of data; in the writing of the manuscript; or in the decision to publish the results.

References

1. Nicholas, A.P. Modelling the Continuum of River Channel Patterns. *Earth Surf. Process. Landf.* **2013**, *38*, 1187–1196. [CrossRef]
2. Lotsari, E.; Lintunen, K.; Kasvi, E.; Alho, P.; Blåfield, L. The Impacts of Near-Bed Flow Characteristics on River Bed Sediment Transport under Ice-Covered Conditions in 2016–2021. *J. Hydrol.* **2022**, *615*, 128610. [CrossRef]
3. Lotsari, E.; Thorndycraft, V.; Alho, P. Prospects and Challenges of Simulating River Channel Response to Future Climate Change. *Prog. Phys. Geogr. Earth Environ.* **2015**, *39*, 483–513. [CrossRef]
4. Aghaji Zare, S.G.; Moore, S.A.; Rennie, C.D.; Seidou, O.; Ahmari, H.; Malenchak, J. Boundary Shear Stress in an Ice-Covered River during Breakup. *J. Hydraul. Eng.* **2016**, *142*, 04015065. [CrossRef]
5. Smith, K.; Cockburn, J.; Villard, P.V. The Effect of Ice Cover on Velocity and Shear Stress in a Riffle-pool Sequence. *Earth Surf. Process. Landf.* **2023**, esp.5557. [CrossRef]
6. Lotsari, E.; Tarsa, T.; Kämäri, M.; Alho, P.; Kasvi, E. Spatial Variation of Flow Characteristics in a Subarctic Meandering River in Ice-Covered and Open-Channel Conditions: A 2D Hydrodynamic Modelling Approach. *Earth Surf. Process. Landf.* **2019**, *44*, 1509–1529. [CrossRef]

7. Lotsari, E.; Kasvi, E.; Kämäri, M.; Alho, P. The Effects of Ice Cover on Flow Characteristics in a Subarctic Meandering River. *Earth Surf. Process. Landf.* **2017**, *42*, 1195–1212. [CrossRef]
8. Demers, S.; Buffin-Bélanger, T.; Roy, A.G. Helical Cell Motions in a Small Ice-Covered Meander River Reach. *River Res. Appl.* **2011**, *27*, 1118–1125. [CrossRef]
9. Schwartz, J.S.; Neff, K.J. Use of River2D Hydrodynamic Model for Stream Restoration Assessment and Design. In Proceedings of the World Environmental and Water Resources Congress 2011: Bearing Knowledge for Sustainability, Palm Springs, CA, USA, 22–26 May 2011; pp. 2593–2602. [CrossRef]
10. Turcotte, B.; Morse, B.; Bergeron, N.E.; Roy, A.G. Sediment Transport in Ice-Affected Rivers. *J. Hydrol.* **2011**, *409*, 561–577. [CrossRef]
11. Kämäri, M.; Alho, P.; Colpaert, A.; Lotsari, E. Spatial Variation of River-Ice Thickness in a Meandering River. *Cold Reg. Sci. Technol.* **2017**, *137*, 17–29. [CrossRef]
12. Robert, A.; Tran, T. Mean and Turbulent Flow Fields in a Simulated Ice-Covered Channel with a Gravel Bed: Some Laboratory Observations. *Earth Surf. Process. Landf.* **2012**, *37*, 951–956. [CrossRef]
13. Wang, J.; Sui, J.-Y.; Karney, B.W. Incipient Motion of Non-Cohesive Sediment under Ice Cover—An Experimental Study. *J. Hydrodyn.* **2008**, *20*, 117–124. [CrossRef]
14. Turcotte, B.; Morse, B.; Pelchat, G. Impact of Climate Change on the Frequency of Dynamic Breakup Events and on the Risk of Ice-Jam Floods in Quebec, Canada. *Water* **2020**, *12*, 2891. [CrossRef]
15. Lindenschmidt, K.-E. RIVICE—A Non-Proprietary, Open-Source, One-Dimensional River-Ice Model. *Water* **2017**, *9*, 314. [CrossRef]
16. Morales-Marín, L.A.; Sanyal, P.R.; Kadowaki, H.; Li, Z.; Rokaya, P.; Lindenschmidt, K.E. A Hydrological and Water Temperature Modelling Framework to Simulate the Timing of River Freeze-up and Ice-Cover Breakup in Large-Scale Catchments. *Environ. Model. Softw.* **2019**, *114*, 49–63. [CrossRef]
17. Shen, H.T. Mathematical Modeling of River Ice Processes. *Cold Reg. Sci. Technol.* **2010**, *62*, 3–13. [CrossRef]
18. Namaee, M.R.; Sui, J. Local Scour Around Two Side-by-Side Cylindrical Bridge Piers under Ice-Covered Conditions. *Int. J. Sediment Res.* **2019**, *34*, 355–367. [CrossRef]
19. Parsapour-Moghaddam, P.; Rennie, C.D. Calibration of a 3D Hydrodynamic Meandering River Model Using Fully Spatially Distributed 3D ADCP Velocity Data. *J. Hydraul. Eng.* **2018**, *144*, 04018010. [CrossRef]
20. Peters, M.; Dow, K.; Clark, S.P.; Malenchak, J.; Danielson, D. Experimental Investigation of the Flow Characteristics beneath Partial Ice Covers. *Cold Reg. Sci. Technol.* **2017**, *142*, 69–78. [CrossRef]
21. Vietz, G.J.; Rutherfurd, I.D.; Stewardson, M.J.; Finlayson, B.L. Hydrodynamics and Sedimentology of Concave Benches in a Lowland River. *Geomorphology* **2012**, *147–148*, 86–101. [CrossRef]
22. Steffler, P.; Blackburn, J. Two-Dimensional Depth Averaged Model of River Hydrodynamics and Fish Habitat. Available online: https://relicensing.pcwa.net/documents/Library/PCWA-L%20452.pdf (accessed on 30 June 2021).
23. Environment and Climate Change Water Survey of Canada. Available online: https://www.canada.ca/en/environment-climate-change/services/water-overview/quantity/monitoring/survey.html (accessed on 1 April 2022).
24. Robert, A. Characteristics of Velocity Profiles along Riffle-Pool Sequences and Estimates of Bed Shear Stress. *Geomorphology* **1997**, *19*, 89–98. [CrossRef]
25. Smith, K. Investigating Ice-Impacted Fluvial Processes in a Small Riffle-Pool Sequence in Southern Ontario. Master's Thesis, University of Guelph, Guelph, ON, Canada, 2022.
26. Crowder, D.W.; Diplas, P. Using Two-Dimensional Hydrodynamic Models at Scales of Ecological Importance. *J. Hydrol.* **2000**, *230*, 172–191. [CrossRef]
27. Sontek Flowtracker 2 Wading Discharge Measurement Instrument. 2019. Available online: https://www.sontek.com/media/pdfs/flowtracker2-brochure-s19-02-1219.pdf (accessed on 9 September 2021).
28. Sontek RiverSuveyor Discharge, Bathymetry and Current Profiling. 2015. Available online: https://info.xylem.com/rs/240-UTB-146/images/riversurveyor-m9-s5-brochure.pdf?_ga=2.67666392.894680228.1648847751-199850675.1648847751 (accessed on 9 September 2021).
29. Parsons, D.R.; Jackson, P.R.; Czuba, J.A.; Engel, F.L.; Rhoads, B.L.; Oberg, K.A.; Best, J.L.; Mueller, D.S.; Johnson, K.K.; Riley, J.D. Velocity Mapping Toolbox (VMT): A Processing and Visualization Suite for Moving-Vessel ADCP Measurements. *Earth Surf. Process. Landf.* **2013**, *38*, 1244–1260. [CrossRef]
30. Waddle, T.; Steffler, P. *Mesh Generation Program for River2D Two Dimensional Depth Averaged Finite Element*; US Geological Survey: Reston, VA, USA, 2002.
31. Alabyan, A.M.; Lebedeva, S.V. Flow Dynamics in Large Tidal Delta of the Northern Dvina River: 2D Simulation. *J. Hydroinform.* **2018**, *20*, 798–814. [CrossRef]
32. Robert, A. *River Processes: An Introduction to Fluvial Dynamics*; Routledge: London, UK, 2003.
33. Waddle, T. *Simulation of Flow Habitat Conditions Under Ice, Cache La Poudre River—January 2006*; Open-File Report; US Geological Survey: Reston, VA, USA, 2007.
34. Church, M.; Hassan, M.A. Size and Distance of Travel of Unconstrained Clasts on a Streambed. *Water Resour. Res.* **1992**, *28*, 299–303. [CrossRef]

35. Lepelletier, T.; Araud, Q. Implicit Formulation for 1D and 2D St Venant Equations—Presentation of the Method, Validation and Applications. In *Advances in Hydroinformatics, Proceedings of the Advances in Hydroinformatics: SimHydro 2017—Choosing The Right Model in Applied Hydraulics, Nice, France, 14–16 June 2017*; Gourbesville, P., Cunge, J., Caignaert, G., Eds.; Springer: Singapore, 2018; pp. 91–103.
36. Beltaos, S.; Prowse, T. River-Ice Hydrology in a Shrinking Cryosphere. *Hydrol. Process.* **2009**, *23*, 122–144. [CrossRef]
37. Mosley, M.P. Analysis of the Effect of Changing Discharge on Channel Morphology and Instream Uses in a Braided River, Ohau River, New Zealand. *Water Resour. Res.* **1982**, *18*, 800–812. [CrossRef]
38. Hassan, M.A.; Radić, V.; Buckrell, E.; Chartrand, S.M.; McDowell, C. Pool-Riffle Adjustment Due to Changes in Flow and Sediment Supply. *Water Resour. Res.* **2021**, *57*, e2020WR028048. [CrossRef]
39. Urroz, G.E.; Ettema, R. Application of Two-Layer Hypothesis to Fully Developed Flow in Ice-Covered Curved Channels. *Can. J. Civ. Eng.* **1994**, *21*, 101–110. [CrossRef]
40. Blanckaert, K.; de Vriend, H.J. Meander Dynamics: A Nonlinear Model without Curvature Restrictions for Flow in Open-Channel Bends. *J. Geophys. Res.* **2010**, *115*, F04011. [CrossRef]
41. Bathhurst, J.C. Distribution of Boundary Shear Stress in Rivers. In *Adjustments of the Fluvial System*; Routledge: Oxfordshire, UK, 2020; pp. 95–116.
42. Tokyay, T.; Sinha, S. Channel Width, Bedform Length and Turbulence: Numerical Investigation of Flow Dynamics over Laboratory-Scale Pool–Riffle Sequences. *Environ. Fluid Mech.* **2020**, *20*, 819–842. [CrossRef]
43. Zabilansky, L.J.; Hains, D.B.; Remus, J.I. Increased Bed Erosion Due to Ice. In *Current Practices in Cold Regions Engineering*; ASCE: Reston, VA, USA, 2006; pp. 1–12. [CrossRef]
44. Kang, S.; Sotiropoulos, F. Flow Phenomena and Mechanisms in a Field-Scale Experimental Meandering Channel with a Pool-Riffle Sequence: Insights Gained via Numerical Simulation. *J. Geophys. Res.* **2011**, *116*, F03011. [CrossRef]
45. Booker, D.J.; Sear, D.A.; Payne, A.J. Modelling Three-Dimensional Flow Structures and Patterns of Boundary Shear Stress in a Natural Pool–Riffle Sequence. *Earth Surf. Process. Landf.* **2001**, *26*, 553–576. [CrossRef]
46. Ministry of the Environment, Conservation and Parks. *Silver Shiner*; Government of Ontario: Toronto, ON, Canada, 2021.
47. Bunt, C. Silver Shiner (Notropis Photogenis) Size-Class Structure, Habitat Utilization, Movement and Persistence in an Urbanized Fragment of a Great Lakes Tributary. *Am. Midl. Nat.* **2016**, *176*, 200–209. [CrossRef]
48. Huusko, A.; Greenberg, L.; Stickler, M.; Linnansaari, T.; Nykänen, M.; Vehanen, T.; Koljonen, S.; Louhi, P.; Alfredsen, K. Life in the Ice Lane: The Winter Ecology of Stream Salmonids. *River Res. Appl.* **2007**, *23*, 469–491. [CrossRef]
49. Sawyer, A.H.; Bayani Cardenas, M.; Buttles, J. Hyporheic Exchange Due to Channel-Spanning Logs. *Water Resour. Res.* **2011**, *47*, W08502. [CrossRef]
50. Church, M. Channel Stability: Morphodynamics and the Morphology of Rivers. In *Rivers—Physical, Fluvial and Environmental Processes*; GeoPlanet: Earth and Planetary Sciences; Springer: Cham, Switzerland, 2015; pp. 281–321. ISBN 978-3-319-17718-2.
51. Salmela, J.; Kasvi, E.; Vaaja, M.T.; Kaartinen, H.; Kukko, A.; Jaakkola, A.; Alho, P. Morphological Changes and Riffle-Pool Dynamics Related to Flow in a Meandering River Channel Based on a 5-Year Monitoring Period Using Close-Range Remote Sensing. *Geomorphology* **2020**, *352*, 106982. [CrossRef]
52. MacWilliams, M.L., Jr.; Wheaton, J.M.; Pasternack, G.B.; Street, R.L.; Kitanidis, P.K. Flow Convergence Routing Hypothesis for Pool-Riffler Maintenance in Alluvial. *Water Resour. Res.* **2006**, *42*, W10427. [CrossRef]
53. Cockburn, J.M.H.; Villard, P.V.; Hutton, C. Assessing Instream Habitat Suitability and Hydraulic Signatures of Geomorphic Units in a Reconstructed Single Thread Meandering Channel. *Ecohydrology* **2016**, *9*, 1094–1104. [CrossRef]
54. Hjulström, F. Studies of the Morphological Activity of Rivers as Illustrated by the River Fyris. Ph.D. Thesis, Uppsala University, Uppsala, Sweden, 1935.
55. Shamloo, H.; Rajaratnam, N.; Katopodis, C. Hydraulics of Simple Habitat Structures. *J. Hydraul. Res.* **2001**, *39*, 351–366. [CrossRef]
56. Jay Lacey, R.W.; Millar, R.G. Reach Scale Hydraulic Assessment of Instream Salmonid Habitat Restoration. *JAWRA J. Am. Water Resour. Assoc.* **2004**, *40*, 1631–1644. [CrossRef]
57. Franca, M.J.; Ferreira, R.M.L.; Lemmin, U. Parameterization of the Logarithmic Layer of Double-Averaged Streamwise Velocity Profiles in Gravel-Bed River Flows. *Adv. Water Resour.* **2008**, *31*, 915–925. [CrossRef]
58. Rickenmann, D.; Recking, A. Evaluation of Flow Resistance in Gravel-Bed Rivers through a Large Field Data Set. *Water Resour. Res.* **2011**, *47*, W07538. [CrossRef]

water

MDPI

Article

Numerical Investigation of the Stress on a Cylinder Exerted by a Stratified Current Flowing on Uneven Ground

Yin Wang [1,2], Ming Xu [2], Lingling Wang [3], Sha Shi [2], Chenhui Zhang [2], Xiaobin Wu [1], Hua Wang [4], Xiahui Xiong [2] and Chunling Wang [5,*]

1 Institute of Disaster Prevention & Mitigation and Water Engineering Safety, Jiangxi Academy of Water Science and Engineering, Nanchang 330029, China
2 School of Hydraulic & Ecological Engineering, Nanchang Institute of Technology, Nanchang 330099, China
3 College of Water Conservancy and Hydropower Engineering, Hohai University, Nanjing 210024, China
4 College of Environment, Hohai University, Nanjing 210024, China
5 Department of Civil Engineering and Architecture, Taizhou University, Taizhou 318000, China
* Correspondence: wangchunling039@163.com

Abstract: In this study, a three-dimensional internal wave (IW)—cylinder—terrain coupled numerical model is established. Based on the large-eddy simulation (LES) method, the IW mechanical characteristics of the cylinder and the flow field evolution around the cylinder over different types of terrains are explored. The similarities and differences in the mechanical characteristics of the cylinders in the environments with and without terrains are compared. The research results show that, when the IWs propagate over terrain, the waveform structures are prone to continuous changes. The intense reverse alternating flow of the upper and the lower water, bounded by the pycnocline, results in huge IWs forces differences between the case without terrains and the cases with terrains. In the case without terrains, the maximum horizontal resultant force on the cylinder is positive, while the resultant forces are negative in the cases with terrain. Compared with the case without terrain, the shallow-water effect caused by the combined action of the terrain and the IWs enhances the flow field strength, making the lower parts of the cylinder suffer larger horizontal forces in the opposite direction to the IW direction. Moreover, the additional vortices produced by the interaction between the IWs and the terrain causes a more complex flow field around the cylinder and the greater forces on the cylinder.

Keywords: internal wave; terrain; force on the cylinder; flow field; numerical simulation

1. Introduction

A vertical density gradient characterizes stratified flow. In various aquatic environments such as oceans, estuaries, and lakes, the fluid density varies with depth due to the changes in temperature, salinity, and other factors, resulting in stable density stratification [1]. Such a stable stratified environment can generate internal waves (IWs) of different amplitudes under small or weak disturbances. The IWs carry a large amount of energy during the propagation. The hydrodynamic characteristics in the IW environment are more complicated than those in a homogeneous fluid due to the opposite flow directions in the upper and lower water layers. The shear flow can induce fatigue failure in most structures [2], which can jeopardize the safety and stability of underwater cylinders in deep seas, near areas, estuaries, and lakes and reservoirs [3]. The development of coastal areas is deepening year by year. With the construction of a large number of bridges across the sea and river near the estuary, the structural safety of cylinders under the action of IWs has become an issue of great concern in the engineering field.

Depression-type IWs widely exist in all kinds of waters [4]. Due to the changeable terrain of coastal areas [5], the continuous change in the wave pattern results in continuous change in the hydrodynamic characteristics during the IWs propagation along the coastal slope. The induced shear instability can easily lead to turbulent mixing [6] and aggravate the shearing effect of the pycnocline [7]. Therefore, the variable terrain environment has a great influence on the propagation and evolution of the IWs and significantly changes the hydrodynamic characteristics of the IWs during propagation.

Existing studies on the forces of IWs on cylinders are mostly carried out on the ideal generalized level-bed terrain, while the continuous and irregular bottom terrains in the near-shore and estuary areas are ubiquitous [8,9]. There are few studies on the response mechanism of the IW forces on cylinders over such non-uniform terrains. Therefore, it is necessary to obtain the hydrodynamic characteristics of the IW environment and the mechanical mechanism of the structure when the IW propagate over different terrains. In the present study, a three-dimensional (3D) IW numerical tank with a large eddy simulation (LES) approach is established to explore the mechanical regularity of the cylinders in such a complex IW-terrain coupled environment, and only estuarine and riverine water are included in the simulation studies. The numerical models and verification are shown in Section 2, the results and analysis are presented in Section 3, and the conclusions are given in Section 4.

2. Numerical Models

2.1. Governing Equations

The Navier–Stokes (N–S) equations are adopted to describe the three-dimensional transient motion of an incompressible viscous fluid, which could be described as:

$$\frac{\partial \rho}{\partial t} + \frac{\partial (\rho u_i)}{\partial x_i} = 0, \tag{1}$$

$$\frac{\partial (\rho u_i)}{\partial t} + \frac{\partial (\rho u_i u_j)}{\partial x_j} = -\frac{\partial p}{\partial x_i} + \frac{\partial}{\partial x_j}\left(\mu \frac{\partial u_i}{\partial x_j}\right) + f_i, \tag{2}$$

where ρ stands for the water density; t stands for the time; i stands for the Cartesian coordinates directions; x_i stands for the spatial coordinate; u_i stands for the flow velocity; p stands for the pressure; μ stands for the kinematic viscosity; and f_i stands for the body force.

2.2. Scalar Transport Equation

In the present research, IWs are excited by the density difference between the upper and lower layers in the two-layer flow system. The convection–diffusion effect is expressed as:

$$\frac{\partial C}{\partial t} + \frac{\partial (u_i C)}{\partial x_j} = \frac{\partial}{\partial x_j}\left(k \frac{\partial C}{\partial x_j}\right) + S \tag{3}$$

where C, ranging from 0 to 1, stands for the scalar volume concentration; the water density ρ can be controlled by the formula: $\rho = C\rho_2 + (1 - C)\rho_1$, ($\rho_1$ and ρ_2 represent the upper layer density and the lower layer density, respectively); k stands for the diffusion coefficient; and S stands for the source term or sink term.

2.3. Turbulence Model

In this paper, the spatial averaging method LES applies a filtering function to separate large-scale and small-scale vortexes to simulate the stratified flow.

The filtering functions for the governing equations above (labeled by an overbar) are expressed as follows:

$$\frac{\partial \overline{u_j}}{\partial x_j} = 0 \tag{4}$$

$$\frac{\partial \overline{u_i}}{\partial t} + \overline{u_j}\frac{\partial \overline{u_i}}{\partial x_j} = -\frac{1}{\rho}\frac{\partial \overline{p}}{\partial x_i} + v\frac{\partial^2 \overline{u_i}}{\partial x_j \partial x_j} + \frac{\partial \tau_{ij}}{\partial x_i} + \overline{f_i} \tag{5}$$

$$\frac{\partial\overline{C}}{\partial t}+\overline{u_j}\frac{\partial\overline{C}}{\partial x_j}=k\frac{\partial^2\overline{C}}{\partial x_j\partial x_j}+\frac{\partial\chi_j}{\partial x_j} \tag{6}$$

where the sub-grid stress tensor $\tau_{ij} = \overline{u_i u_j} - \overline{u_i}\,\overline{u_j}$, the Sub-Grid Scale (SGS) flux $\chi_j = \overline{C u_j} - \overline{C}\,\overline{u_j}$. In this study, the turbulent eddy viscosity ν_t is a function of the filtered scale and the strain rate tensor, which can be described as:

$$\nu_t = (C_s\Delta)^2\left(2\overline{S_{ij}S_{ij}}\right)^{\frac{1}{2}} \tag{7}$$

where Δ stands for the filtered scale; the filtered scale $\overline{S_{ij}} = \frac{1}{2}\left(\frac{\partial^2\overline{u_i}}{\partial x_j} + \frac{\partial^2\overline{u_j}}{\partial x_j}\right)$; C_s stands for the Smagorinsky constant. In the IWs environment, C_s, changing with space-time, can be calculated by a dynamic procedure [10].

2.4. Establishment of a Numerical Tank

In the current study, the three-dimensional numerical terrain models are established based on a single-cylinder (SC, without terrain) tank model. The dimensions of this numerical model are 4.0 m long (X) × 0.3 m high (Y) × 0.3 m wide (Z), see Figure 1. A SC with a diameter D = 0.05 m and a length h_1 = 0.2 m is placed on the centerline in the spread direction (Z). The cylinder bottom center is located at (x, y, z) = (2.0, 0, 0.15) m. By using the gravity collapse approach to motivate the IWs [11], the numerical tank is separated into the IW generation region (x = 0–0.23 m) and the IW propagation region (x = 0.23–4 m), where Δh is the thickness difference in the pycnocline between the two regions, and η_0 is the initial amplitude.

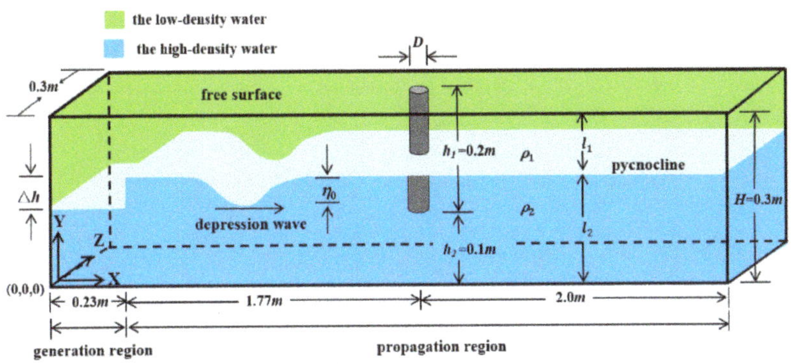

Figure 1. Sketch map of the single cylinder (SC) numerical model.

In the stratified fluid environment, the upper layer is defined as the lower-density water layer with density ρ_1 = 0.998 g/cm^3 and depth l_1 = 0.075 m, and the lower layer is defined as the higher-density water layer with density ρ_2 = 1.017 g/cm^3 and depth l_2 = 0.225 m, respectively. Therefore, the total water depth of the tank H is kept at 0.3 m.

The generation and propagation of IWs of depression are simulated by the LES method [12]. The governing equations are discretized by the finite-volume method. The PISO algorithm is employed to ensure mass conservation and obtain the pressure field by coupling the velocity-pressure terms [13,14]. The second-order central differential scheme, the second-order upwind scheme, and the second-order implicit scheme are adopted to discretize the diffusion term, the convection term, and the time step, respectively.

Nonslip solid-rigid walls are chosen to define the numerical tank left wall (upstream boundary), the sidewalls, the bottom, and the cylinder periphery. A "Sommerfeld radiation" boundary condition is used to define the tank's right wall (downstream boundary) [15] to avoid the reflection of IWs. The model top boundary is specified by the rigid-lid approximation [16] to neglect the surface waves. Surface waves are usually very small compared to the IWs [17], so the rigid-lid approximation is reasonable for the water surface.

2.5. Numerical Model Verification

2.5.1. Verification by Physical Model Test

A physical model test conducted by Chen [18] is applied to verify the numerical model. The physical model tank with dimensions of length × width × depth = 12 m × 0.5 m × 0.7 m is illustrated in Figure 2. The thicknesses of the upper freshwater layer are 0.1 m with a density of 0.998 g/cm^3, and the thicknesses of the lower saline layer are 0.4 m with density of 1.030 g/cm^3. In the initial phase, there is a thickness difference controlled by a gate in the pycnocline between the IW generation region and the IW propagation region, which is similar to Figure 1 above. By lifting the gate, a leading IW of depression is motivated by the gravity collapse (caused by the thickness difference) method and propagates toward the tank's left end [19], as shown in Figure 2. Two ultrasonic probes fixed in the tank top were adopted to capture the IWs spatial distribution in the propagation region of the numerical tank. A triangular bar was fixed between the probes, and the basis length *Lw* was equal to the wavelength.

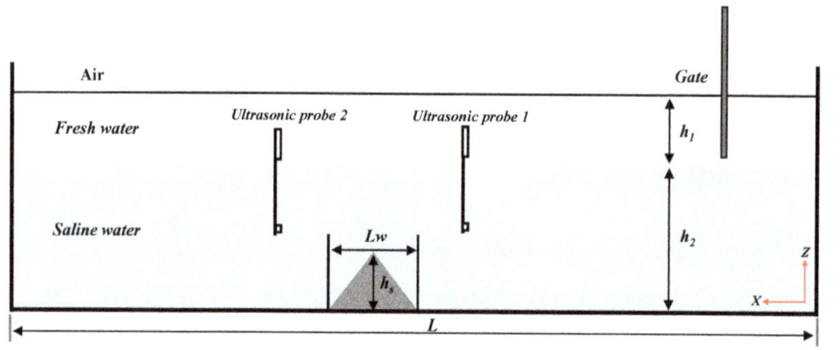

Figure 2. Schematic diagram of the physical model test [18].

The evolution of IWs excited by the current numerical simulation is compared with the physical model results of Chen [18] for verification. The comparison results of the verification, see Figure 3, presents that the simulation results agree well with the physical model results. Consequently, in this study, it is reasonable to adopt the numerical model for studying the IW interactions with cylinders and terrain.

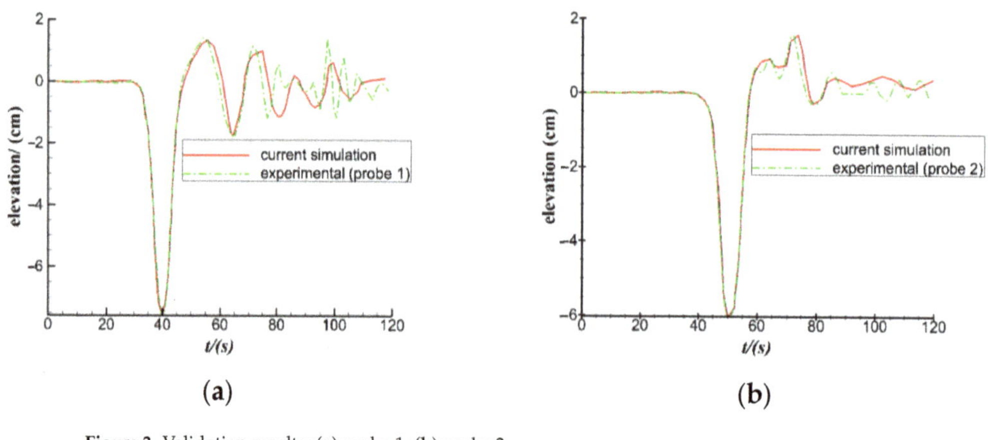

Figure 3. Validation results: (**a**) probe 1, (**b**) probe 2.

2.5.2. Grid Independence Test

The grid-size convergence analysis of the single cylinder (SC) without terrain is carried out in this section. Three cases of different grid types are set according to the differences in the number of nodes on the cylinder periphery, as illustrated in Figure 4. The details of the calculation cases are shown in Table 1.

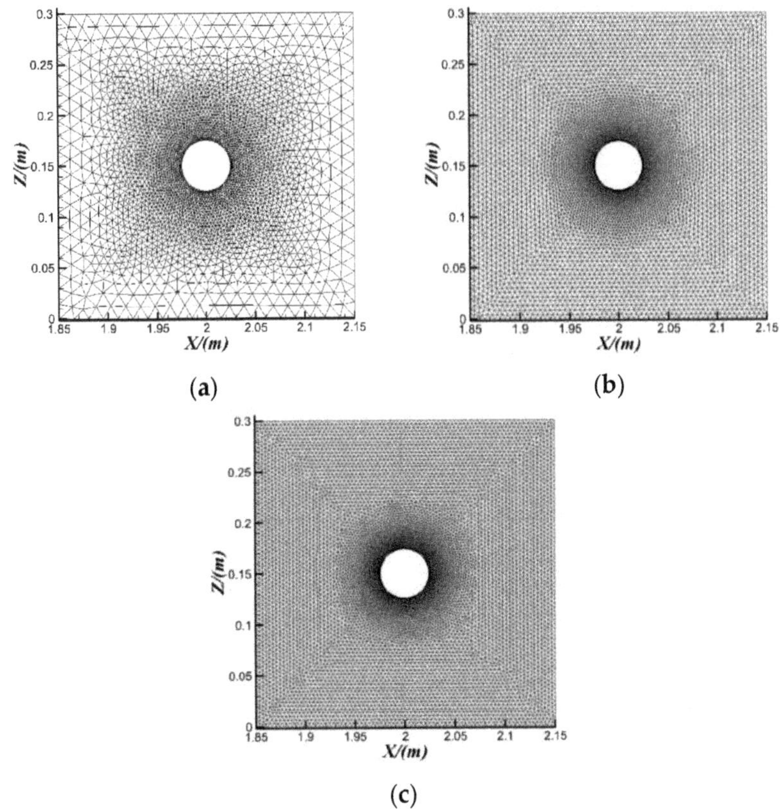

Figure 4. Detail view of the unstructured mesh around a cylinder periphery: (**a**) T1, (**b**) T2, (**c**) T3.

Table 1. The cases of grid-size independence test, $\eta_0/H = 0.0575$.

No.	Case	Δt (s)	$C_{Fn\text{-}max}$	Elements Number
1	T_1 (low density)	0.02	0.0810	525,454
2	T_2 (moderate density)	0.01	0.0857	2,384,640
3	T_3 (high density)	0.006	0.0862	3,318,278

In the current study, C_{F_n} is used to define the dimensionless horizontal total force on the cylinder, which is calculated by the following formula:

$$C_{F_n} = \frac{F_n}{\rho g A H} \tag{8}$$

where, F_n stands for the horizontal total force on the cylinder; g stands for the gravitational acceleration; A stands for the cylinder windward surface area (cylinder frontal side); and H is the total water depth of the tank.

The numerical results of the three different grid densities are illustrated in Figure 5. The force trend of the horizontal total force in the low-density grid is different from that in the medium-density grid, and the maximum horizontal force difference in the horizontal force in the two cases is about 0.5%. At the same time, the maximum horizontal force difference between medium-density grid and high-density grid is only approximately 0.05%, and C_{Fn} in T_2 almost coincides with that in T_3 over time. Therefore, the grid independence test results show the convergence of the grids. Consequently, the moderate gird density (case T_2) is sufficiently fine enough to discretize the computational domain, which can be adopted in the rest numerical simulations.

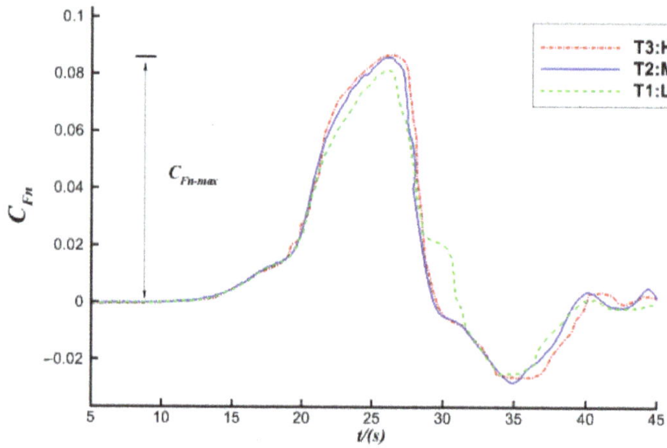

Figure 5. Numerical model verification results of the IWs forces on a cylinder of different mesh densities, $\eta_0/H = 0.0575$.

3. Result and Analysis

A nonuniform bottom is very common, such as bottom-step terrain [20], flat-top-knoll terrain [21], and flat-top-platform terrain [22]. To further explore the IW mechanical characteristics for the cylinder over terrain, one case of SC and three cases of different types of terrain are selected for comparative analysis, as shown in Figure 6a–d, including the SC model (Case N_1), the bottom-step terrain model (Case N_2), the flat-top-knoll terrain model (Case N_3), and the flat-top-platform model (Case N_4).

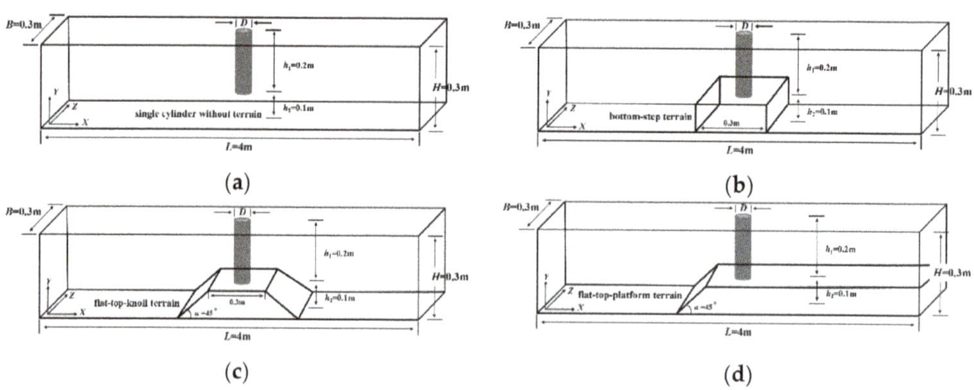

Figure 6. Sketch map of numerical tank for four cases: (**a**) N_1, (**b**) N_2, (**c**) N_3, (**d**) N_4.

In this study, the stratified characteristic parameters (the water depth ratio l_1/l_2 of the two water layers, and the density difference $\Delta\rho$ between the two water layers) and the dimensionless IW amplitude η_0/H are consistent: the upper water depth l_1 = 0.075 m, the lower water depth l_2 = 0.225 m, making l_1/l_2 = 0.33; $\Delta\rho$ = 0.019 g/cm^3, η_0/H = 0.0575. The diameter of the cylinder is 0.05 m, the length of the cylinder h_1 = 0.2 m, and the coordinates of the center of the cylinder bottom are (2, 0.1, 0.15) *XYZ*. The height difference between the cylinder bottom and the tank bottom is h_2 = 0.1 m. The lengths of the terrain platform in Cases N_2, N_3, and N_4 are 0.3 m, 0.3 m, and 2.15 m, respectively, and the slope angles α in Cases N_3 and N_4 are 45°. Schematic diagrams of numerical tank for the four cases are shown in Figure 6a–d, and the cases introduction is presented in Table 2.

Table 2. Case introduction for different terrains.

No.	Case	h_1/h_2	η_0/H	$C_{Fn\text{-}max}$
1	Single cylinder (N_1)	0.33	0.0575	0.0857
2	Bottom-step terrain (N_2)	0.33	0.0575	−0.0683
3	Flat-top-knoll terrain (N_3)	0.33	0.0575	−0.0692
4	Flat-top-knoll terrain (N_4)	0.33	0.0575	−0.0753

3.1. Coupled Influence of Terrain and IWs on the Forces on the Cylinder

Figure 7 shows the graphs of C_{Fn} on the cylinder vs. time *t* for the SC case and the three terrain cases. The figure presents that the peak value of C_{Fn} in the cylinder of the N_1 case is significantly larger than that of the other three terrain cases, indicating that the bottom topography significantly changes the IW forces on the cylinder and makes the IW forces peak value $C_{Fn\text{-}max}$ change from positive to negative. Obvious negative peaks of C_{Fn} appear in the three terrain cases N_2, N_3, and N_4 when the IWs are close to the front edge of the terrain. The reduction in the lower layer depth of IW due to the topographical factors causes a shallow-water effect, and the strength of the fluid flow field in the lower layer around the bottom terrain is enhanced. Moreover, the negative forces applied to the cylinder over the terrain result in the negative peak value of C_{Fn}. As shown in Figure 8, the negative forces on the cylinder in the lower layer in cases N_2, N_3, and N_4 are much greater than those in case N_1, which can also explain the difference in the peak value between N_1 and N_2–N_4 in Figure 7. As a result, the shallow-water effect enhances the strength of the flow field around the cylinder in the lower layer, thereby causing a greater negative force on the lower parts of the cylinder.

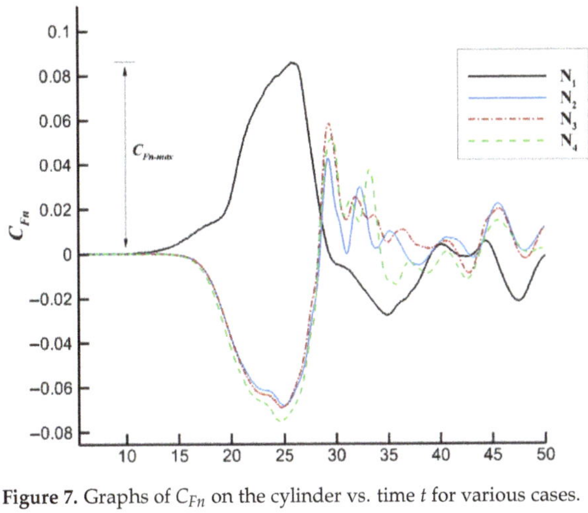

Figure 7. Graphs of C_{Fn} on the cylinder vs. time *t* for various cases.

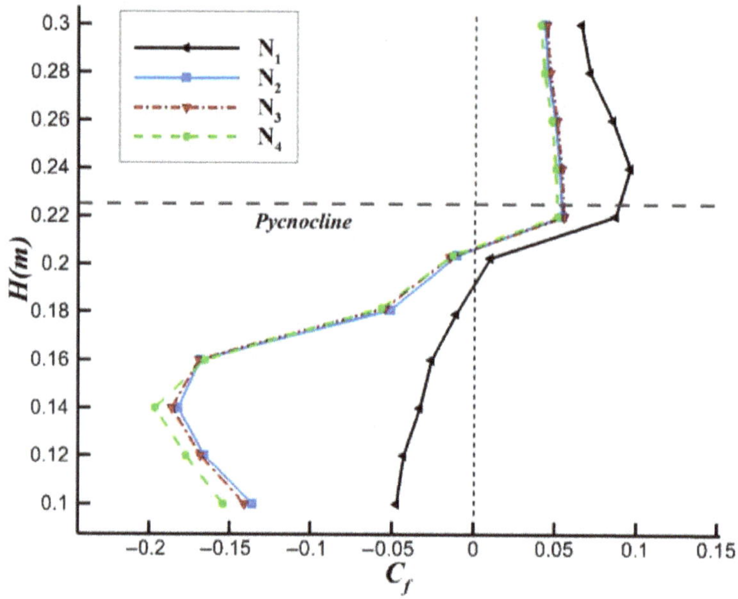

Figure 8. Comparison diagram of layered horizontal resultant force under four working conditions, $C_{Fn} = C_{Fn\text{-}max}$.

3.2. Comparison of the Vertical Distribution of the Force on the Cylinder in Different Cases

The calculation results corresponding to the time $C_{Fn} = C_{Fn\text{-}max}$ of all the cases (as shown in Figure 8, $t \approx 25$ s, the time of the most unfavorable forces on the cylinder) are selected for analysis. By further analyzing the stress mechanism of the cylinder over different terrains, the cylinder body is divided into ten parts along the vertical direction, and a section is taken at a separation of 0.02 m. The parts in the upper layer are defined as "upper parts", and the parts in the lower layer are called "lower parts". The pycnocline between the upper parts and lower parts is at the depth of 0.225 m. The vertical distribution of the non-dimensional horizontal forces C_f at various water depths in all the cases are shown in Figure 8. The calculation expression of C_f is similar to Equation (8) for C_{Fn}. C_f on the upper parts of the cylinder is positive, and the value in case N_1 condition is significantly greater than that in the other three cases. C_f on the lower parts of the cylinder becomes negative and the values in cases N_2, N_3, and N_4 are much greater than that in case N_1.

The pressure coefficient of the cylinder periphery at a certain depth section is C_{Py}, and the expression can be defined as follows:

$$C_{Py} = \frac{2 \times \left(P_y - P_{oy}\right)}{\rho_y U_{\max y}^2} \tag{9}$$

where P and $U_{\max}$ are the point pressure on the cylinder periphery and the maximum velocity, respectively; P_o is the point hydrostatic pressure on the cylinder periphery; and ρ is the fluid density. The subscript y denotes a certain depth y. In this paper, the values of y are 0.26 m and 0.14 m, which are the center heights of the upper and lower layers, respectively. The schematic diagram for the definition of the circumferential angle is illustrated in Figure 9, where the angel degrees ranging from 90° to 270° form the cylinder windward surface area (frontal side), and the angel degrees ranging from 270° to 90° from the lee side.

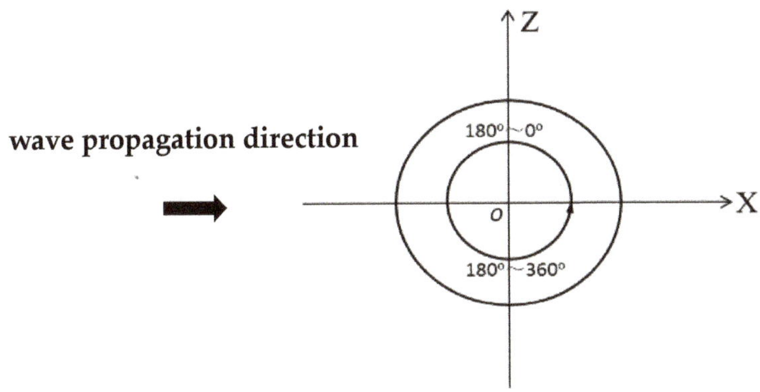

Figure 9. Schematic diagram for the definition of circumferential angle.

Figure 10a,b shows the distribution of the pressure coefficient C_p on the cylinder at Plane$_{y=0.26}$ ($P_{0.26}$ for short, y = 0.26 m) and Plane$_{y=0.14}$ ($P_{0.14}$ for short, y = 0.14 m) sections.

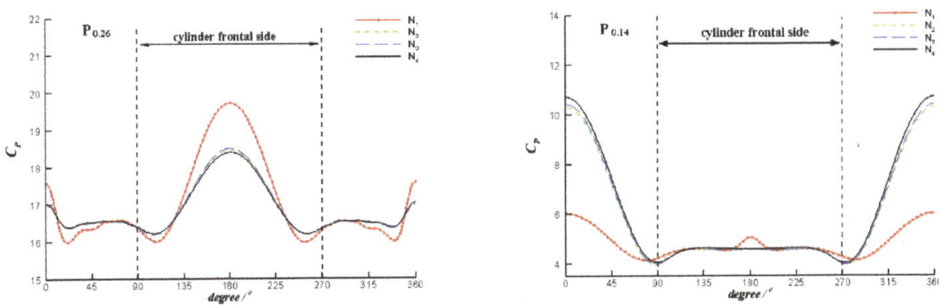

(a) (b)

Figure 10. Comparison of the pressure distribution between N1 and N_2–N_4. (**a**) y = 0.26 m, (**b**) y = 0.14 m.

As shown in the Figure 10a of the pressure distributions in the upper layer (y = 0.26 m), the maximum pressure points on the periphery are all at the former stagnation points, and the lee sides of the cylinders are all immersed in the low-pressure areas. The pressure on the frontal sides is greater than that on the lee side, so the forces on the upper parts of the cylinder are generally positive. Moreover, the pressure on the frontal sides in case N_1 is significantly greater than that in cases N_2–N_4, resulting in the forces on the upper parts in case N_1 being significantly greater than those in cases N_2–N_4.

As shown in Figure 10b, the pressure distributions in the lower layer (y = 0.14 m), the high-pressure areas all appear near the cylinder lee sides. The pressure on the cylinder frontal sides is less than that on the lee sides, resulting in negative horizontal resultant forces on the lower parts (see Figure 8). Moreover, the differences in pressure between the frontal sides and the lee sides in cases N_2–N_4 are much greater than those in case N_1. Therefore, the reverse horizontal forces on the cylinder in cases N_2–N_4 are much greater than those in case N_1.

Combined with the vorticity contours, the pressure distribution characteristics of the cylinders are further analyzed below. The propagation direction of the IWs is defined as the positive direction. When the IWs pass through the cylinder, the flow direction in the upper water is the same as the wave direction, so that the frontal side of the upper parts is impacted by the IWs, while the vortexes are formed near the lee sides (as shown in Figure 11). Therefore, the pressure on the frontal side increases, while the pressure on the lee side immersed in the vortexes decreases. The pressure difference between the frontal side and lee side leads to a positive horizontal total force on the upper parts, and the maximum pressure point is at the former stagnation point. Moreover, Figure 11 shows that the vortices around the cylinder in the SC case are obviously different from those in the other three terrain cases, indicating that the existence of terrain weakens the horizontal forces on the upper parts to a certain extent.

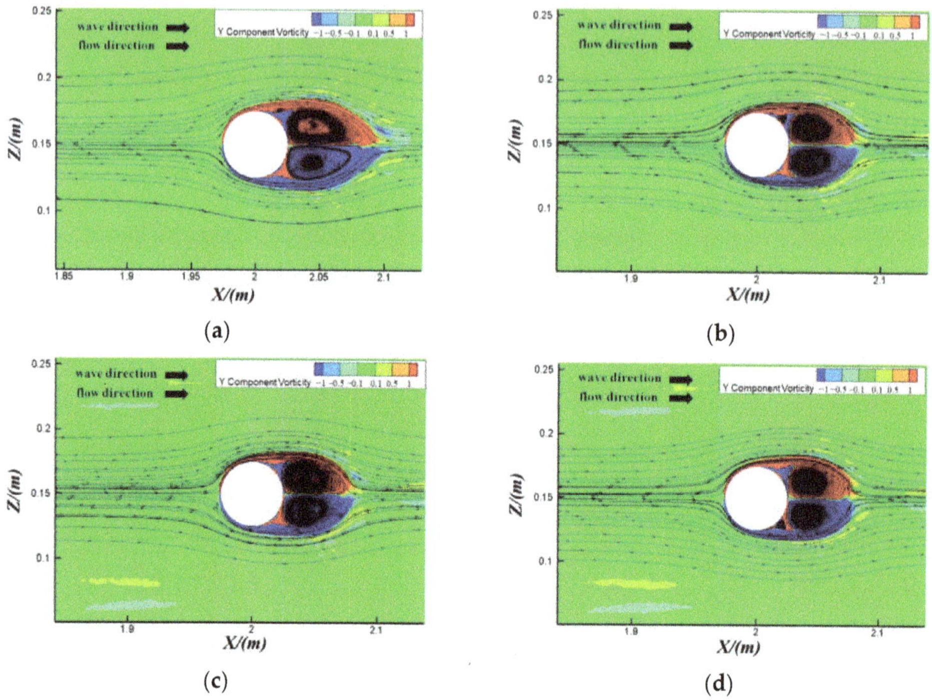

Figure 11. Plots of vorticity around the cylinder at $P_{0.26}$: (**a**) N_1, (**b**) N_2, (**c**) N_3, (**d**) N_4, when $C_{Fn} = C_{Fn\text{-}max}$.

In the lower layer, the wave propagation direction is opposite to the flow propagation direction, and the vortices of the low-pressure region are formed near the frontal side of the cylinder (see Figure 12), thus reducing the pressure on the frontal side. The pressure difference between the frontal side and lee side results in the reverse horizontal resultant force, as shown in Figure 12. In addition, combined with Figure 13, the shallow-water effect occurs when IWs propagate to the bottom terrain [23,24], which strengthens the flow field near the terrain in the lower layer and makes the lower parts subject to the impact of the additional reverse wave fluid. Additionally, although in the four cases the vortices all appear near the frontal sides of the lower parts, the vortices in the SC case are significantly different from those in the other three terrain cases due to the influence of the topographic factors, resulting in the difference between the horizontal resultant forces on the lower parts in the terrain cases and the no terrain case.

In summary, the existence of bottom terrains weakens the forces on the upper parts of the cylinder in the upper layer fluid, and due to the shallow-water effect in the lower layer, the lower parts of the cylinder are subject to greater horizontal resultant forces in the reverse wave direction.

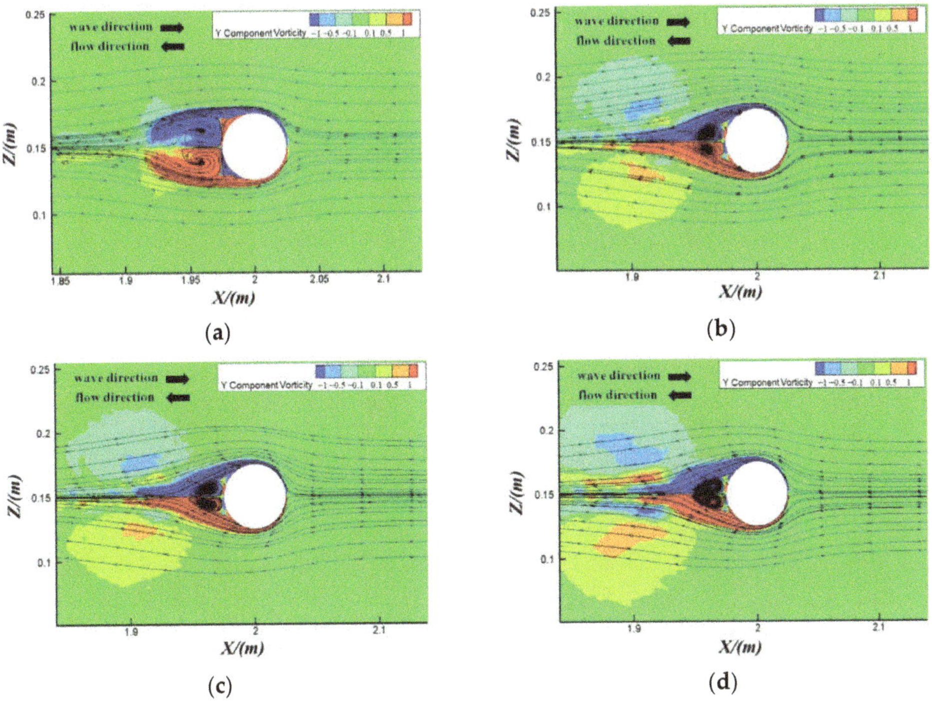

Figure 12. Plots of vorticity around cylinder at P_{014}: (**a**) N_1, (**b**) N_2, (**c**) N_3, (**d**) N_4, when $C_{Fn} = C_{Fn\text{-}max}$.

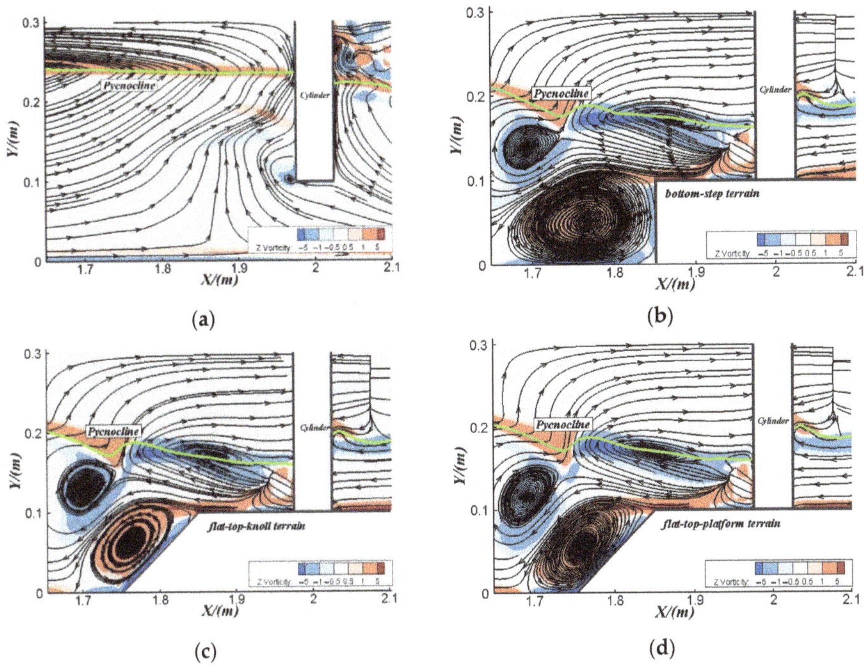

Figure 13. Plots of the flow field in Z direction, when $C_{Fn} = C_{Fn\text{-}max}$: (**a**) N_1, (**b**) N_2, (**c**) N_3, (**d**) N_4.

3.3. Variation in the Flow Field around the Cylinder under Different Cases

By comparing the flow field distributions of the SC case (N_1) with the terrain cases (N_2–N_4) in Figure 13a–d, it is found that, when the IWs reach the bottom terrain, vortexes can be found near the frontal side of the terrain, but no obvious vortex appears in case N_1. The vortexes are generated by the coupled effects of the IWs, the topography, and the shallow-water effect. Compared with the SC case, when the IWs propagate over the terrain, the interactions between the IWs and the terrain make the flow field around the cylinder more complex and changeable, especially near the lower parts of the cylinder. As a result, the complex hydrodynamic environment caused by the IWs and the terrains compels the cylinder to experience larger forces.

3.4. Influence of the Amplitudes on the IWs Forces and the Flow Field over the Terrain

On the base of the above research, the terrain is a key consideration to explore the hydrodynamic characteristics in IW environment, but the IWs amplitudes also have a big effect on the forces on and the flow field around the cylinder [25,26]. Therefore, more calculation results of the influence of the amplitudes on IWs forces are presented to study the research further in the following. The cases introduction is presented in Table 3. The case S stands for the SC case and the case F stands for the flat-top-platform terrain case, where the subscripts 1-5 indicate the different amplitudes ranging from 0.0275 to 0.0674.

Table 3. Case introduction for different amplitudes.

No.	Case	h_1/h_2	η_0/H	$C_{Fn\text{-}max}$	$R_{Fn\text{-}max}$
1	S_1	0.33	0.0275	0.0245	17.6%
2	F_1	0.33	0.0275	0.0202	
3	S_2	0.33	0.0384	0.0428	17.5%
4	F_2	0.33	0.0384	−0.0353	
5	S_3	0.33	0.0494	0.0664	13.9%
6	F_3	0.33	0.0494	−0.0572	
7	S_4	0.33	0.0575	0.0857	12.1%
8	F_4	0.33	0.0575	−0.0753	
9	S_5	0.33	0.0674	0.132	34.5%
10	F_5	0.33	0.0674	−0.0864	

3.4.1. Influence on the IWs Forces

The comparison of C_{Fn} vs. time t between the case S (SC) and the case F (flat-top-platform terrain) under different amplitudes conditions are shown in Figure 14. On the whole, with the increase of the IW amplitude, the forces on the single cylinder increase, and the duration curve of the IW forces on cylinder are all similar in all the cases. Under the small amplitude condition of η_0/H = 0.0275 (see Figure 14a), the IW forces peaks $C_{Fn\text{-}max}$ for the case S_1 and the case F_1 are both positive. $C_{Fn\text{-}max}$ for the cases F_2–F_5 turn negative as the amplitude increases, while $C_{Fn\text{-}max}$ for the cases S_2–S_5 still keep positive (see Figure 14b–e). Meanwhile, a percentage parameter $R_{Fn\text{-}max}$ is applied here to specify the differences of $C_{Fn\text{-}max}$ between the cases S and the cases F, and the expression can be defined as follow:

$$R_{Fn-max} = \frac{(C_{Fn-max})_S - \left|(C_{Fn-max})_F\right|}{(C_{Fn-max})_S} \tag{10}$$

where $(C_{Fn-max})_S$ and $(C_{Fn-max})_F$ are the IW forces peaks on the cylinder for the case S and the case F, respectively.

As can be seen from the Table 3, $R_{Fn\text{-}max}$ decreases as the IW amplitudes η_0/H increase from 0.0275 to 0.0575 but it sharply increases to 34.5% when η_0/H reaches 0.0674. This can be explained by the flow field and density distribution characteristics as illustrated in Figures 15e and 16e. When IW with large amplitude propagates to the bottom terrain, the interaction between the IW and the terrain is intensified and the shallow-water effect occurs, which strengthens the flow field strength near the terrain in the lower layer and influences the force on the cylinder. A phenomenon of "elevation" simultaneously occurs because of the "blockage" effect caused by the bottom topography, which directly reflects the influence of the existing terrain on the flow field near the cylinder.

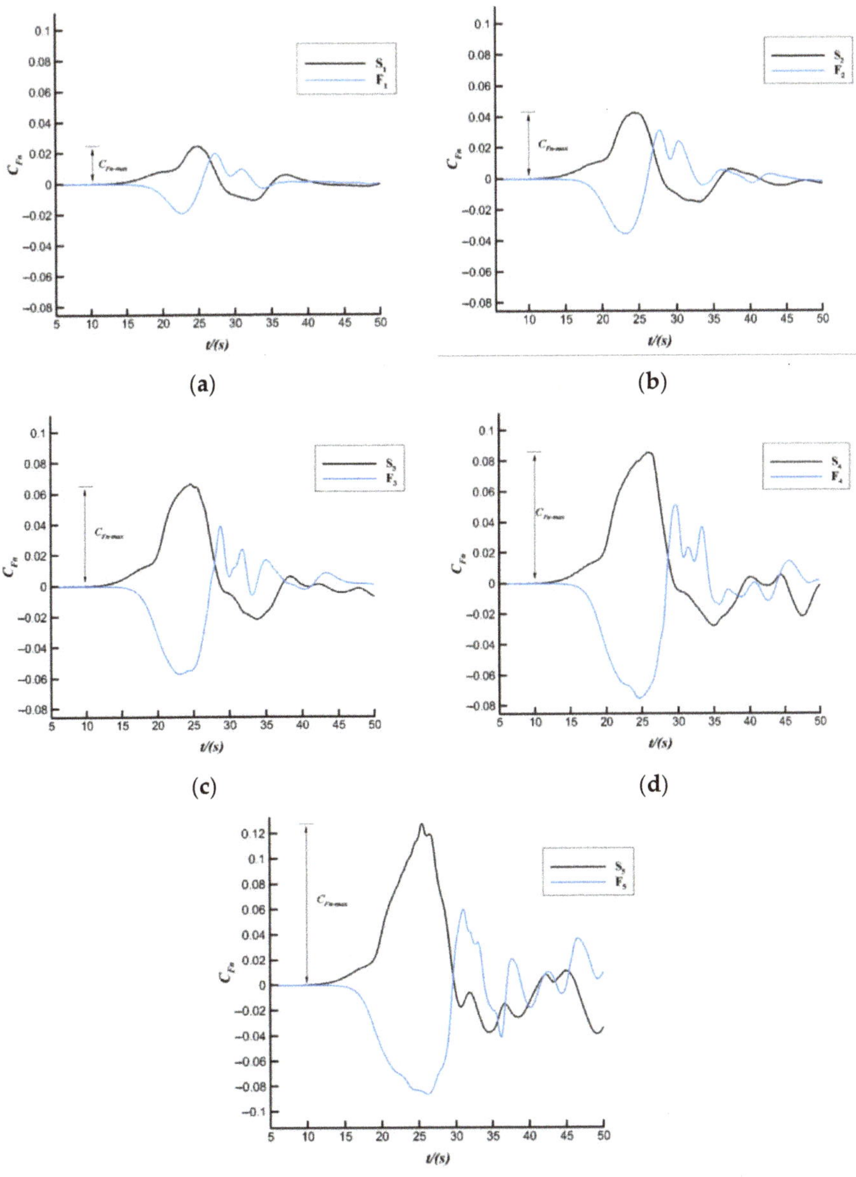

Figure 14. Comparison of C_{Fn} vs. time t between the S case and the F case: (**a**) η_0/H = 0.0275, (**b**) η_0/H = 0.0384, (**c**) η_0/H = 0.0494, (**d**) η_0/H = 0.0575, (**e**) η_0/H = 0.0674.

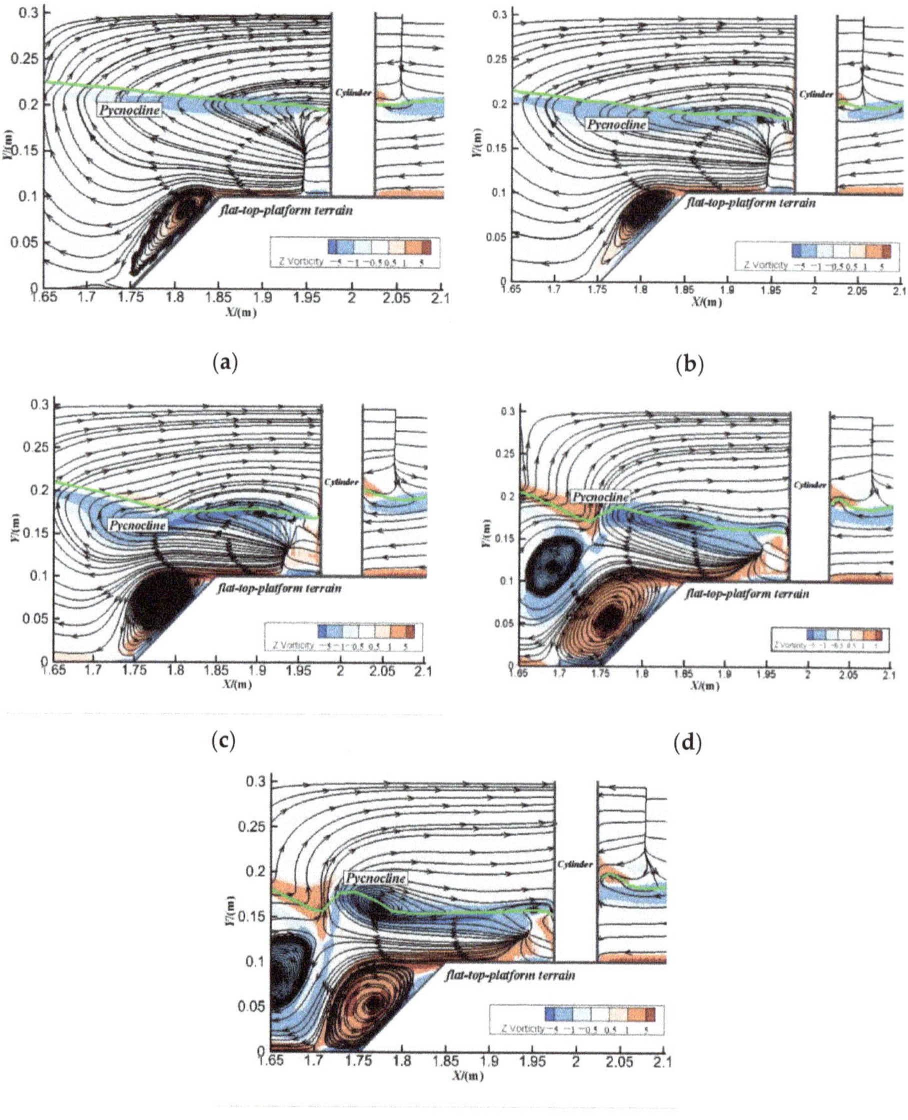

Figure 15. Plots of the flow field in Z direction, when $C_{Fn} = C_{Fn\text{-}max}$: (**a**) $\eta_0/H = 0.0275$, (**b**) $\eta_0/H = 0.0384$, (**c**) $\eta_0/H = 0.0494$, (**d**) $\eta_0/H = 0.0575$, (**e**) $\eta_0/H = 0.0674$.

3.4.2. Influence on the Flow Field

The flow field distribution characteristics of different amplitudes conditions are plotted in Figure 15a–e. With the increase of the wave amplitude, the interaction between the IW and the terrain as well as the flow field intensity are both enhanced. The vortices can be found on the bank slope in all the cases, but the size of the vortex is obviously different when the amplitude changes. It is obvious that the vortex size increases with the amplitude, and more than one vortex appears when the amplitude exceeds 0.0575, which can be found in Figure 15d,e.

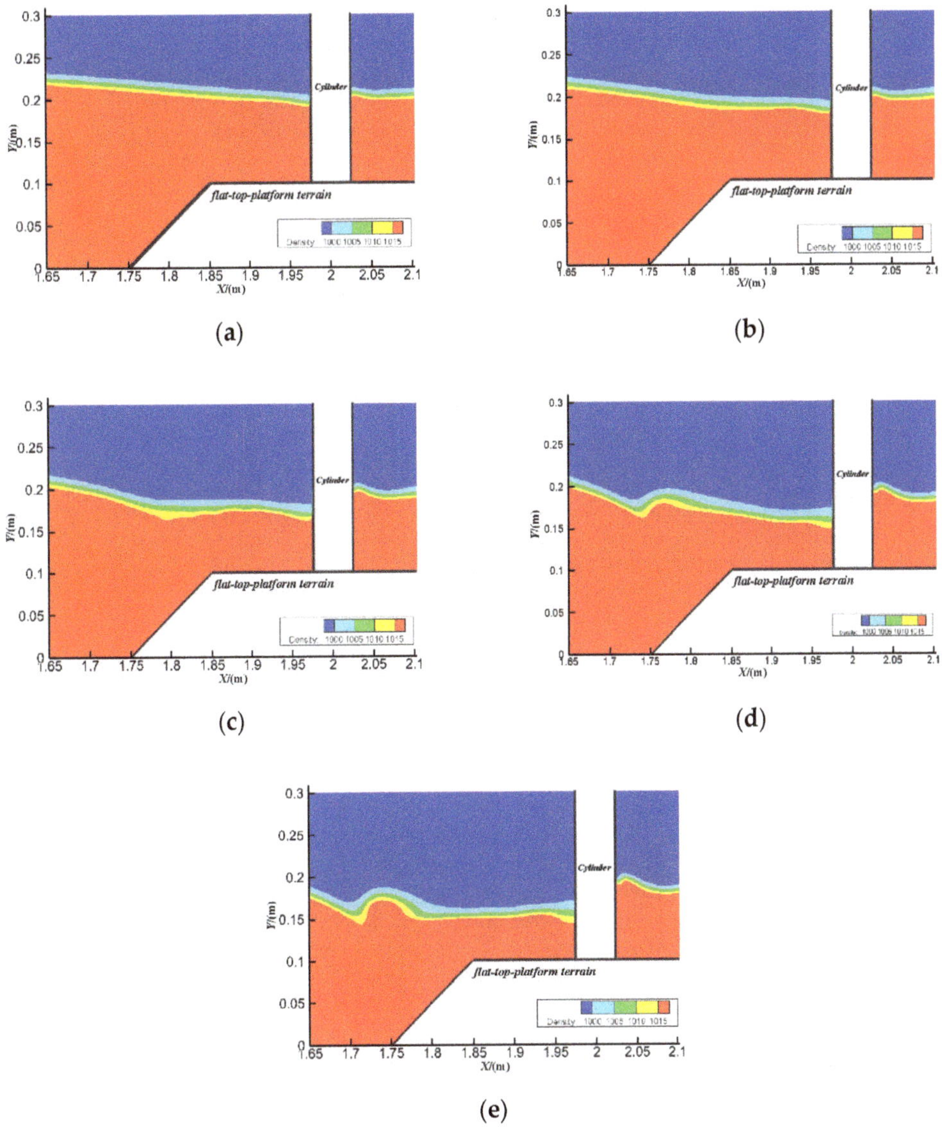

Figure 16. Plots of the density distribution in Z direction, when $C_{Fn} = C_{Fn\text{-}max}$: (**a**) $\eta_0/H = 0.0275$, (**b**) $\eta_0/H = 0.0384$, (**c**) $\eta_0/H = 0.0494$, (**d**) $\eta_0/H = 0.0575$, (**e**) $\eta_0/H = 0.0674$.

The density distribution characteristics plotted in Figure 16a–e also give evidence to present the interaction between the IWs and the terrain. With the increase of IW amplitude, the IW pattern is more strongly disturbed by the terrain. IW propagating over the bank slope is partially reflected, causing a remarkable "blockage" near the terrain and a "elevation" phenomenon in the reverse wave propagation direction as illustrated in Figure 16e. Therefore, the increase of the interaction strength between the IWs and the terrain could not only cause greater horizontal forces on the lower parts of the cylinder, but also make the flow field around the terrain more complex, further affecting the mechanical characteristics of the cylinder.

4. Conclusions

A 3D numerical tank model of a flume with internal waves and terrain is set up to research the mechanics characteristics of the cylinder in an IW-terrain coupled environment. The main described are as follows:

(1) The topographic factors of the terrain significantly affect the IW forces on the cylinder. There is a strong distinction between the SC case and the three terrain cases: in the SC case, the maximum resultant forces on the cylinder are positive, and the maximum resultant forces are negative in the terrain cases.
(2) Compared with the SC case, the shallow-water effect caused by the IW-terrain coupled environment enhances the strength of the flow field around the cylinder, so that the lower parts of the cylinder are subjected to larger forces in the reverse wave direction.
(3) Compared with the SC case, when the IWs propagate over the terrain, the interactions between the IWs and the terrain make the flow field around the cylinder more complex and changeable. As a result, the complex hydrodynamic environment compels the cylinder to experience larger forces.
(4) A percentage parameter $R_{Fn\text{-}max}$ is applied in this research to specify the differences of $C_{Fn\text{-}max}$ between the SC case and the terrain case. $R_{Fn\text{-}max}$ decreases as the IW amplitude increases when the amplitude is relatively small, but it sharply increases when amplitude is large enough. It is can be explained by the shallow-water effect. When IWs with large amplitude propagate to the bottom terrain, the interaction between the IW and the terrain is intensified and the shallow-water effect occurs, which strengthens the flow field strength near the terrain in the lower layer.
(5) With the increase of IW amplitude, the interaction between the IW and the terrain is enhanced. Vortices can be found on the bank slope in all the cases, but the size of the vortices is obviously different when amplitude changes. The vortex size increases with the amplitude, and more than one vortex appears when the amplitude is large enough.
(6) With the increase of the IW amplitude, the IW pattern is more strongly disturbed by the terrain. IW propagating over the bank slope is partially reflected, causing a "blockage" near the terrain and a "elevation" in the reverse wave propagation direction. Therefore, the intensification of the interaction strength between the IWs and the terrain could not only cause greater horizontal forces on the lower parts of the cylinder, but also make the flow field around the terrain more complex.

Author Contributions: Methodology, L.W.; Software, Y.W., C.Z., X.X. and C.W.; Validation, M.X., S.S., C.Z., X.X. and C.W.; Writing—original draft, Y.W.; Writing—review & editing, Y.W., M.X., L.W., X.W., H.W. and C.W.; Funding acquisition, S.S. All authors have read and agreed to the published version of the manuscript.

Funding: The work was supported by the Sci-tech Program by Jiangxi Provincial Education Department of China (Grant No. GJJ180953), the National Natural Science Foundation of China (Grant Nos. 52109090, 52009087, 51879086), the China Postdoctoral Science Foundation Funded Project (Grant No. 2022M721426), the Belt and Road Special Foundation of the State Key Laboratory of Hydrology-Water Resources and Hydraulic Engineering (Grant No. 2018492011), and the Jiangxi Natural Science Foundation of China (Grant No. 20192ACBL20041).

Data Availability Statement: The data that support the findings of this study are available from the corresponding author upon reasonable request.

Conflicts of Interest: The authors declare no conflict of interest.

References

1. Mei, Y.; Wang, J.; Huang, S.; Mu, H.; Chen, X. Experimental investigation on the optical remote sensing images of internal solitary waves with a smooth surface. *Acta Oceanol. Sin.* **2019**, *38*, 124–131. [CrossRef]
2. Wang, X.; Zhou, J.-F.; Wang, Z.; You, Y.-X. A numerical and experimental study of internal solitary wave loads on semi-submersible platforms. *Ocean Eng.* **2018**, *150*, 298–308. [CrossRef]
3. Xu, W.; Lin, Z.Y.; You, Y.X.; Yu, R. Numerical Simulations for the Load Characteristics of Internal Solitary Waves on a Vertical Cylinder. *J. Ship* **2017**, *21*, 1071–1085.

4. Colosi, J.A.; Kumar, N.; Suanda, S.H.; Freismuth, T.M.; MacMahan, J.H. Statistics of Internal Tide Bores and Internal Solitary Waves Observed on the Inner Continental Shelf off Point Sal, California. *J. Phys. Oceanogr.* **2018**, *48*, 123–143. [CrossRef]
5. Moum, J.N.; Farmer, D.M.; Smyth, W.D.; Armi, L.; Vagle, S. Structure and Generation of Turbulence at Interfaces Strained by Internal Solitary Waves Propagating Shoreward over the Continental Shelf. *J. Phys. Oceanogr.* **2003**, *33*, 2093–2112. [CrossRef]
6. Timothy, W. Internal Solitons on the Pycnocline: Generation, Propagation, and Shoaling and Breaking over a Slope. *J. Fluid Mech.* **1985**, *159*, 19–53.
7. Chen, M.; Chen, K.; You, Y.X. Experimental investigation of internal solitary wave forces on a semi-submersible. *Ocean Eng.* **2017**, *141*, 205–214. [CrossRef]
8. Jia, T.; Liang, J.J.; Li, X.-M.; Sha, J. SAR Observation and Numerical Simulation of Internal Solitary Wave Refraction and Reconnection Behind the Dongsha Atoll. *J. Geophys. Res. Oceans* **2018**, *123*, 74–89. [CrossRef]
9. Chen, L.; Zheng, Q.; Xiong, X.; Yuan, Y.; Xie, H.; Guo, Y.; Yu, L.; Yun, S. Dynamic and Statistical Features of Internal Solitary Waves on the Continental Slope in the Northern South China Sea Derived From Mooring Observations. *J. Geophys. Res. Oceans* **2019**, *124*, 4078–4097. [CrossRef]
10. Germano, M.; Piomelli, U.; Moin, P.; Cabot, W.H. A dynamic subgrid-scale eddy viscosity model. *Phys. Fluids* **1991**, *3*, 1760–1765. [CrossRef]
11. Michallet, H.; Barthelemy, E. Experimental study of large interfacial solitary waves. *Fluid Mech.* **1998**, *366*, 159–177. [CrossRef]
12. Camassa, R.; Choi, W.; Michallet, H.; Rusås, P.-O.; Sveen, J.K. On the realm of validity of strongly nonlinear asymptotic approximations for internal waves. *J. Fluid Mech.* **2006**, *549*, 1–23. [CrossRef]
13. Li, X.; Ren, B.; Wang, G.Y.; Wang, Y.X. Numerical simulation of hydrodynamic characteristics on an arx crown wall using volumn of fluid method based on BFC. *J. Hydrodyn.* **2001**, *23*, 767–776. [CrossRef]
14. Chen, Y.X. Flow simulation of car air conditioner duct based on simple algorithm. *Mech. Eng.* **2008**, *11*, 111–112.
15. Cao, Y.; Ye, Y.T.; Liang, L.L.; Zhao, H.; Jiang, Y.; Wang, H.; Wang, J. Uncertainty analysis of two-dimensional hydrodynamic model parameters and boundary conditions. *J. Hydroelectr. Eng.* **2018**, *37*, 47–61. (In Chinese)
16. Forgia, G.L.; Sciortino, G. Free-surface effects induced by internal solitons forced by shearing currents. *Phys. Fluids* **2021**, *33*, 072102. [CrossRef]
17. Zhu, H.; Wang, L.L.; Avital, E.J.; Tang, H.W.; Williams, J.J.R. Numerical Simulation of Shoaling Broad-Crested Internal Solitary Waves. *J. Hydraul. Eng.* **2017**, *143*, 04017006. [CrossRef]
18. Chen, C.-Y. Amplitude decay and energy dissipation due to the interaction of internal solitary waves with a triangular obstacle in a two-layer fluid system: The blockage parameter. *J. Mar. Sci. Technol.* **2009**, *14*, 499–512. [CrossRef]
19. Chen, C.Y.; Hsu, J.R.C.; Chen, C.W.; Chen, H.H.; Kuo, C.F. Generation of internal solitary wave by gravity collapse. *J. Mar. Sci. Technol.* **2007**, *15*, 1–7. [CrossRef]
20. Talipova, T.; Terletska, K.; Maderich, V.; Brovchenko, I.; Jung, K.T.; Pelinovsky, E.; Grimshaw, R. Internal solitary wave transformation over a bottom step: Loss of energy. *Phys. Fluids* **2013**, *25*, 032110. [CrossRef]
21. Cheng, C.Y.; Hsu, J.R.C.; Chen, M.H. An investigation on internal solitary waves in a two-layer fluid:Propagation and reflection from steep slopes. *Ocean. Eng.* **2007**, *34*, 171–184. [CrossRef]
22. Cheng, M.-H.; Hsu, J.R.-C.; Chen, C.-Y. Laboratory experiments on waveform inversion of an internal solitary wave over a slope-shelf. *Environ. Fluid Mech.* **2011**, *11*, 353–384. [CrossRef]
23. Shroyer, E.L.; Moum, J.N.; Nash, J. Observations of Polarity Reversal in Shoaling Nonlinear Internal Waves. *J. Phys. Oceanogr.* **2009**, *39*, 691–701. [CrossRef]
24. Sutherland, B.R.; Barrett, K.J.; Ivey, G.N. Shoaling internal solitary waves. *J. Geophys. Res. Ocean.* **2013**, *118*, 4111–4124. [CrossRef]
25. Wei, G.; Du, H.; Xu, X.; Zhang, Y.; Qu, Z.; Hu, T.; You, Y. Experimental investigation of the generation of large-amplitude internal solitary wave and its interaction with a submerged slender body. *Sci. China Phys. Mech. Astron.* **2014**, *57*, 301–310. [CrossRef]
26. Zou, P.; Bricker, J.D.; Uijttewaal, W. The impacts of internal solitary waves on a submerged floating tunnel. *Ocean Eng.* **2021**, *238*, 109762. [CrossRef]

water

MDPI

Article

Analysis of Flow Characteristics around a Square Cylinder with Boundary Constraint

Zhun Xu *, Shiqiang Wu, Xiufeng Wu, Wanyun Xue, Fangfang Wang, Ang Gao and Weile Zhang

State Key Laboratory of Hydrology-Water Resources and Hydraulic Engineering, Nanjing Hydraulic Research Institute, Nanjing 210098, China
* Correspondence: xuzhun_nanjing@foxmail.com

Abstract: Based on the two-dimensional hydrodynamic model of the finite volume method and structured multigrid, the flow characteristics around a square cylinder with boundary constraint are analysed. The gap ratio G/D (G is the distance from the cylinder to the channel boundary, and D is the side length of the square cylinder) does not change the four flow patterns. Under the laminar vortex street phase, the boundary constraint only reduces the scale of the vortices. The vortex centres are pressed toward the boundaries of the channel, and a low velocity zone is formed near the boundary, but the law of vorticity attenuation along the flow direction is not changed. The flow pattern classification map shows that the boundary constraint increases the Reynolds number required to generate the turbulence flow pattern, and the range of the Reynolds number in the flow pattern of the laminar vortex street has a maximum increase range. The correlation between the time-averaged drag coefficient or the vortex shedding frequency and Reynolds number under different gap ratios indicates that the resistance of the square cylinder and the vortex shedding frequency increase accordingly with the strengthening of the boundary constraint. When $G/D < 3.5$, the increase is particularly obvious. Meanwhile, the correlation characteristics between the resistance or the vortex shedding frequency of the square cylinder and the Reynolds number are unrelated to the boundary constraint strength.

Keywords: flow around a square cylinder; boundary constraint; wake flow pattern; resistance strength; vortices shedding frequency

1. Introduction

Flow around a square cylinder is a classic problem in fluid mechanics. When fluids such as air and water flow through square components such as high-rise buildings, bridge piers, and floodgate pillars, they are affected by the cylinder, and flow patterns with different characteristics are produced. The characteristics are related to the Reynolds number of fluids, the shape and size of the cylinder, the space size, etc. The pulsation and resonance effects caused by fluid separation and vortex shedding of the flow around a cylinder can affect the safety of the building itself. Therefore, the complex flow characteristics around a square cylinder have always been a hotspot and a challenge in academic research, and they provide important references for related practical applications.

Flow around a cylinder is a typical phenomenon of the flow around a blunt body. The development of the flow around a cylinder is closely related to the boundary layer of the cylinder. Because of the retarding action of the boundary mutation of the cylinder, the velocity and pressure gradient near the cylinder are obvious. Under large gradients, the thickness of the fluid boundary layer increases and separates from the cylinder boundaries, forming a backflow in the downstream direction of the cylinder. The above process is the origin of the flow around a cylinder. The separation and shedding of the boundary layer, the flow changes of the shear layer, and the flow pattern distribution of the wake are coupled with each other to form a complete flow around a cylinder. The Reynolds number is a key factor in determining the flow pattern around a cylinder. At different Reynolds numbers, the flow around a cylinder exhibits completely different fluid characteristics. According to the flow characteristics under different Reynolds numbers, the flow around a square cylinder can usually be divided into four types of flow pattern classifications, namely L1, L2, L3, and T. The four types of flow pattern classifications and their main characteristics are shown in Table 1 (Williamson, 1988 [1]; Wang, et al., 2001 [2]); Behara and Mittal, 2010 [3]).

Table 1. An overview of the flow pattern classifications of the flow around a square cylinder.

Flow Pattern Classification		Flow Characteristics
L1	Stable attached vortex phase	Stable and symmetrical attached vortices formed by the separated flow behind a cylinder (Taneda, 1956 [4]; Coutanceau and Bouard, 1977 [5]).
L2	Unstable attached vortex phase	The attached vortices behind a cylinder are asymmetric, fluctuating sinusoidally.
L3	Laminar vortex street phase	Attached vortex shedding appearing behind a cylinder and forming periodic laminar vortex streets (Thoman and Szewczyk, 1969 [6]; Collins and Dennis, 1973 [7]).
T	Turbulence phase	Vortex streets are unstable, attached vortex shedding appearing non-sinusoidally, and turbulence is formed behind a cylinder.

A physical model test and numerical simulation are the main methods for the study of the flow around a cylinder. Using a tracer (Coutanceau and Defaye, 1991 [8]), aluminium powder (Taneda, 1956 [4]), soap liquid film (Wu, et al., 2004 [9]; Couder, 1981 [10]; Chomaz, 2001 [11]; Beizaie and Gharib, 1997 [12]), PIV (Kim, et al., 2015 [13]; Palau-Salvador et al., 2010 [14]) and other tools, physical model tests have been widely used to study the flow around a cylinder, but these tests have been limited in terms of their scale effect correction, boundary layer restoration, and measurement accuracy. Compared with the physical model test, the numerical simulation has the characteristics of accurate data extraction, no scaling effect, and low cost. Thus, it is widely used in the study of the flow around a cylinder. However, the mutation of boundaries of the flow around a cylinder, the complexity of the local flow pattern, and the strong turbulence characteristics demand high requirements in terms of the accuracy of the numerical simulation results. Considering the calculation efficiency while improving the calculation accuracy is an important indicator of the excellence of the numerical simulation method of the flow around a cylinder. The turbulence model and grid meshing are the primary methods of optimizing the numerical simulation method of the flow around a cylinder.

To reduce the influence of boundary conditions, the distance between the cylinder and the boundaries is usually increased, or the boundaries are set as slip boundaries during numerical simulation (Sen et al., 2011 [15]; Tong et al., 2022 [16]). However, the influence of boundary conditions on the flow around a cylinder is unavoidable in the real environment. Bridge piers, submarine cables, oil and gas pipelines, radiators, grids etc. are all typical examples of the flow around a cylinder constrained by boundaries (Yang et al., 2021 [17]; Prsic et al., 2016 [18]; Chen and Wu, 2019 [19]). The gap ratio G/D provides a quantitative basis for the constraint degree of the boundaries on the cylinder (Lei, 2000 [20]). According to the size of the gap ratio, the constraint effect of the boundaries on the flow around a cylinder can be subdivided into several stages (Nishino et al., 2007 [21]; Lei et al., 2000 [22]). When the gap is relatively large, for example, when the G/D of the flow around a cylinder is greater than 0.8, the boundaries have no effect on the vortex shedding and the wake flow pattern around a cylinder (Khabbouchi et al., 2013 [23]). As the boundaries of the channel narrow, for example, when the G/D is less than 0.8, the boundary constraint effect begins to increase significantly, and the shedding of the vortices of the flow around a cylinder is inhibited by the boundaries (Prsic et al., 2016 [18]; Durao et al., 1991 [24]). Meanwhile, the separation angle, frequency and vorticity intensity of the resistance, lift intensity, and vortex shedding of the cylinder change with the variation of the G/D (Yang et al., 2021 [18]; Lei et al., 2000 [22]; Khabbouchi et al., 2013 [23]; Martinuzzi et al., 2003 [25]), and the distribution and stability of the wake behind the cylinder are affected (Nishino et al., 2007 [22]; Martinuzzi et al., 2003 [25]; Bailey et al., 2002 [26]; Bearman and Zdravkovich, 1978 [27]; Grass et al., 1983 [28]).

Physical model test tools, such as a wind tunnel or a PIV, can trace the vortex shedding frequency and wake structure of the flow around a cylinder under boundary constraint (Khabbouchi et al., 2013 [23]; Bearman and Zdravkovich, 1978 [27]; Shi et al., 2010 [29]; Zhou et al., 2022 [30]; Price et al., 2002 [31]), but numerical simulation is still required to obtain the exact values of physical quantities such as force conditions. The confinement effect of the boundaries can amplify the local gradient of physical quantities such as velocity and pressure around the cylinder, which demands higher requirements for the error control of the numerical simulation.

The selection of the turbulence model and grid meshing method is closely related to the calculation accuracy of the flow around a cylinder with boundary constraint. Direct numerical simulation (DNS) can accurately restore various fluid characteristics of the flow around a cylinder (Zhang et al., 2020 [32]; Trias et al., 2015 [33]; Dong et al., 2006 [34]), but the high calculation cost constrains the application of DNS in the numerical simulation of the flow around a cylinder with boundary constraint (Behera and Saha, 2019 [35]; Ji et al., 2019 [36]). Large eddy simulation (LES) has been widely used in the numerical simulation of the flow around a cylinder (Rodi et al., 1997 [37]; Voke, 1997 [38]; Kahil et al., 2019 [39]); it has mainly been used to accurately restore the transient flow pattern of the flow around a cylinder because calculation costs are reduced significantly. In the numerical simulation of the flow around a cylinder with boundary constraint, LES can accurately restore the separation angle, the vortex shedding frequency, the resistance, and lift intensity of the cylinder (Prsic et al., 2016 [18]; Rodi, 1997 [40]; Grigoriadis et al., 2003 [41]). The grid resolution is closely related to the simulation accuracy of the flow around a cylinder with boundary constraint. The O-H meshing strategy based on a structured multigrid divides the computational domain of the flow around a cylinder into the high-resolution area near the cylinder and the ordinary resolution area in other areas. This strategy not only accurately restores the fluid characteristics and the wake flow pattern near the cylinder, but also considers the calculation efficiency (Khan and Ibrahim, 2019 [42]) and is the preferred meshing strategy for the numerical simulation of the flow around a cylinder with boundary constraint (Prsic et al., 2016 [18]; Kanaris et al., 2011 [43]).

Unlike cylinders, square cylinders have a discontinuous boundary shape with several right-angle mutations. Different boundary layer separation mechanisms lead to differences in the vortex shedding laws between cylinders and square cylinders (Norberg, 1993 [44]). Among these, abrupt boundaries, such as right angles, can have a significant impact on the flow pattern, especially the generation of vortices. At present, the research on the flow around a cylinder with boundary constraint has mainly used cylinders as the research object (Prsic et al., 2016 [18]; Chen and Wu, 2019 [19]; Khabbouchi et al., 2013 [23]), and there have been few research results related to the flow around a square cylinder (Yang et al., 2021 [17]). However, in practical application, it is impossible for the cylinder to be of an absolutely smooth construction as structures with edge angles widely exist. Meanwhile, the research has mainly focused on the force condition of the cylinder itself and the relationship between the local flow pattern near the cylinder and boundary constraint (Khabbouchi et al., 2013 [23]). There have been few studies on the distribution and development law of the wake of the flow around a cylinder by boundary constraint (Prsic et al., 2016 [18]). The wake of the flow around a cylinder has a wider range of influence and this also has research value.

In this study, a square cylinder was taken as the research object. Based on the finite volume method and a structured multigrid, a two-dimensional mathematical model was established to simulate the flow around a square cylinder with boundary constraint in a rectangular channel. The boundary constraint effect was studied under four flow patterns: laminar, transitional flow pattern, vortex street, and turbulence. The resistance of the cylinder, the vortex shedding frequency caused by the flow around a square cylinder, and the flow pattern of the wake behind the square cylinder are studied, and the mechanism of the boundary constraint interfering with the flow around a square cylinder is discussed.

2. Numerical Simulation Method for Flow around a Square Cylinder

2.1. Mathematical Model

According to the flow characteristics around a square cylinder, a two-dimensional mathematical model was developed. The governing equations were modified as follows:

$$\text{continuity equation}: \frac{\partial \rho}{\partial t} + \frac{\partial \rho u_i}{\partial x_i} = 0 \quad (1)$$

$$\text{momentum equation}: \frac{\partial \rho u_i}{\partial t} + \frac{\partial}{\partial x_i}(\rho u_i u_j) = -\frac{\partial p}{\partial x_i} + \frac{\partial}{\partial x_j}\left[\mu\left(\frac{\partial u_i}{\partial x_j} + \frac{\partial u_j}{\partial x_i}\right)\right] + f_i \quad (2)$$

Here ρ is the fluid density, μ is the dynamic viscosity, u is the velocity vector, p is the scalar pressure, f is the inertia force, x is the cartesian coordinate, and i and j are the x direction unit vector and the y direction unit vector, respectively.

The finite volume method was used to discretize the governing equation, the quadrilateral multigrid was used to discretize the two-dimensional model, and the PISO calculation method was used to couple the pressure and velocity.

The inlet was set as the velocity inlet boundary with steady velocity, and the outlet was set as the free outflow boundary. A no-slip wall boundary was used for the plane boundary of the model.

For the initial condition, the computational domain was set as a stationary flow field, and each physical quantity was assigned as 0.

The transient model was used for the numerical simulation. The convergence accuracy of each physical quantity was set to 1.0×10^{-6}.

2.2. Computational Domain and Boundary Conditions

In the actual application scenarios of the flow around a cylinder, such as bridge piers, high-rise buildings, wind turbines, etc., the geometric shapes in the horizontal direction are similar to those in the vertical direction (Blackburn and Henderson, 1999 [45]; Sadeque et al., 2008 [46]; Yuce and Kareem, 2016 [47]; Zaid et al., 2019 [48]), so the characteristics of the flow around a cylinder in the horizontal direction are representative. Meanwhile, the velocity distribution of the approaching flow, such as water flow and wind, are uniform and symmetrical on the scale of the cylinder. Therefore, the computational domain was set as a two-dimensional rectangular channel with length L and width H. The two ends of the channel were the inlet and outlet. A square cylinder with a side length of D was set at the axis of the channel. The distance between the square cylinder and the inlet and outlet of the channel were $\frac{1}{4}$L and $\frac{3}{4}$L, respectively. The length L of the computational domain and the side length D of the square cylinder, as well as their positions, were fixed. To keep the flow pattern near a square cylinder free from the influence of the inlet and outlet boundary, it was necessary to keep the cylinder and boundaries at a greater distance. Thus, we set L = 30 D, and the cylinder was $\frac{1}{4}$L and $\frac{3}{4}$L from the inlet and outlet boundaries, respectively. The width H of the computational domain can be adjusted in the range of H = 2 D~12 D during the calculation. In this study, D was set to 0.4 m. The flow around a square cylinder was simulated under different Reynolds numbers by adjusting the velocity at the inlet boundary. The schematic diagram of the computational domain and boundary conditions is shown in Figure 1, where x is the flow direction, and y is the width direction of the rectangular channel.

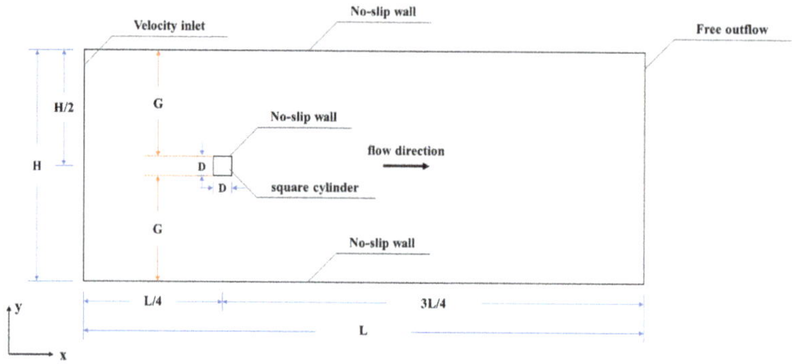

Figure 1. Schematic diagram of the computational domain and boundary conditions.

2.3. Computational Grids

Structured grids were used to fit the rectangular computational domain with square inner boundaries. Different grid sizes were set in different parts of the computational domain. In the area close to the square cylinder, the grid density was relatively high, while in the other area the density was relative. The computational domain was divided into nine zones with different grid sizes. The location of these nine zones is shown in Figure 2.

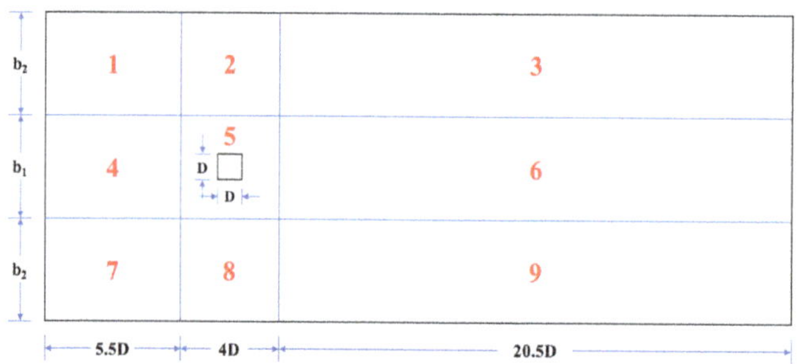

Figure 2. The location of nine zones for grid generation.

Here, when $H > 4D$, $b_1 = 4D$ and $H \leq 4D$, $b_1 = H$, only zones 4, 5, and 6 exist. The grid sizes in the nine zones are shown in Table 2, where grid sizes are adjusted by variable δ. The computational grid distribution of the computational domain is shown in Figure 3.

Table 2. Grid sizes in nine zones.

Zone	Grid Length	Grid Width
1	3δ	3δ
2	δ	3δ
3	3δ	3δ
4	3δ	δ
5	δ	δ
6	3δ	δ
7	3δ	3δ
8	δ	3δ
9	3δ	3δ

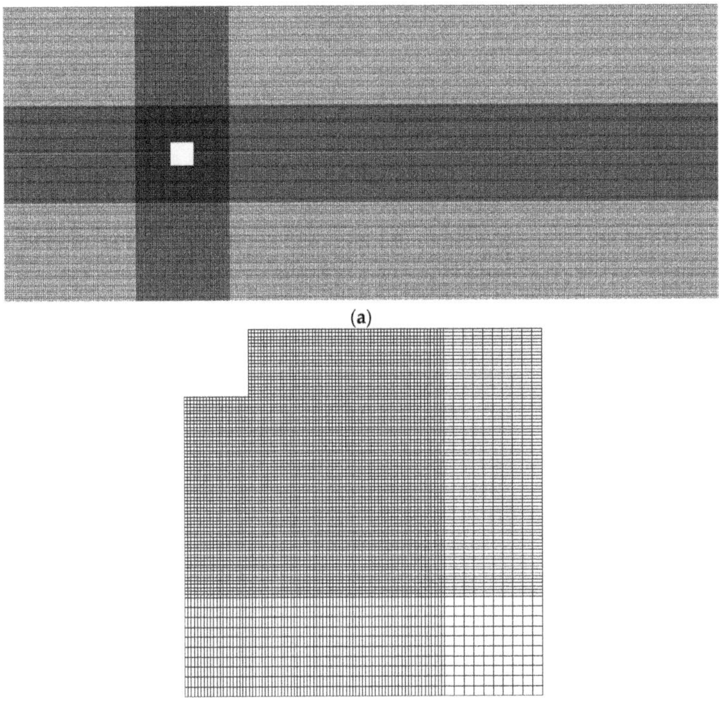

Figure 3. Computational grids distribution of the computational domain; (**a**) total computational grids distribution; (**b**) detailed computational grids distribution close to the square cylinder.

To optimize the simulation in terms of accuracy and computational cost, the influence of the grid resolution on the simulation results was examined. A specific channel width (H = 12 D) was set, and the simulation results under different grid resolutions were compared. The time-averaged drag coefficient $\overline{C_d}$ and the Strouhal number S_t were important parameters to describe the fluid characteristics around a square cylinder.

The time-averaged drag coefficient $\overline{C_d}$ is a dimensionless coefficient that reflects the resistance of a square cylinder in the fluid and is modified as follows:

$$\overline{C_d} = \frac{1}{T}\int_0^T \frac{2F_d}{\rho U^2 D}dt \quad (3)$$

Here, $\overline{C_d}$ is the time-averaged drag coefficient, F_d is the resistance of a square cylinder, that is, the pressure difference between the stoss side and slip slope of a square cylinder, U is the streamwise velocity, D is the side length of the square cylinder and the characteristic length defining the Reynolds number, T, is the period that includes a certain variation period of the drag coefficient. The Strouhal number S_t is a dimensionless coefficient describing the periodic unsteady flow, which, in this study, was used to reflect the shedding frequency of the vortex from the flow around a square cylinder.

$$S_t = \frac{f_s D}{U} \quad (4)$$

Here, S_t is the Strouhal number, f_s is the shedding frequency of the vortex from the flow around a square cylinder.

By adjusting the velocity at the inlet boundary, the Reynolds number R_e of the flow around a square cylinder was set. The Reynolds number was modified as follows:

$$R_e = \frac{UD}{v} \tag{5}$$

Here R_e is the Reynolds number and v is the kinematic viscosity of the fluid.

The flow around a square cylinder under different grid resolutions (δ/D = 0.125, 0.075, 0.050, 0.025, 0.0125, 0.0075) was simulated where R_e = 40 and 100, and the results were compared. The time-averaged drag coefficient $\overline{C_d}$ and the Strouhal number S_t of the flow around a square cylinder under different grid resolutions are shown in Table 3.

Table 3. Comparison of $\overline{C_d}$ and S_t of the flow around a square cylinder under different grid resolutions.

Grids	δ/D	Total Elements	Re = 40	Re = 1000	
			$\overline{C_d}$	$\overline{C_d}$	S_t
G1	0.125	5444	1.49	1.88	0.117
G2	0.075	15,194	1.54	2.02	0.120
G3	0.050	33,636	1.59	2.17	0.121
G4	0.025	134,544	1.64	2.29	0.121
G5	0.0125	535,076	1.67	2.39	0.122
G6	0.0075	1,487,034	1.69	2.42	0.122

It can be seen from Table 3 that, with the increase in the grid density, the simulation results of the flow around a square cylinder tended to converge. When δ/D was less than 0.0125, with the further reduction of δ/D, the change range of $\overline{C_d}$ and S_t was very small. When R_e = 40, replacing the grid G5 with the finest grid G6 resulted in only a 0.9% change in $\overline{C_d}$. When R_e = 1000, replacing the grid G5 with the finest grid G6 resulted in only a 1.4% change in $\overline{C_d}$ and 0.2% in S_t. Though grid G6 ($\delta/D = 0.0075$) provided the finest simulation results, the computational cost of this grid resolution was too high. Grid G5 ($\delta/D = 0.0125$) was acceptable. Thus, $\delta/D = 0.0125$ was set to define the grid sizes.

2.4. Model Analysis

A specific channel width (H = 12 D) was selected, the numerical results under different velocity inlet conditions were compared, and the rationality of this calculation model was analysed.

Using the flow around a square cylinder under the conditions of R_e = 40, 200, 1000, and 22,000, the earlier research results were compared with the calculation results in this study. The time-averaged drag coefficient $\overline{C_d}$ and the Strouhal number S_t of the flow around a square cylinder under different Reynolds numbers between the present and earlier results are compared in Table 4.

It can be seen from Table 4 that when R_e = 40, 200, 1000, and 22,000, both $\overline{C_d}$ and S_t in this study were in good agreement with the earlier numerical and physical results, indicating that the calculation model established in this study had good effectiveness and accuracy under different R_e values.

Table 4. Comparison of $\overline{C_d}$ and S_t of the flow around a square cylinder under different R_e values between the present and earlier results.

R_e	Source	$\overline{C_d}$	S_t
40	Sen et al. [15]	1.67	-
	Lan et al. [49]	1.72	-
	Present	1.67	-
200	Lan et al. [49]	1.49	0.143
	Gera et al. [50]	-	0.145
	Okajima [51]	1.50	0.141
	Present	1.49	0.141
1000	Lan et al. [49]	2.23	0.122
	Okajima [51]	2.10	0.120~0.130
	Wang [52]	2.40	0.124
	Present	2.39	0.122
22,000	Trias et al. [33]	2.19	0.132
	Lyn [53]	2.18	0.134
	Bouris et al. [54]	2.10	0.135
	Fureby et al. [55]	2.11–2.30	0.126~0.138
	Cao et al. [56]	-	0.130
	Bearman et al. [57]	2.05	0.122
	Lee [58]	2.18	0.132
	Present	2.06	0.124

3. Characteristics of the Flow around a Square Cylinder under Boundary Constraint

3.1. Feature of Wake Flow Patterns of the Flow around a Square Cylinder

The flow pattern distributions around a square cylinder in the four classifications L1, L2, L3, and T (Table 1) are shown in Figure 4.

When R_e was low, the flow around a square cylinder formed a closed wake composed of two attached vortices behind the square cylinder (Classification L1, Classification L2), as shown in Figure 4a,b. Here, the positions of the attached vortices were stable, and they did not appear to be shedding from the triangular area behind the square cylinder and moving downstream. The action range of the vortices was limited to the triangular area behind the square cylinder and the flow field in most other areas of the channel was still laminar flow. When R_e exceeded a specific value, the attached vortices behind the square cylinder appeared to be shedding, and the vortices moved freely in the downstream direction (Classifications L3 and T), as shown in Figure 4c,d. Here, the area behind the square cylinder was the area affected by the eddy current, and its action range was much larger than that of Classifications L1 and L2.

3.2. Boundary Constraint Effect in Vortex Street Flow

Under the laminar vortex street flow pattern (Classification L3), the regular shedding of the attached vortices had an impact on the change of the flow pattern within a specific range behind the square cylinder, so Classification L3 was chosen for study. Quantitative research was carried out on the distribution and motion law of the vortices and the influence of the boundary constraint on the wake flow pattern around a square cylinder.

A dimensionless coefficient gap ratio G/D was defined, which characterized the constraint degree of the boundaries to the flow around a square cylinder. G/D was modified as follows:

$$G/D = \frac{H-D}{2D} \quad (6)$$

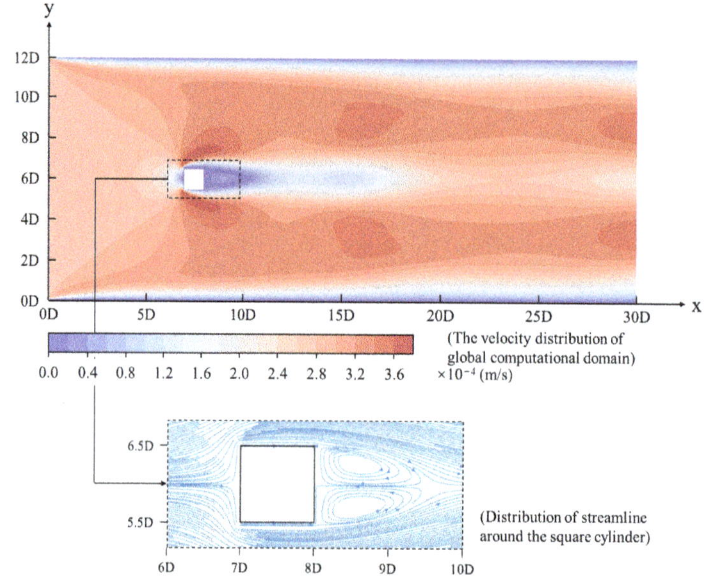

(**a**)

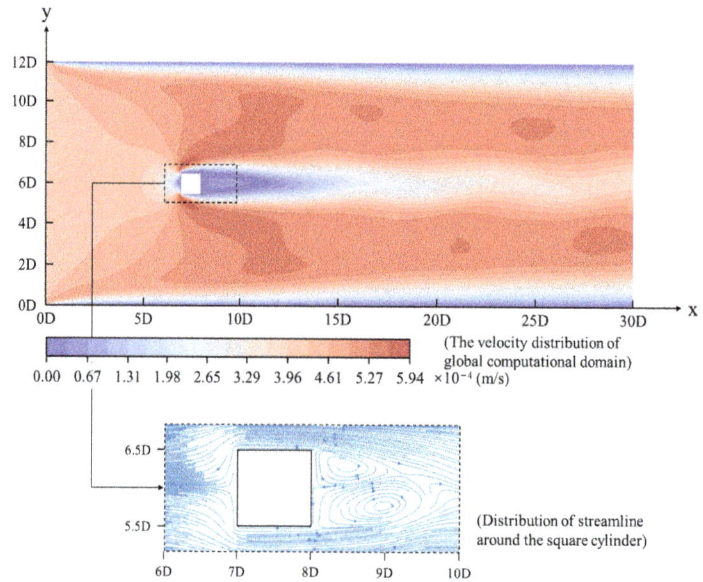

(**b**)

Figure 4. *Cont.*

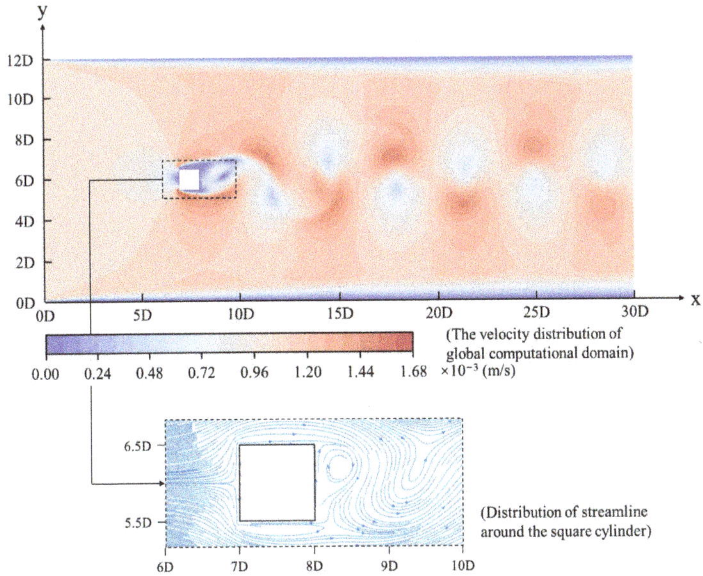

(c)

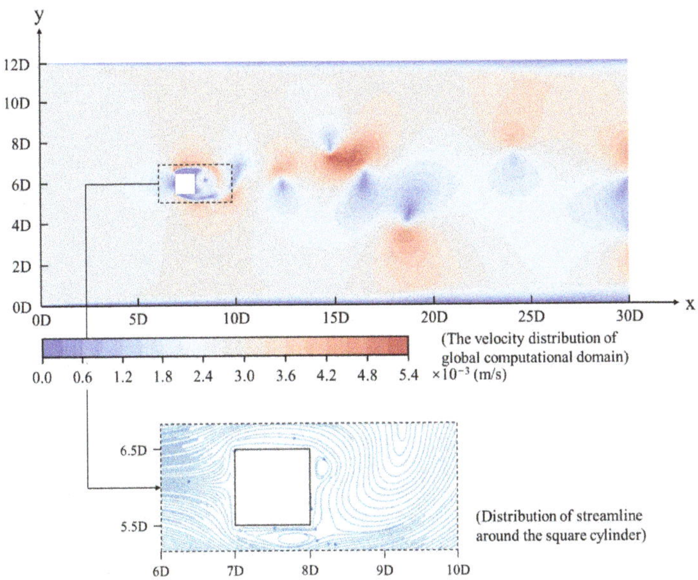

(d)

Figure 4. Different classifications of flow distribution around a square cylinder; (**a**) Classification L1 (R_e = 40); (**b**) Classification L2 (R_e = 95); (**c**) Classification L3 (R_e = 400); and (**d**) Classification T (R_e = 1000).

Here G/D is the gap ratio, H is the width of the rectangular channel, D is the side length of the square cylinder, G is the distance from the side of the square cylinder to the plane boundary of the channel (Figure 2).

According to Section 2.1, when $R_e = 400$, the vortex street can form behind the square cylinder. The flow pattern around a square cylinder, the trajectory of the vortex centres, and the average distance of the vortex centres $\overline{d_x}$ along the flow direction when $G/D = 5.5$, 3.5, 2.5, 1.5, 1.0, and 0.5 are shown in Figures 5–7, respectively.

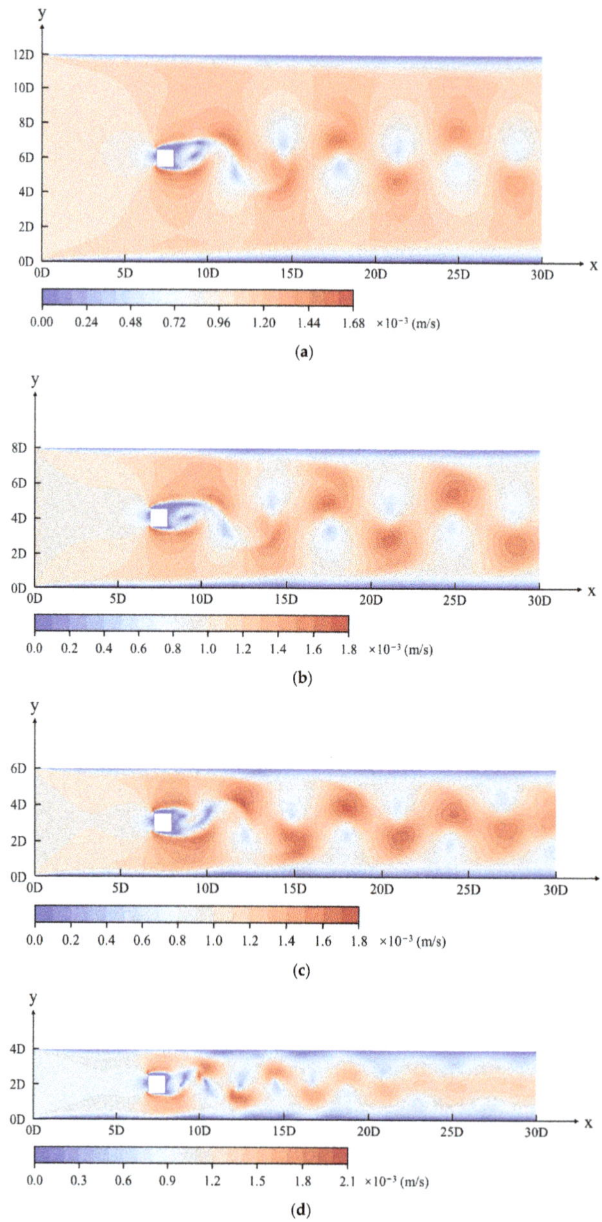

Figure 5. *Cont.*

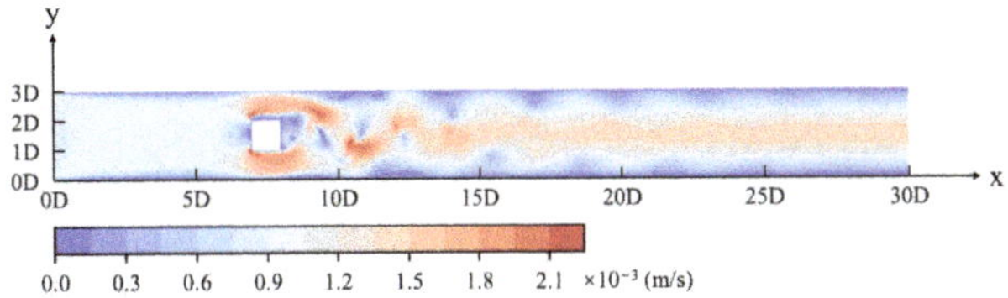

(e)

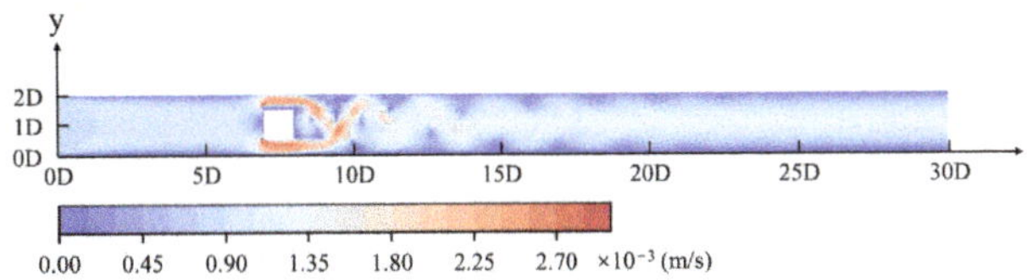

(f)

Figure 5. Flow pattern around a square cylinder under different boundary constraint degrees when $R_e = 400$. (**a**) $G/D = 5.5$; (**b**) $G/D = 3.5$; (**c**) $G/D = 2.5$; (**d**) $G/D = 1.5$; (**e**) $G/D = 1.0$; and (**f**) $G/D = 0.5$.

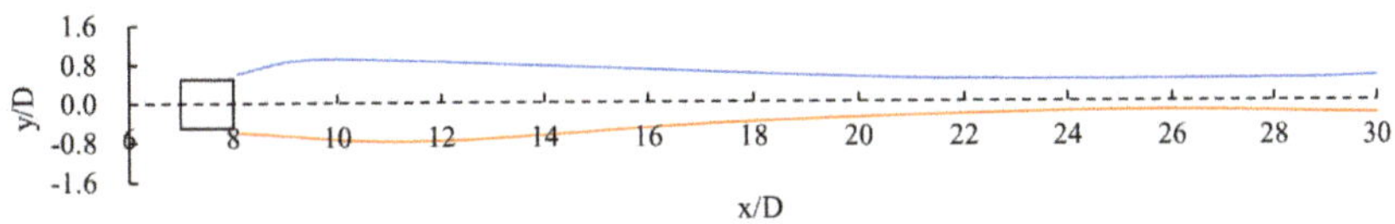

(a)

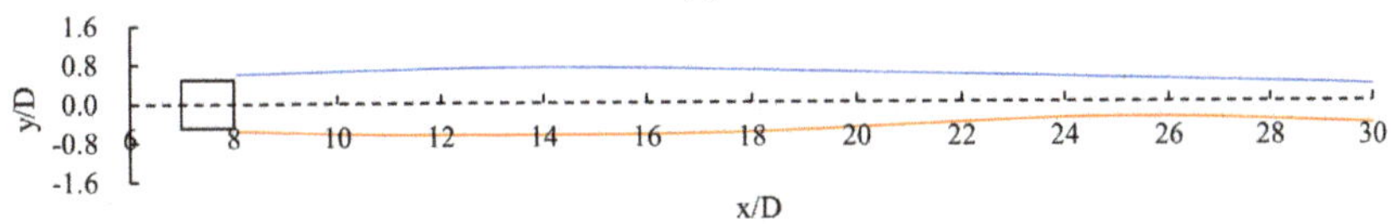

(b)

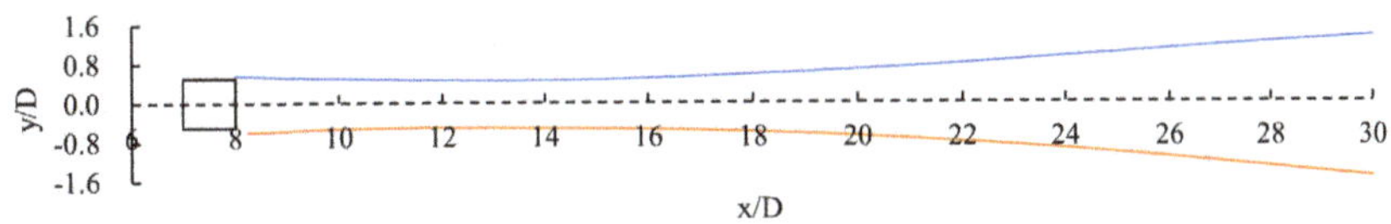

(c)

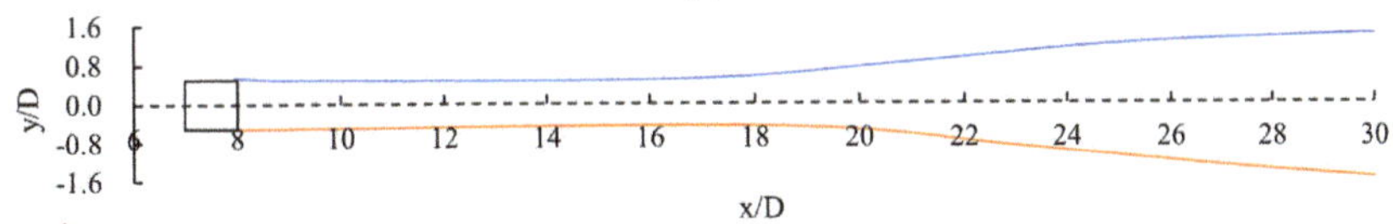

(d)

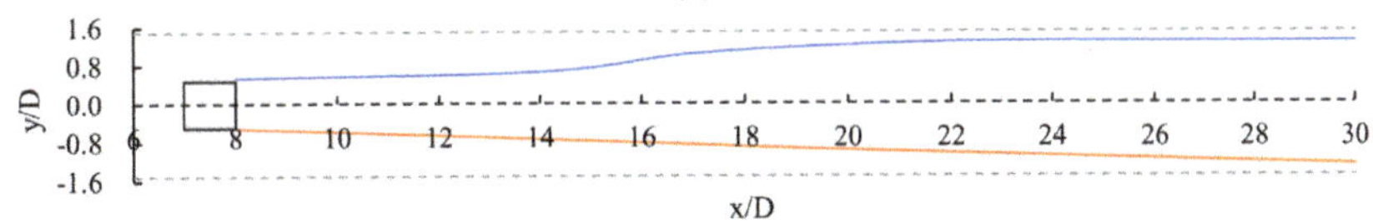

(e)

Figure 6. *Cont.*

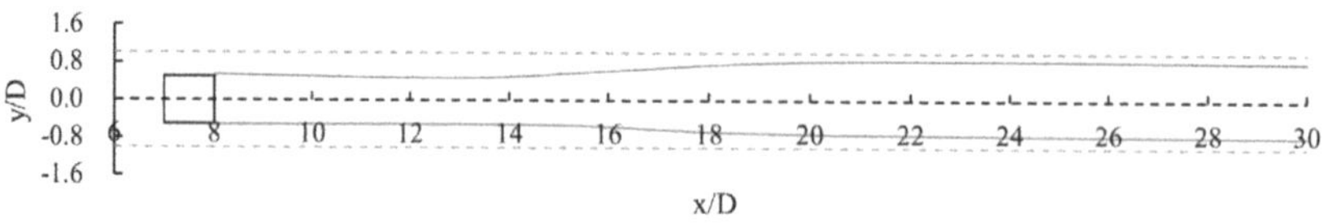

(f)

Figure 6. Trajectory of the vortex centres of the flow around a square cylinder under different boundary constraint degrees when $R_e = 400$. (**a**) G/D = 5.5; (**b**) G/D = 3.5; (**c**) G/D = 2.5; (**d**) G/D = 1.5; (**e**) G/D = 1.0; and (**f**) G/D = 0.5.

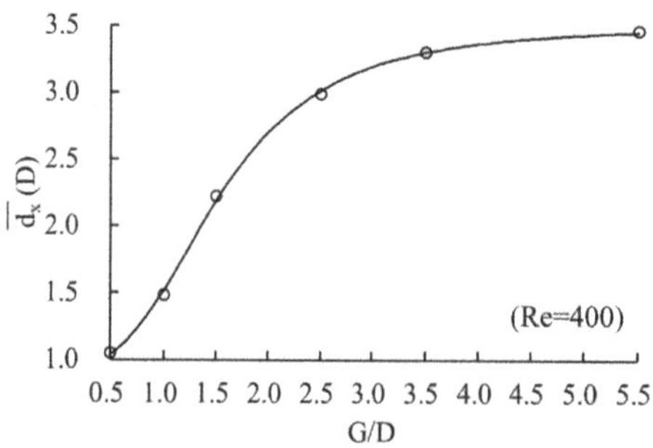

Figure 7. Average distance of the vortex centres of the flow around a square cylinder $\overline{d_x}$ along the flow direction under different boundary constraint degrees when $R_e = 400$.

The constraint degree of the channel boundaries will affect the distribution of the vortex street and the movement of the vortices. When the gap of the channel is large ($G/D \geq 3.5$), the flow pattern of the post-cylinder wake is similar, and the constraint degree of boundaries is weak. Here, the two trajectories of the vortex centres were closer to the axis of the channel. When the gap was small ($G/D < 3.5$), the constraint degree of the boundaries became more obvious, the channel boundaries began to interfere with the shape and distribution of the vortex street, and the degree of interference increased along the flow direction. After extrusion by the constrained boundary, the volume of the vortices decreased, and the two trajectories of the vortex centres gradually migrated to both sides of the channel along the flow direction. As the constraint degree was further enhanced ($0.5 < G/D < 2.5$), the vortices were extruded by the channel boundaries, and its size was further reduced. The vortex centres moved a certain distance near the axis and then migrated to the two sides of the channel. Then, they reached the channel boundaries and moved downstream along them. A low velocity area was formed near the boundaries. The greater the constraint degree of the boundaries, the more forward the starting point of the trajectory migration of the vortex centres. The constraint effect of the boundaries reduced the distance between adjacent vortices in the wake along the flow direction. When $G/D \geq 3.5$, the distance between adjacent vortex centres along the flow direction did not decrease significantly, but when $G/D < 3.5$, the effect of the boundary constraint was increasingly strengthened, and the amplitude reduction of $\overline{d_x}$ increased significantly.

The variation of vortex vorticity Ω along the flow direction with different boundary constraints is shown in Figure 8. Ω was modified as follows:

$$\Omega = 0.5 \times \left(\frac{\partial u_i}{x_j} - \frac{\partial u_j}{x_i} \right) \tag{7}$$

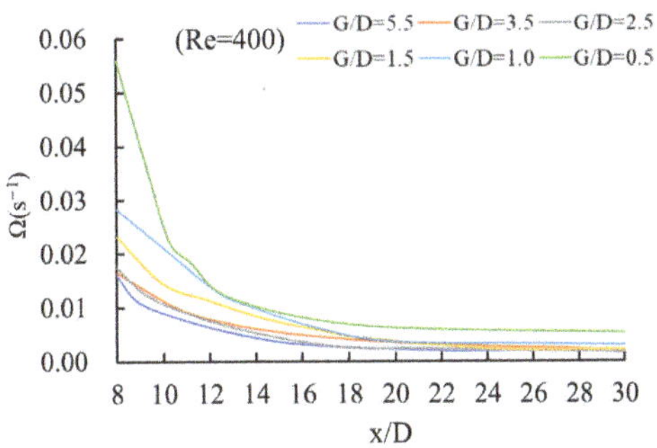

Figure 8. Variation of the vorticity of the flow around a square cylinder along the flow direction under different boundary constraint degrees.

Here, Ω is vortex vorticity.

The vorticity at the initial position of the wake was modified as Ω_0, and the vorticity reaching the outlet of the channel was modified as Ω_1. The relationship between Ω_1/Ω_0 and G/D is shown in Figure 9.

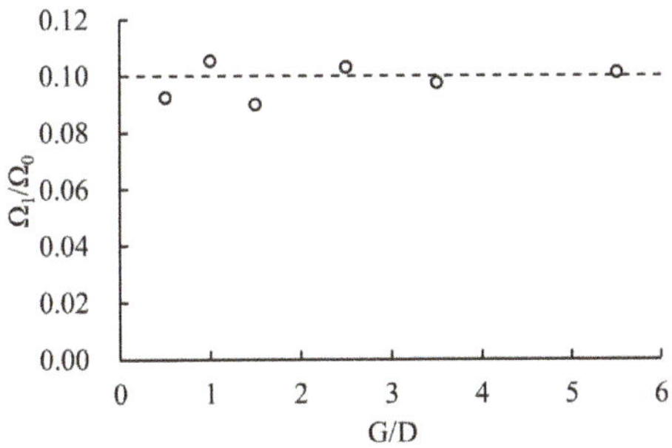

Figure 9. Relationship between Ω_1/Ω_0 and G/D in the vortex street flow pattern.

The vorticity of the vortex street around a square cylinder attenuated along the flow direction, and the boundary constraint affecting only the law of the initial vorticity attenuated along the flow direction. When the vortex moved to the area beyond 18 D behind the square cylinder, the vorticity tended to be stable, and there was no obvious attenuation along the flow direction. Under different boundary constraint degrees, the vorticity attenuated to 10% of the initial vorticity. For the flow around a square cylinder, the final attenuation degree of the vorticity of the vortex street was unrelated to the distribution, shape, and motion law of the vortex street caused by the boundary constraints.

3.3. Fluid Characteristics of the Flow around a Square Cylinder

The time-averaged drag coefficient $\overline{C_d}$ and the Strouhal number S_t were used as the evaluation indexes of the fluid characteristics of the four flow pattern classifications L1, L2, L3, and T in the flow around a square cylinder. The range of the boundary gap ratio G/D was 0.5 to 5.5. The relationship between $\overline{C_d}$ and G/D, R_e when R_e was between 0.005 and 2000 is shown in Figure 10. The relationship between S_t and G/D, R_e when R_e was between 150 and 2000 is shown in Figure 11.

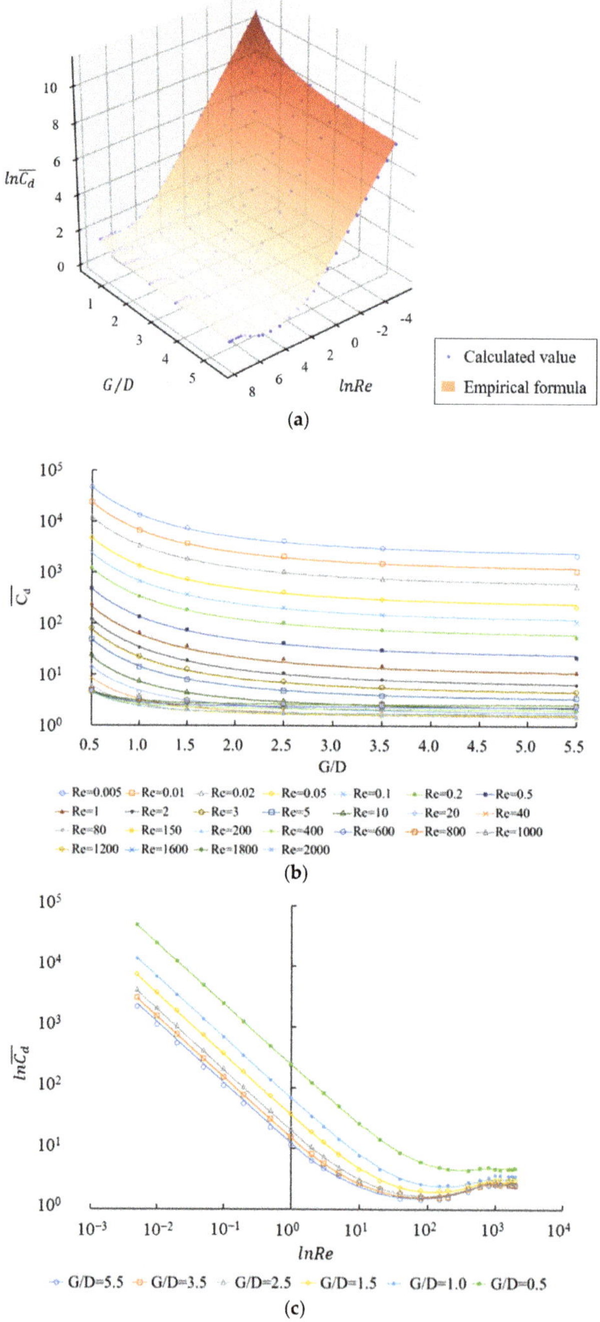

Figure 10. Relationship of $\overline{C_d}$ nd G/D, R_e. (**a**) $ln\overline{C_d}$ versus $lnRe$ and G/D; (**b**) $\overline{C_d}$ versus G/D; and (**c**) $ln\overline{C_d}$ versus $lnRe$.

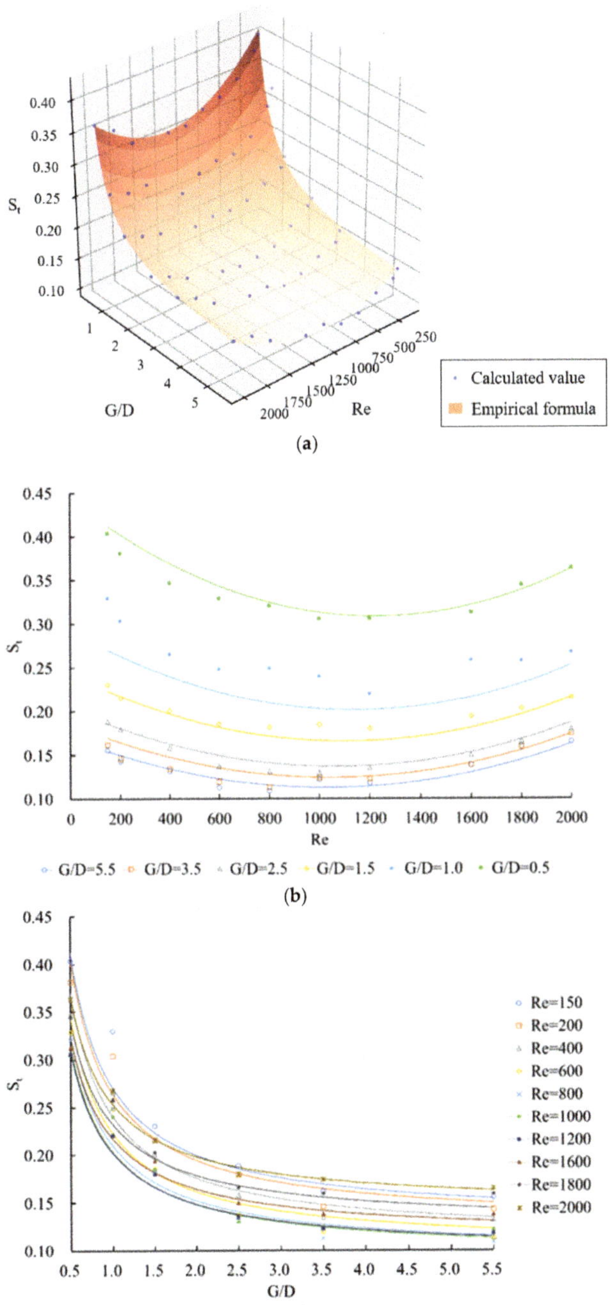

Figure 11. Relationship of $\overline{C_d}$ and G/D, R_e. (**a**) S_t versus Re and G/D; (**b**) S_t versus G/D; and (**c**) S_t versus Re.

The relationship between $\overline{C_d}$ and R_e in the flow around a square cylinder was similar to that of the flow around a cylinder. The relationship between $\overline{C_d}$ and G/D, R_e in Figure 10 is expressed by Equation (8a–c):

$$\overline{C_d} = exp(a) + exp(b)(G/D)^{-2} \tag{8a}$$

$$a = \begin{cases} 2.35 - lnRe & (Re \leq 2.46) \\ 2.07 - 0.77lnRe + 0.09(lnRe)^2 & (2.46 < Re \leq 860) \\ 0.934 & (Re > 860) \end{cases} \tag{8b}$$

$$b = \begin{cases} 4.06 - lnRe & (Re \leq 12.18) \\ 4.94 - 1.62lnRe + 0.12(lnRe)^2 & (12.18 < Re \leq 880) \\ -0.569 & (Re > 880) \end{cases} \tag{8c}$$

Under the four different flow patterns, $\overline{C_d}$ values were all related to the boundary constraint. After the boundary constraint degree was increased, the resistance of the square cylinder was significantly enhanced. In Figure 10a,b and Equation (8a–c), it can be seen that, though R_e was different, the law of $\overline{C_d}$ changing with G/D was consistent, and all flow patterns conformed to the negative square relationship. When $G/D \geq 3.5$, the boundaries were far away from the square cylinder. Its constraint effect was small, and $\overline{C_d}$ tended to $exp(a)$ with the increase in G/D (see Equation (8a–c)). $exp(a)$ is the drag coefficient of the flow around a square cylinder without the constraint effect. When $G/D < 3.5$, the constraint effect of the boundaries was significantly enhanced, and $\overline{C_d}$ grew exponentially with the decrease in G/D. For example, in a turbulence flow pattern, when R_e was 1200 and when $G/D \geq 3.5$, the boundaries were far away from the square cylinder, and $\overline{C_d}$ was close to the value of $exp(a)$ without a constraint effect. Between them, when $G/D = 3.5$, then $\overline{C_d} = 2.59$, which was only 1.8% larger than $exp(a)$ ($exp(a) = 2.54$). Here, the impact of the boundary constraint on $\overline{C_d}$ was minimal. When $G/D < 3.5$, as the boundary constraint was strengthened, $\overline{C_d}$ increased. After $G/D < 1.0$, the distance between the cylinder and the boundaries was smaller than the width of the cylinder, and $\overline{C_d}$ increased significantly. When $G/D = 0.5$, $\overline{C_d}$ reached 4.81.

The relationship between $\overline{C_d}$ and R_e was not affected by the boundary constraint degree. It can be seen in Figure 10a,b that under each G/D, the law of the change of $\overline{C_d}$ with R_e can be divided into three stages. When $R_e \leq 6$, $lnRe$ was linearly related to $ln\overline{C_d}$, and $\overline{C_d}$ decreased with the increase in R_e. When $6 < R_e \leq 1100$, $ln\overline{C_d}$ had a quadratic relation with $lnRe$. $\overline{C_d}$ decreased first and then increased as R_e increased. After $R_e > 1100$, $\overline{C_d}$ did not change with R_e and became a specific value.

The relationship between S_t and G/D, R_e in Figure 11 can also be expressed by Equation (9a–c):

$$S_t = a + b(G/D)^{-1} \tag{9a}$$

$$a = 0.14 - 10^{-4}Re + 5 \times 10^{-8}Re^2 \tag{9b}$$

$$b = 0.15 - 6 \times 10^{-5}Re + 2 \times 10^{-8}Re^2 \tag{9c}$$

It can be seen from Equation (9a–c) that the relationship between S_t and G/D conformed to the hyperbolic functional relationship at each R_e. In Figure 11a,b, when $G/D < 3.5$, the boundary constraints increased the shedding frequency of vortices, and when $G/D < 1.5$, the shedding frequency of vortices increased significantly, and the amplification of S_t increased significantly.

Like $\overline{C_d}$, the relationship between S_t and R_e was also unaffected by the boundary constraint degree. S_t and R_e had a quadratic relation, which is shown in Figure 11a,b and Equation (9a–c). Under each G/D, S_t first decreased and then increased with the elevation of R_e, and reached the minimum value when $R_e = 1100$.

3.4. Flow Pattern Division of Flow around a Square Cylinder

The four flow patterns (L1, L2, L3, T) around a square cylinder under different R_e were all affected by the boundary constraint. When G/D was 0.5~5.5, the dichotomy method was used to locate the R_e range corresponding to each wake flow pattern of the flow around a square cylinder under a different boundary constraint degree, and the flow pattern classification map of the flow around a square cylinder is shown in Figure 12.

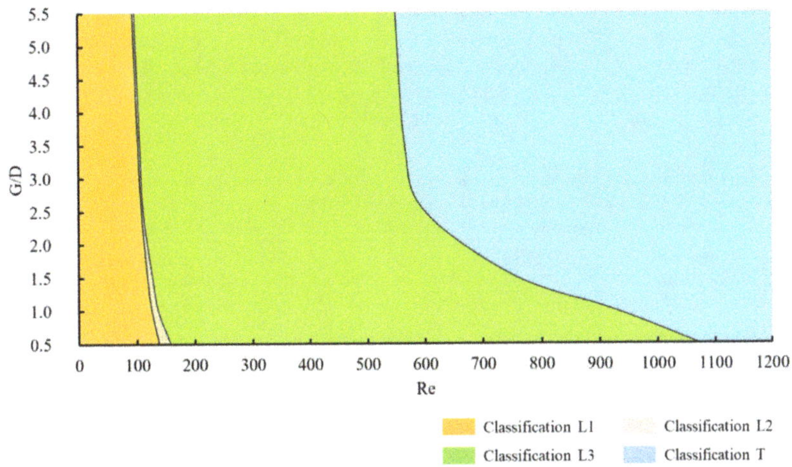

Figure 12. Flow pattern classification map of the flow around a square cylinder under boundary constraint.

The increase in the boundary constraint led to the appearance of Classification T (turbulence phase), which required a larger R_e, and the R_e ranges corresponding to classifications L1 (stable attached vortex phase), L2 (unstable attached vortex phase), and L3 (laminar vortex street phase) also expanded. When $G/D < 2.5$, the increase in the constraint degree resulted in a significant expansion of the R_e range corresponding to Classification L3. As seen in Figures 4d, 5 and 6, after being constrained by boundaries, the steady state of the staggered distribution and regular movement downstream of the vortex streets in the original turbulence flow pattern was destroyed, and vortices were shed from the square cylinder and randomly diverged downstream to both sides of the boundaries. The vortices of the flow around a square cylinder were constrained by the channel boundaries. The divergence of vortices to both sides was blocked, and they could only move downstream. The relative stability of the flow around a square cylinder was maintained, which caused the appearance of Classification T, requiring a larger R_e. Thus, the boundary constraint had a rectification effect on the flow around a square cylinder, which improved the stability of the flow pattern and moderated the occurrence of turbulence.

4. Conclusions

In the present study, numerical simulation was used to study the flow characteristics around a square cylinder under different boundary constraints. The major results are summarized as follows:

(1) The flow pattern around a square cylinder can be divided into four types under different Reynolds numbers: L1 (stable attached vortex phase), L2 (unstable attached vortex phase), L3 (laminar vortex street phase), and T (turbulence phase). The boundary constraint does not change the flow pattern around a square cylinder.

(2) In the flow pattern of a laminar vortex street, the boundary constraint squeezes the vortex shapes, pushes the vortex centres to the channel boundaries, and reduces the distance between the vortices along the flow direction, but the boundary con-

straint cannot change the attenuation law and magnitude of the vorticity along the flow direction.

(3) When the constraint degree of boundaries exceeds a specific level ($G/D < 3.5$), the time-averaged drag coefficient of the cylinder and the shedding frequency of the attached vortices can increase significantly.

(4) The variation law of the drag coefficient of the flow around a square cylinder and the shedding frequency of attached vortices with the Reynolds number will not be affected by the boundary constraint degree.

(5) The boundary constraint rectifies the wake flow pattern around a square cylinder, and the appearance of turbulence requires a larger Reynolds number. With the increase of boundary constraint degree, the expansion of the of Reynolds number range corresponding to the flow pattern of the laminar vortex street is the most significant.

Author Contributions: Conceptualization, Z.X.; methodology, Z.X.; software, Z.X.; validation, Z.X., F.W., A.G. and W.Z.; formal analysis, Z.X.; investigation, Z.X.; resources, S.W. and X.W.; data curation, Z.X.; writing—original draft preparation, Z.X.; writing—review and editing, Z.X., S.W. and W.X.; visualization, Z.X.; supervision, S.W. and X.W.; project administration, S.W. and X.W.; funding acquisition, S.W. and X.W. All authors have read and agreed to the published version of the manuscript.

Funding: This research was funded by the National Natural Science Foundation of China, grant number 51909169; and the Research Foundation for Basic Research (Natural Science Foundation) of Jiangsu Province, China, grant number SBK2019042181.

Institutional Review Board Statement: Not applicable.

Informed Consent Statement: Not applicable.

Data Availability Statement: Not applicable.

Conflicts of Interest: The authors declare no conflict of interest.

References

1. Williamson, C.H.K. Defining a universal and continuous Strouhal–Reynolds number relationship for the laminar vortex shedding of a circular cylinder. *Phys. Fluids* **1988**, *31*, 2742–2744. [CrossRef]
2. Wang, Y.L.; Liu, Y.Z.; Miao, G.P. Three-Dimensional Numerical Simulation of Viscous Flow around Circular Cylinder. *J. Shanghai Jiaotong Univ.* **2001**, *2001*, 1461–1469.
3. Behara, S.; Mittal, S. Wake transition in flow past a circular cylinder. *Phys. Fluids* **2010**, *22*, 114104. [CrossRef]
4. Taneda, S. Experimental investigation of the wakes behind cylinders and plates at low Reynolds numbers. *J. Phys. Soc. Jpn.* **1956**, *11*, 302–307. [CrossRef]
5. Coutanceau, M.; Bouard, R. Experimental determination of the main features of the viscous flow in the wake of a circular cylinder in uniform translation. Part 1. Steady flow. *J. Fluid Mech.* **1977**, *79*, 231–256. [CrossRef]
6. Thoman, D.C.; Szewczyk, A.A. Time-dependent viscous flow over a circular cylinder. *Phys. Fluids* **1969**, *12*, II-76–II-86. [CrossRef]
7. Collins, W.M.; Dennis, S.C.R. Flow past an impulsively started circular cylinder. *J. Fluid Mech.* **1973**, *60*, 105–127. [CrossRef]
8. Coutanceau, M.; Defaye, J.R. Circular cylinder wake configurations: A flow visualization survey. *Apply Mech. Rev.* **1991**, *44*, 255–305. [CrossRef]
9. Wu, M.H.; Wen, C.Y.; Yen, R.H.; Weng, M.C.; Wang, A.B. Experimental and numerical study of the separation angle for flow around a circular cylinder at low Reynolds number. *J. Fluid Mech.* **2004**, *515*, 233–260. [CrossRef]
10. Couder, Y. The observation of a shear flow instability in a rotating system with a soap membrane. *J. De Phys. Lett.* **1981**, *42*, 429–431. [CrossRef]
11. Chomaz, J.M. The dynamics of a viscous soap film with soluble surfactant. *J. Fluid Mech.* **2001**, *442*, 387–409. [CrossRef]
12. Beizaie, M.; Gharib, M. Fundamentals of a liquid (soap) film tunnel. *Exp. Fluids* **1997**, *23*, 130–140. [CrossRef]
13. Kim, N.; Kim, H.; Park, H. An experimental study on the effects of rough hydrophobic surfaces on the flow around a circular cylinder. *Phys. Fluids* **2015**, *27*, 85113. [CrossRef]
14. Palau-Salvador, G.; Stoesser, T.; Fröhlich, J.; Kappler, M.; Rodi, W. Large eddy simulations and experiments of flow around finite-height cylinders. *Flow Turbul. Combust.* **2010**, *84*, 239–275. [CrossRef]
15. Sen, S.; Mittal, S.; Biswas, G. Flow past a square cylinder at low Reynolds numbers. *Int. J. Numer. Methods Fluids* **2011**, *67*, 1160–1174. [CrossRef]
16. Tong, B.; Zhu, B.; Zhou, B.K. Numerical Simulation of Flow around Square Cylinder. *Chin. Q. Mech.* **2002**, *23*, 77–81.

17. Yang, F.; Zhou, Z.; Tang, G.; Lu, L. Steady flow around a square cylinder near a plane boundary. *Ocean Eng.* **2021**, *222*, 108599. [CrossRef]
18. Prsic, M.A.; Ong, M.C.; Pettersen, B.; Myrhaug, D. Large Eddy Simulations of flow around a circular cylinder close to a flat seabed. *Mar. Struct.* **2016**, *46*, 127–148. [CrossRef]
19. Chen, L.F.; Wu, G.X. Boundary shear flow past a cylinder near a wall. *Appl. Ocean Res.* **2019**, *92*, 101923. [CrossRef]
20. Lei, C.; Cheng, L.; Kavanagh, K. Re-examination of the effect of a plane boundary on force and vortex shedding of a circular cylinder. *J. Wind Eng. Ind. Aerodyn.* **1999**, *80*, 263–286. [CrossRef]
21. Nishino, T.; Roberts, G.T.; Zhang, X. Vortex shedding from a circular cylinder near a moving ground. *Phys. Fluids* **2007**, *19*, 25103. [CrossRef]
22. Lei, C.; Cheng, L.; Armfield, S.W.; Kavanagh, K. Vortex shedding suppression for flow over a circular cylinder near a plane boundary. *Ocean Eng.* **2000**, *27*, 1109–1127. [CrossRef]
23. Khabbouchi, I.; Guellouz, M.S.; Nasrallah, S.B. A study of the effect of the jet-like flow on the near wake behind a circular cylinder close to a plane wall. *Exp. Therm. Fluid Sci.* **2013**, *44*, 285–300. [CrossRef]
24. Durao, D.F.G.; Gouveia, P.S.T.; Pereira, J.C.F. Velocity characteristics of the flow around a square cross section cylinder placed near a channel wall. *Exp. Fluids* **1991**, *11*, 341–350. [CrossRef]
25. Martinuzzi, R.J.; Bailey, S.C.C.; Kopp, G.A. Influence of wall proximity on vortex shedding from a square cylinder. *Exp. Fluids* **2003**, *34*, 585–596. [CrossRef]
26. Bailey, S.C.C.; Kopp, G.A.; Martinuzzi, R.J. Vortex shedding from a square cylinder near a wall. *J. Turbul.* **2002**, *3*, 3. [CrossRef]
27. Bearman, P.W.; Zdravkovich, M.M. Flow around a circular cylinder near a plane boundary. *J. Fluid Mech.* **1978**, *89*, 33–47. [CrossRef]
28. Grass, A.J.; Raven, P.W.J.; Stuart, R.J.; Bray, J.A. The influence of boundary layer velocity gradients and bed proximity on vortex shedding from free spanning pipelines. *J. Energy Resour. Technol.* **1984**, *106*, 70–78. [CrossRef]
29. Shi, L.L.; Liu, Y.Z.; Wan, J.J. Influence of wall proximity on characteristics of wake behind a square cylinder: PIV measurements and POD analysis. *Exp. Therm. Fluid Sci.* **2010**, *34*, 28–36. [CrossRef]
30. Zhou, J.; Qiu, X.; Li, J.; Liu, Y. Vortex evolution of flow past the near-wall circular cylinder immersed in a flat-plate turbulent boundary layer. *Ocean Eng.* **2022**, *260*, 112011. [CrossRef]
31. Price, S.J.; Sumner, D.; Smith, J.G.; Leong, K.; Païdoussis, M.P. Flow visualization around a circular cylinder near to a plane wall. *J. Fluids Struct.* **2002**, *16*, 175–191. [CrossRef]
32. Zhang, Z.; Ji, C.; Alam, M.M.; Xu, D. DNS of vortex-induced vibrations of a yawed flexible cylinder near a plane boundary. *Wind Struct.* **2020**, *30*, 465–474. [CrossRef]
33. Trias, F.X.; Gorobets, A.; Oliva, A. Turbulent flow around a square cylinder at Reynolds number 22,000: A DNS study. *Comput. Fluids* **2015**, *123*, 87–98. [CrossRef]
34. Dong, S.; Karniadakis, G.E.; Ekmekci, A.; Rockwell, D. A combined direct numerical simulation–particle image velocimetry study of the turbulent near wake. *J. Fluid Mech.* **2006**, *569*, 185–207. [CrossRef]
35. Behera, S.; Saha, A.K. Characteristics of the flow past a wall-mounted finite-length square cylinder at low Reynolds number with varying boundary layer thickness. *J. Fluids Eng.* **2019**, *141*, 1–17. [CrossRef]
36. Ji, C.N.; Zhang, Z.M.; Alam, M.M.; Xu, D. Three-dimensional DNS of vortex-induced vibrations of an inclined flexible cylinder near a plane boundary. In Proceedings of the 2019 World Congress on Advances in Structural Engineering and Mechanics, Jeju Island, Republic of Korea, 17–21 September 2019.
37. Rodi, W.; Ferziger, J.H.; Breuer, M.; Pourquié, M. Status of large eddy simulation: Results of a workshop. *Trans. -Am. Soc. Mech. Eng. J. Fluids Eng.* **1997**, *119*, 248–262. [CrossRef]
38. Voke, P.R. Flow past a square cylinder: Test case LES2. In *Direct and Large-Eddy Simulation II*; Springer: Dordrecht, The Netherlands, 1997; pp. 355–373. [CrossRef]
39. Kahil, Y.; Benhamadouche, S.; Berrouk, A.S.; Afgan, I. Simulation of subcritical-Reynolds-number flow around four cylinders in square arrangement configuration using LES. *Eur. J. Mech.-B/Fluids* **2019**, *74*, 111–122. [CrossRef]
40. Rodi, W. Comparison of LES and RANS calculations of the flow around bluff bodies. *J. Wind Eng. Ind. Aerodyn.* **1997**, *69*, 55–75. [CrossRef]
41. Grigoriadis, D.G.E.; Bartzis, J.G.; Goulas, A. LES of the flow past a rectangular cylinder using the immersed boundary concept. *Int. J. Numer. Methods Fluids* **2003**, *41*, 615–632. [CrossRef]
42. Khan, N.B.; Ibrahim, Z.; Bin Mohamad Badry, A.B.; Jameel, M.; Javed, M.F. Numerical investigation of flow around cylinder at Reynolds number = 3900 with large eddy simulation technique: Effect of spanwise length and mesh resolution. *Proc. Inst. Mech. Eng. Part M J. Eng. Marit. Environ.* **2019**, *233*, 417–427. [CrossRef]
43. Kanaris, N.; Grigoriadis, D.; Kassinos, S. Three dimensional flow around a circular cylinder confined in a plane channel. *Phys. Fluids* **2011**, *23*, 64106. [CrossRef]
44. Norberg, C. Flow around rectangular cylinders: Pressure forces and wake frequencies. *J. Wind Eng. Ind. Aerodyn.* **1993**, *49*, 187–196. [CrossRef]
45. Blackburn, H.M.; Henderson, R.D. A study of two-dimensional flow past an oscillating cylinder. *J. Fluid Mech.* **1999**, *385*, 255–286. [CrossRef]
46. Sadeque, M.A.; Rajaratnam, N.; Loewen, M.R. Flow around cylinders in open channels. *J. Eng. Mech.* **2008**, *134*, 60–71. [CrossRef]

47. Yuce, M.I.; Kareem, D.A. A numerical analysis of fluid flow around circular and square cylinders. *J. -Am. Water Work. Assoc.* **2016**, *108*, E546–E554. [CrossRef]
48. Zaid, M.; Yazdanfar, Z.; Chowdhury, H.; Alam, F. Numerical modeling of flow around a pier mounted in a flat and fixed bed. *Energy Procedia* **2019**, *160*, 51–59. [CrossRef]
49. Lan, X.J.; Zhao, W.W.; Wan, D.C.; Zou, L. Numerical simulation of low-Re flow around a square cylinder by MPS method. *Ocean Eng.* **2002**, *40*, 83–92. [CrossRef]
50. Gera, B.; Sharma, P.K. CFD analysis of 2D unsteady flow around a square cylinder. *Int. J. Appl. Eng. Res.* **2010**, *1*, 602.
51. Okajima, A. Strouhal numbers of rectangular cylinders. *J. Fluid Mech.* **1982**, *123*, 379–398. [CrossRef]
52. Wang, J.C. Direct Numerical Simulation of Medium Reynolds Flow Past Square Cylinder. Master's Thesis, China Ship Research and Development Academy, Beijing, China, 2016.
53. Lyn, D.A. Phase-averaged turbulence measurements in the separated shear layer region of flow around a square cylinder. In Proceedings of the 23rd Congress of the International Association for Hydraulic Research, Ottawa, ON, Canada, 21–25 August 1989.
54. Bouris, D.; Bergeles, G. 2D LES of vortex shedding from a square cylinder. *J. Wind Eng. Ind. Aerodyn.* **1999**, *80*, 31–46. [CrossRef]
55. Fureby, C.; Tabor, G.; Weller, H.G.; Gosman, A.D. Large eddy simulations of the flow around a square prism. *AIAA J.* **2000**, *38*, 442–452. [CrossRef]
56. Cao, Y.; Tamura, T. Large-eddy simulations of flow past a square cylinder using structured and unstructured grids. *Comput. Fluids* **2016**, *137*, 36–54. [CrossRef]
57. Bearman, P.W.; Obasaju, E.D. An experimental study of pressure fluctuations on fixed and oscillating square-section cylinders. *J. Fluid Mech.* **1982**, *119*, 297–321. [CrossRef]
58. Lee, B.E. The effect of turbulence on the surface pressure field of a square prism. *J. Fluid Mech.* **1975**, *69*, 263–282. [CrossRef]

water

MDPI

Review

Technical Challenges of Safety Emergency Drawdown for High Dam and Large Reservoir Project

Xueyu Zheng [1,2], Luchen Zhang [2], Jiaxiu Yang [1,*], Shuaiqun Du [1], Shiqiang Wu [2] and Shaoze Luo [2]

1 Power China Guiyang Engineering Corporation Limited, Guiyang 550081, China; dusq66@163.com (S.D.)
2 State Key Laboratory of Hydrology-Water Resources and Hydraulic Engineering, Nanjing Hydraulic Research Institute, Nanjing 210029, China
* Correspondence: yjs6308@126.com; Tel.: +86-18985190890

Abstract: With the development of a 300 m high dam and large reservoir construction, the emergency drawdown capacity of cascade reservoirs, especially high dams, has become a hot issue of concern to all sectors of society while giving play to huge comprehensive benefits. Based on a thorough investigation of the current situation of drawdown facilities for high dams and large reservoirs with a height of 200 m or more in the world, this paper finds that drawdown facilities currently face difficulties such as insufficient drawdown capacity, poor safety and stability of high head structures, extremely high lift hoisting equipment, and high difficulty in high head water seal technology. It is pointed out that the key technologies that need to be urgently addressed for a deep drawdown of high dams and large reservoirs are the pressure-bearing capacity of gates and the capacity limit of hoists. As a result, the elevation of the bottom tunnel of the drawdown building cannot be arranged and the orifice is limited, and the drawdown depth and discharge capacity are limited.

Keywords: high dam and large reservoir; security; technical for reservoir drawdown; drawdown capacity; deep drawdown theory

1. Introduction

High dams and large reservoirs play an irreplaceable role in ensuring social and economic development and national water resource security, and their security is related to national major public security. With the intensification of the impact of human activities, global climate change, and the frequent occurrence of natural disasters, the problems of reservoir dam-break and landslide dam-break have become increasingly prominent [1]. Effectively ensuring the safety of hydropower projects and cascade hydropower stations in the basin is of far-reaching and significant significance to ensuring national energy security and the safety of human life and property. The Basin Hydropower Emergency Plan and Requirements, which is being prepared by the National Development and Reform Commission and the Energy Administration of the People's Republic of China, clearly stipulates the principles of emergency work for basin hydropower emergencies and the requirements for emergency dispatching and security and requires that cascade hydropower projects should have the emergency venting capacity and should regularly repair and maintain the flood discharge and venting facilities. When the cascade hydropower project encounters such emergencies as excessive inflow flood in the upstream river channel, inflow flood in the upstream cascade or reservoir landslide, a flood caused by the dangerous situation of the backwater building itself, and war and terrorist attacks on the project, the reservoir capacity shall be adjusted in time according to the requirements of the government emergency management department. The requirements for emergency drawdown capacity of high dams and large reservoirs are written into the specification, which shows the importance and necessity of emergency venting capacity of reservoir dams.

The operation and maintenance experience of water conservancy and hydropower projects shows that the drawdown facilities play a significant role in reducing the incidence of engineering accidents, reducing the impact of accidents and disasters, and improving the safe operation of the project. For example, in China, after an earthquake, the Zipingpu concrete face dam was damaged. The upstream water level of the dam was reduced in time through the flushing and venting tunnels and the flood discharge tunnel, which created construction conditions for the repair of the dam's concrete slab and water stop system. And Seepage occurred downstream of the Qizishan dam with the flow rate reached 1170 L/s. From March to July 2002, the emptying tunnel was used to withdraw the reservoir for treatment. The operation has been normal so far, which has played an important role in the rapid inspection and treatment of basic leakage. Others, such as Sayanshushensk in Russia, Kolbrien and Zillergrund in Austria, Zezier and Sacred Maria in Switzerland, and El Ataza in Spain, also occurred on the heel. In the case of cracking, the emptying of the reservoir provides favorable conditions for engineering maintenance.

According to incomplete statistics, there are relatively few venting facilities for rigid dams in China, and high dams generally have bottom outlets for flood discharge, such as Xiaowan, Guangzhao, Ertan, etc. There are many examples of setting up emptying facilities for rock-fill dams. Medium and low dams generally meet the emptying requirements by combining spillway tunnels and diversion tunnels; high dams are generally equipped with special emptying facilities. Investigation of the relevant parameters of the deep-hole arc gates for hydropower projects at home and abroad shows that the domestic leader in various indicators is the second-level concrete double-curved arch dam of Dongjiang Hydropower Station. The maximum dam height is 157 m, the working gate of the emptying tunnel has a water head of 123.40 m, and the size of the orifice is 6.4 m × 7.5 m (width × height). The total horizontal static water pressure is 5743 kN, and the maximum flow is 1625.00 m^3/s. The leading foreign index is India's Tarbela. The earth-rock dam has an inclined core, and the maximum dam height is 143 m. The water head of the working door of the emptying tunnel is 135.60 m, the orifice size is 4.88 m × 7.3 m (width × height), the total horizontal static water pressure is 4697 kN, and the maximum flow rate is 1260.00 m^3/s. As the layout depth of the vent hole increases, the size of the orifice of the arc gate decreases and the flow rate decreases. The layout depth of the prior art venting buildings is generally about 120 m, and the small orifice size results in limited leakage, which can only be vented during specific periods of the dry season, and the venting depth is limited and can only be used for routine maintenance and overhaul of the project.

In short, the scientific and reasonable setting of venting facilities will not only provide the dam with maintenance conditions but also improve the emergency response capability of the dam, avoid major disasters, and ensure the safety of people's lives and property. However, the emergency response capacity of the reservoir dam emptying facility designed according to the existing specifications is insufficient and does not fully meet the requirements of national emergency management. The main manifestations are the inability to withdraw all the time, the limited drawdown volume, the insufficient drawdown depth, the limited drawdown storage capacity, and so on.

2. Method

On the basis of researching a large amount of engineering data, the research method in this article is shown in Figure 1:

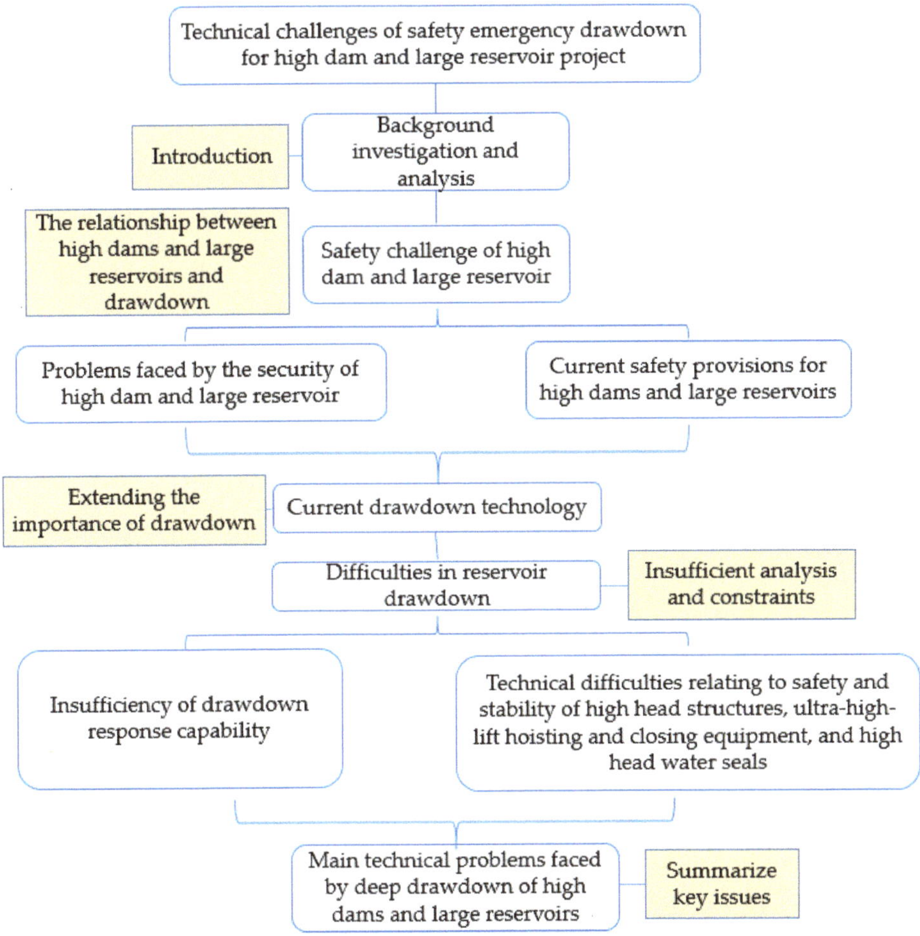

Figure 1. Method flowchart.

3. Safety Challenge of High Dam and Large Reservoir

3.1. Problems Faced by the Security of High Dam and Large Reservoir

Global climate change and the increase in extreme weather, earthquakes, landslides, debris flows, and other natural disasters lead to the damage of flood discharge facilities, dam overtopping, dam collapse, and other accidents in some hydropower projects, which seriously affect the safety of people's lives and property and public safety, e.g., the failure of the dam in Zhumadian, China, in 1975; the landslide surge generated by the Cascade Mountains in the United States in 1966; the swell generated by the Nosteyuko River in Canada in 1983, which caused severe damage, including the collapse of the downstream barrier dam; the formation of the Hongshiyan barrier lake, caused by the Ludian earthquake in Yunnan, China, in 2014; the Baige barrier lake on the Jinsha River, caused by a landslide in the Tibetan Autonomous Region of China in 2018; and the dam failure in the Sardoba Reservoir in Uzbekistan in 2020. All these examples caused a huge loss of life and property and undermined the safety of local people.

Large reservoirs with high dams are generally the leading reservoirs in a basin, and their structural design is usually based on natural floods and check floods. However, in recent years, with the emergence of non-natural extreme flooding, it has become clear that most high dam reservoirs are lacking in emergency response capacity. When a major incident occurs in a cascade reservoir group, there are severe potential safety hazards which may even trigger a disastrous chain reaction through successive reservoirs. Therefore, safety provision in terms of emergency emptying capacity in cascade reservoirs has become a matter of widespread concern; the improvement of the emergency capacity of high dam and large reservoir projects is now an urgent priority requiring immediate attention.

3.2. Current Safety Provisions for High Dams and Large Reservoirs

At present, the safety guarantee measures of high dam and large reservoir projects are mainly to deal with sudden adverse conditions such as non-standard floods, landslide surges, earthquakes, etc., by setting up relevant water withdrawal structures and comprehensively evaluating whether the overall operation status of the building is safe and supplemented by special underwater detection, regular maintenance, and other work. The traditional underwater detection method cannot meet the underwater detection and maintenance requirements of high dams and large reservoirs due to its low efficiency, high cost, high safety risk, and limited coverage of underwater work within 60 m. There are a few projects in China that attempt to use underwater robots or manned submersibles to detect the high dam reservoir [2], but the existing deep-water defect detection and intelligent operation underwater robots (ROV) and miniaturized manned diving equipment [3] can only adapt to the water depth of 100 m, far from meeting the requirements for the repair and maintenance of 300 m high dam. In addition, the existing underwater detection means are mainly to inspect the surface of the building, and the types of NDT equipment that can be carried are also extremely limited. The acquired detection data can not only reach the depth of comprehensive evaluation of the building's operation status but also cannot play the role of emergency elimination. Therefore, where large reservoir capacity, wide water area, and deep-water levels prevail, existing methods of detection cannot meet operational safety needs, and emptying facilities take on particular importance.

Years of operational and maintenance experience also show that emptying facilities play a significant role in reducing the occurrence and impact of accidents and improving the assurance of a project's operational safety. For example, the deep outlets in the Zipingpu Dam, Zhushuqiao Panel Dam, Qiezi Mountain, and Meishan Reservoir projects in China play an important role in facilitating rapid inspection and the treatment of problems. When damage occurs, the reservoir can be emptied promptly, and maintenance conditions can be provided to ensure the continuing safety of the project. In 2018, China's "11.03" Baige barrier lake emergency rescue took "about 1.3 billion m^3 of storage capacity to absorb dam break flood" in emergency treatment measures [4,5], which successfully averted the danger. The arch dam of Kölnbrein in Austria, with cracks in the heel, was also reinforced by a number of drainage reservoirs. However, the early Baiyun Hydropower Station, Sanbanxi Hydropower Station, Zhushuqiao Panel Dam, and other projects in China cannot be overhauled or maintained because of a lack of drainage facilities; consequently, problems with these dams cannot be adequately addressed, and their operation continues to be compromised [6,7]. The Oroville Dam in the United States is also not equipped with venting facilities [8]. In February 2017, due to the large inflow, the reservoir water level rose sharply, and there was no time to discharge. The spillway was damaged, resulting in major losses. So, high dams and large reservoirs lack effective means to respond to extreme events and still face great hidden dangers. With the continuous development of dam height, the existing drawdown technology has been difficult to cope with the challenges faced by high dams and large reservoirs.

4. Current Drawdown Technology

Emptying facilities are defined in this paper as structures that can lower the reservoir water level. These principally include spillways, spillway tunnels, and discharge cavities in embankment dams, surface outlets, mid-level outlets, low-level (bottom) outlets, and other drainage structures in concrete dams. These can be categorized by function, type, and layout. In terms of function, specialized emptying facilities include emptying tunnels and low-level (bottom) outlets, that is, those facilities intended for emptying the water in the reservoir as far as possible. In terms of type, current usage includes more pressure holes, no pressure holes with pressure, and full pressure holes; in terms of layout on the facade, there are "dragon head up" type, "dragon tail down" type as well as steep groove type, deep-hole type/straight line type, and swirling vertical shaft type.

Current emptying facilities in a representative worldwide range of dams over 200 m in height are summarized in Table 1.

Table 1. Characteristics of the emptying facilities in a representative selection of high dam projects.

No.	Name	Country	Type	Height (m)	Capacity (bn·m^3)	Facilities for Emptying or Lowering Water Levels
1	ShuangJiangKou	China	ER	314	31.15	1 emptying outlet
2	Nurek [9]	Tajikistan	TE	300	105	none
3	LiangHeKou	China	ER	295	101.54	1 emptying outlet
4	XiaoWan	China	VA	294.5	150	2 low-level emptying outlets
5	Inguri [10]	Georgia	VA	272	11.1	7 low-level outlets
6	NuoZhaDu	China	ER	261.5	237.03	none
7	Chicoasen [11]	Mexico	TE	261	16.1	none
8	Deriner	Turkey	VA	249	19.69	8 bottom outlets
9	Sayano-Shushenskoe Dam [12]	Russia	VA/PG	245	313	1 guide bottom outlet
10	ChangHe Dam	China	ER	240	10.75	1 discharge outlet
11	La Esmeralda Dam [13]	Colombia	ER	237	8.2	spillway
12	Oroville [14]	America	TE/ER	234	43.62	none
13	El Cajón Dam [15]	Honduras	VA	234	57	3 discharge outlets
14	Karun-4 Dam [16]	Iran	VA	222	21.9	2 discharge outlets
15	Bhakra Dam [17]	India	PG	226	96.2	16 deep outlets in the dam
16	Hoover Dam [18]	America	VA	221	348	none
17	JiangPingHe	China	CFRD	219	13.66	1 discharge outlet
18	Daniel-Johnson Dam	Canada	PG	214	1418.52	low-level discharge outlet
19	Keban Dam	Turkey	PG	210	300	no
20	Almendra Dam	Spain	VA	202	26.49	2 bottom discharge outlets

Note: The dam type adopts the code of International Commission on dams: PG—gravity dam; VA—arch dam; ER—rock-fill dam; TE—earth dam; CFRD—Concrete faced rock-fill dam.

Amongst high dams above 200 m, early concrete dams were generally equipped with deep outlets for drawdown. The construction and development of high dams and large reservoirs have led to increased attention being given to drawdown facilities. Concrete dams are increasingly equipped with dedicated bottom outlets, used to withdraw the reservoir water below the deep outlet. More than 90% of China's earth-rock dam and rock-fill dam projects are equipped with emptying tunnels as dedicated drawdown facilities. By contrast, the majority of projects in other countries do not have dedicated emptying facilities but combine these with flood discharge, water diversion, or sand discharge structures to withdraw water levels when necessary. Traditionally, withdrawal facilities have been designed primarily to meet the maintenance requirements of high dams and large reservoirs, which is a normal emergency emptying method. However, the Jinsha River, Yarlung Zangbo River, and Niulan River barrier lake accidents that have occurred in China in recent years demonstrate that hydropower and water conservancy projects struggle to meet the challenge of emptying a large number of reservoirs in a short time. The emergency capacity of traditional drawdown facilities is insufficient, and the safety emergency drawdown technology of high dam and large reservoir projects needs innovation.

At present, some exploratory studies have been carried out on the layout of new deep drawdown facilities. Deng Gang et al. [19] proposed an inverted "U" shaped drawdown tunnel and set a vent gate at the downstream side of the top of the emptying tunnel. Since the inlet elevation is higher than the outlet elevation, when the reservoir water level is higher than the top elevation of the emptying tunnel and the reservoir water needs to be emptied, lift the vent gate, and the reservoir water can automatically flow out of the outlet and be emptied under its own weight. Although this structure can reduce the water pressure borne by the vent gate, the vent tunnel is arranged in an inverted "U" shape, and the drawdown of high dams and large reservoirs requires a large tunnel diameter, which increases the construction difficulty. In addition, the hydraulic characteristics under high-water heads are extremely complex, and the safety of this structure needs to be verified.

Yang Jiaxiu et al. [20] proposed a rapid deep drawdown technology for high dams and large reservoirs based on the combined water retaining of multistage gates. This technology provides the reverse hydraulic thrust by filling water at the downstream side of the single-stage gate so that the hydraulic thrust of the single-stage gate can meet the conventional design requirements. The total waterhead is shared by the "graded water retaining" of the multistage gate so that the drawdown facilities are arranged at a deeper underwater position. Under the condition of large discharge and high-water pressure, the dynamic water operating head of each gate and the flow velocity in each channel can still be at the normal level. At the same time, the technical idea of "channel layering + relay drawdown" is adopted to achieve the sustainability of drawdown depth in theory. This technology can rapidly empty high dams and large reservoir projects in the flood season and dry season. This technology meets the drawdown requirements of high dam and large reservoir project safety, cascade joint operation management, health diagnosis, upgrading and reconstruction, engineering accident rescue, disaster prevention, and emergencies and can effectively improve the safety of the operation and maintenance of the high dam and large reservoir project and the anti-risk ability under various major natural disasters and emergencies. This structure theoretically realizes the deep safety emergency venting of high dams and large reservoirs, and its engineering practice application in 300 m high dams is still in progress.

5. Difficulties in Reservoir Drawdown

5.1. Insufficiency of Drawdown Response Capability

The emptying capacity of the reservoir is evaluated by the drawdown ratio of the waterhead and the drawdown ratio of storage capacity [21]. Define the drawdown ratio of waterhead δ_H and storage capacity venting rate δ_V, The higher the value of δ_H and δ_V, the higher the drawdown capacity. δ_H and δ_V can be calculated by using Formulas (1) and (2).

$$\delta_H = \frac{Z_0 - Z_1}{H} \times 100\% \tag{1}$$

$$\delta_V = \frac{V_0 - V_1}{V_0} \times 100\% \tag{2}$$

where:

δ_H—Drawdown ratio of waterhead;
Z_0—Normal water level, m;
Z_1—Drawdown level, m;
H—Dam water head, m;
δ_V—Drawdown ratio of storage capacity;
V_0—Storage capacity corresponding to normal water level, m^3;
V_1—Storage capacity corresponding to emptying water level, m^3.

Using formulas (1) and (2), the following observations can be made regarding the emptying capacity of representative high dam and large reservoir projects in China: at present, the largest reduction rate of emptying head is Shuibuya Hydropower Station with 62.44%, while the reduction rate of most projects is only 30~50%, and the largest reduction rate of emptying capacity is Goupitan Hydropower Station with 96.75%, while most are in the range of 70~90% (Figure 2). Even for the high dam after the emptying storage decreasing rate is as high as 90%, an empty head lower rate can reach 50%. Only the remaining reservoir water sinks to the depth of an a-level one-hundred-meter high dam, which still has more than 100 million cubic meters that need to be filled with water (Figure 3). The same large storage capacity is equal to the (1) type of hydropower project. This emptying capacity is still drastically insufficient, to the extent that conditions do not even allow for adequate maintenance, and emergency safety provisions are seriously compromised.

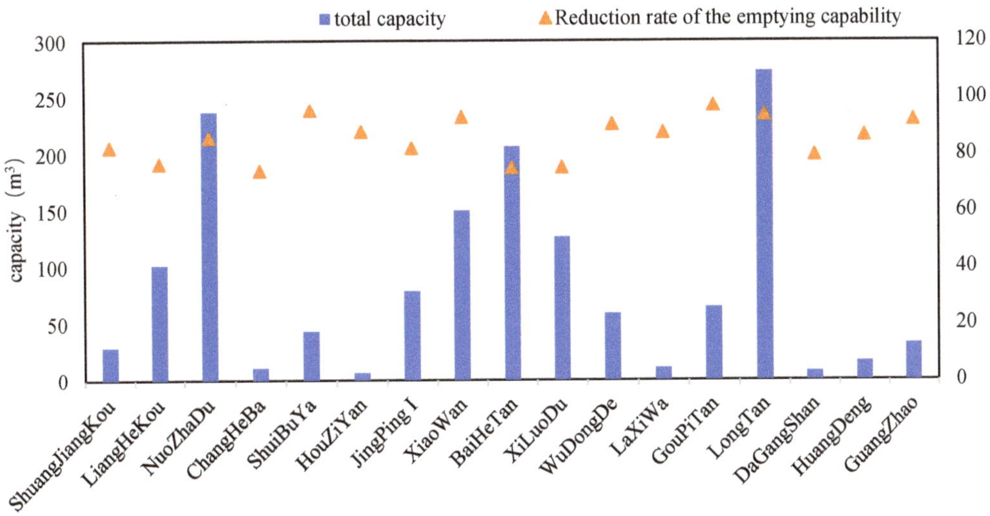

Figure 2. Reduction rate of the drawdown capability of 17 dams in China.

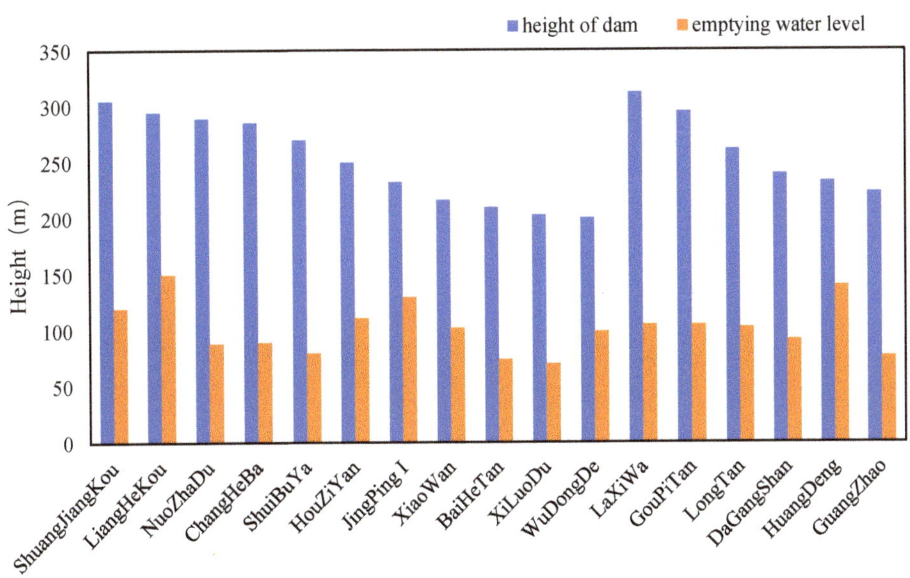

Figure 3. Drawdown water head capacity of 17 dams in China.

From the treatment of dammed lakes such as Jinsha River, Yarlung Zangbo River, and Niulan River in recent years, the frequent occurrence of serious and special natural disasters has put forward new requirements for an emergency response of hydropower and water conservancy projects. The most effective way to reduce the impact of accidents is to empty the storage capacity after disasters. Traditional water conservancy and hydropower projects, even if equipped with drawdown facilities, are basically based on the maintenance needs of high dams and large reservoirs. They are a normal drawdown means. When faced with emergencies, they often cannot withdraw the reservoir in a timely and effective manner, which cannot meet the emergency needs of emergencies. Therefore, the safety emergency drawdown technology needs to emerge at the historic moment. On the premise of meeting the requirements of safe operation and maintenance of high dams and large reservoirs, further improve the corresponding emergency capacity of the project, give full play to the disaster prevention and mitigation capacity of hydropower projects, and reduce the secondary disaster risk.

In addition to the above indicators, the emergency drawdown capacity can also be evaluated according to the drawdown time, drawdown capacity, and orifice area. According to the statistics of the drawdown time of the high dam and large reservoir project (Figure 4), it can be seen that the drawdown time is generally 20 to 40 days, some of which are 4.25 days short, and some of which are 102 days long. The average drawdown time per 100 million m^3 of storage capacity is 1.86 days, the average drawdown time per 1 m of water level drop is 3.78 days, and the drawdown can only be carried out in the dry season when the upstream inflow is small. In the flood season, because the upstream inflow is large, the drawdown flow is small, and the drawdown cannot be carried out. Because the time of the emergency cannot be predicted, the emergency response can only be carried out according to the principle of "the reservoir water shall be as low as possible, the drawdown shall be as fast as possible, and the loss shall be as small as possible". The existing 20-day or even 40-day emergency response time is relatively long. Therefore, the current drawdown time cannot meet the emergency venting requirements.

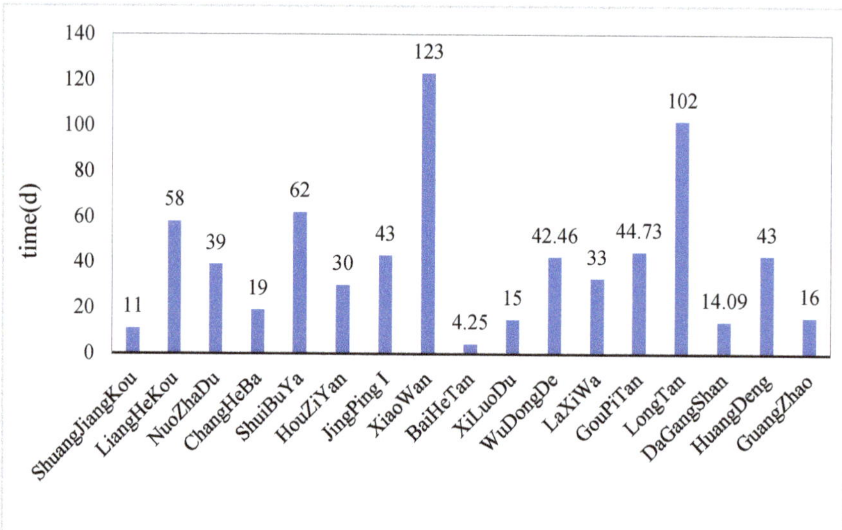

Figure 4. The drawdown time of 17 dams in China.

According to the statistics of the drawdown capacity and orifice area of the high dam and large reservoir project (Figure 5), the existing high dam and large reservoir project has a huge storage capacity. Because the discharge of the single hole of the concrete dam is small, with an average of 1300 m^3/s, it has to increase the discharge by arranging multiple bottom holes to achieve the purpose of emptying the reservoir. However, when the local material dam is equipped with a special emptying reservoir, the maximum discharge is 2000 m^3/s, which is the emptying tunnel of Lianghekou Hydropower Station. At present, the area of the vent hole is 41 m^2 on average, and the largest is the vent hole of Lianghekou Hydropower Station, up to 84 m^2. With the construction of a 300 m high dam, the demand for drawdown discharge is increasing, and the area of the vent hole also needs to be increased accordingly, which brings unprecedented challenges to drawdown technology.

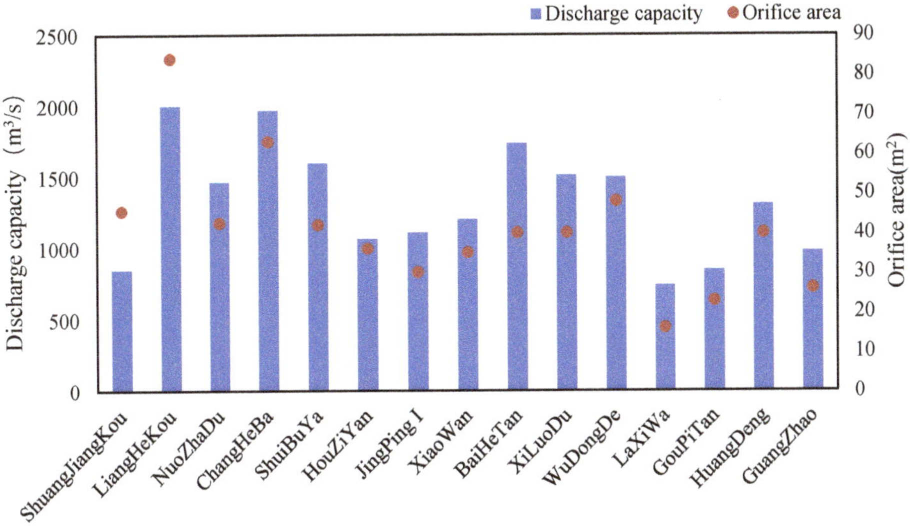

Figure 5. Discharge capacity and orifice area of 17 dams in China.

To sum up, at present, the area of a single hole of high dam and large reservoir projects is relatively small, the drawdown discharge is relatively small, the drawdown time is relatively long, and it can only be vented during the dry season. It is feasible for drawdown under normal operation, but if extreme conditions (earthquake, excessive flood, military strike, and cascade engineering disasters in the basin) require drawdown in flood season or rapid drawdown, it cannot meet the requirements. The lack of drawdown and emergency capacity is an important problem for the development of drawdown facilities in high dams and large reservoirs.

5.2. Technical Difficulties Relating to Safety and Stability of High Head Structures, Ultra-High-Lift Hoisting and Closing Equipment, and High Head Water Seals

In comparison with high dams, either the inlet of the discharge tunnel floor in conventional dams is higher or the gate well is located behind the dam's seepage prevention curtain. In addition, the structure bears less external water pressure, and the main load-bearing components of the gate shaft lining are only used as a safety reserve in the initial support and the secondary lining. However, for high dam and large reservoir projects, the intake floor is set deep, the gate well is located before the anti-seepage curtain, and external water pressure and surrounding rock deformation pressure are larger. When conventional structural mechanics or the boundary value method is used for calculation, the lining section size and the amount of reinforcement are larger. When the external water exceeds a certain head, the excavation section size and lining structure calculated are distinctly uneconomical and unreasonable [22]. In addition, during the operation of the circular hydraulic tunnel, the reinforced concrete lining is under the influence of internal water pressure, and its circum-axial stress is generally expressed as tensile stress. Under the action of a high-water head, the concrete may exceed its tensile strength and crack, leading to the seepage of internal water [23]. Therefore, the following points represent the most important design challenges for the safety and stability of emptying structures: consideration of the interaction between surrounding rock and lining, accurate analysis of the stability and safety of surrounding rock and lining structure, realistic simulation of damage and cracking behavior of hydraulic tunnel lining under pressure, and the economical and rational design of the reinforced concrete lining.

When the reservoir is drawn down, water is discharged by opening the upper spillway or surface outlet, spillway tunnel, and other gates. After the water level drops further, the emptying hole or the bottom outlet's arc gate is opened.

It can be seen from Table 2 that the current maximum water-retaining head for the bottom outlet gate of Xiaowan Hydropower Station is 160 m. This figure represents the highest design level. That is to say, the maximum pressure of the gate is 160 m static pressure; however, the operating water head is less than 160 m. The highest water pressure of 124,000 kN is found in the emptying and repair gate of the Lianghekou Hydropower Station; this represents a worldwide maximum.

Table 2. Technical characteristics of the main domestic and international emptying gates.

No.	Name	Gate Name	Orifice Dimensions (Width × Height/m × m)	Orifice Area (m^2)	Design Head (m)	Total Water Pressure (kN)
1	Rogun	Accident gate of deep spillway tunnel	5 × 7.5	37.5	146	52,330
		Working gate of deep spillway tunnel	5 × 6	30	146	40,245
2	ShuangJiangKou	Emptying accident gate	7 × 9	63	120	71,383
		Emptying radial working gate	7 × 6.5	45.5	120	52,112
3	LiangHeKou	Emptying check gate	8 × 10	80	120	124,000
		Emptying radial working gate	8 × 10.5	84	80	96,750
4	PuBuGou	Emptying accident gate	7 × 9	63	124	84,900
		Emptying working gate	6.5 × 8	52	126.28	121,870
5	LaXiWa	Emergency sluice gate for drain bottom outlet	4 × 9	36	132	48,694
		Sluice gate for bottom outlet	4 × 6	24	132	46,096
6	JinPing a	Emergency sluice gate for emptying bottom outlet	5 × 12	60	133	87,200
		Sluice gate for bottom outlet	5 × 6	30	133	52,120
7	ShuiBuYa	Emptying accident gate	5 × 11	55	152	84,530
		Emptying working gate	6 × 7	42	154	102,200
8	XiaoWan	Emergency sluice gate for drain bottom outlet	5 × 12	60	160	106,300
		Sluice gate for bottom outlet	5 × 7	35	163	110,000
9	ChangHeBa	Emptying gate	14 × 11.5	161	40	98,460

The extreme pressure borne by gates in high dams and large reservoirs necessitates larger hoisting equipment and higher water seal pressure. For high-head arc gates, swing hydraulic presses are generally used as hoisting equipment. By increasing the orifice area to improve discharge volume, the hydraulic press capacity exceeds the limit and the ultra-high lift and high-capacity opening and closing equipment volume. The outer side of the cylinder is too large, making it difficult to manufacture and transport, and the wire rope is easily damaged during the opening and closing of the gate. The head of the water seal of the gate changes in accordance with the water level of the reservoir area, and the stress state of the water seal varies greatly in response to the large water level fluctuations in the emptying tunnel. Research has been carried out on a number of domestic super-high head gates (including Dongfeng, Hongjiadu, Guangzhao, Xiaowan, and Laxiwa) to test water seal materials and types, to address the problem of water seal behavior under the action of high-water heads. When the water level is low, the contact stress between the water seal head and the water seal seat plate is low, and the risk of leakage is high. When the water level is high, the contact stress between the water seal head and the water seal seat plate becomes too large, leading to compression, deformation, and possible failure, of the water seal. The water seal is under unidirectional force: there is no water on the downstream side. Under the action of high-water levels, crevice jets are easily created, leading to "crevice cavities", cavitation or wear of the sluice panel and embedded parts, and vibration of the sluice, adversely affecting the normal operation of the gate and the safety of the structure.

When the water pressure of the gate cannot be increased, there is a conflict between the gate orifice area and emptying depth, namely, as emptying depth increases, the gate orifice area can only be reduced. Concurrently, discharge volume will decrease, which may damage the gate or prevent it from opening by excess pressure. When the discharge volume is less than the incoming flow of the reservoir, the water level of the reservoir cannot continue to decrease, and it cannot be emptied. Therefore, the pressure-bearing capacity of the gate, the capacity of the gate hoist, and the water seal performance in high dam and large reservoir projects, both domestically and internationally, have reached the limits of technology. The safety design of the metal structure is a further industry technical problem facing discharge outlets.

6. Main Technical Problems Faced by Deep Venting of High Dams and Large Reservoirs

China's Lancang River, Jinsha River, Nu River, and Yarlung Zangbo River basins are planning to build a number of 300 m high dams and large reservoirs, mostly cascade "leading" power stations in the basin. Some of the projects are located in the western frontier and minority areas, with complex political and social environments. The valleys with deep overburden and high unloading and steep slopes in the dam areas are ecologically fragile. The project area is cold, at a high altitude, and has a high seismic intensity, and the construction and operation conditions are complex. The technical difficulties of ultra-high dam construction technology, low-pressure flood discharge and energy dissipation, risk control, and emergency treatment have exceeded the scope of existing experience and specifications.

At present, the effective drawdown means of water conservancy projects are mainly to set up bottom holes, and the control equipment is a hydraulic gate. Due to the limitation of the pressure-bearing capacity of the gate and the capacity of the hoist, the drawdown facilities cannot be arranged deeper and the orifice cannot be larger, resulting in the limited drawdown depth and capacity, and the water cannot be discharged in the flood season, which cannot meet the requirements of emptying the reservoir at any time. This is a technical problem in the industry that needs to be solved urgently. However, the understanding of dam emptying at home and abroad is not uniform, and some are still quite different. On the one hand, it is suggested to set up deep drawdown facilities, which must be completely vented; on the other hand, due to the consideration of the project economy and lack of understanding of drawdown, it is recommended not to set up venting facilities. However, for ultra-high dam projects, it is necessary to set up deep drawdown facilities as an important engineering means to control complex political, social, engineering, and other risks.

To sum up, the main technical problems faced by the construction of high dam deep drawdown facilities are as follows:

(1) Determination of drawdown safety control standard: The long operation time of the high dam project, the large range of water level fluctuation, and the large discharge and flow rate will have an impact on the earth-rock dam body, the reservoir bank slope, and the rigid dam abutment and dam foundation. How to discharge water safely and quickly is the premise of studying the deep release of the high dam and large reservoir.

(2) Study of ultra-deep drawdown technology above 180 m water head: A key problem of the existing drawdown facilities is the contradiction between the drawdown depth and the discharge. The hydraulic thrust borne by the gate of the drawdown facility is proportional to the drawdown depth and the gate orifice area. With the increase in the drawdown depth being limited by the level of metal structure manufacturing technology, the hydraulic thrust borne by the gate extends to the maximum value of the current hydraulic thrust, the gate orifice area S decreases, and the discharge Q decreases. When the discharge is less than the inflow, the water increases instead of decreasing and cannot be emptied. Therefore, the first core problem to be solved for the emptying of high dams and large reservoirs is how to bear the high-water head of the gate. The development of ultra-deep drawdown technology above 180 m water head is the key to realizing the deep drawdown of high dams and large reservoirs.

(3) Study on drawdown intelligent control system: In order to realize the deep safe drawdown of high dams and large reservoirs, it is necessary to ensure the safe and stable operation of the drawdown system under various working conditions, and at the same time, it can be maintained frequently to avoid the situation that it cannot be opened when it is necessary to open. Therefore, the development of an intelligent control platform, the formation of an intelligent control process and emergency plan, and the realization of an intelligent reservoir drawdown control are important conditions to ensure the safe drawdown of high dams and large reservoirs.

(4) Drawdown effect evaluation: The effect of drawdown is directly related to the strength of the emergency response capability of high dams and large reservoirs. The traditional drawdown usually takes the drawdown depth, period, and drawdown rate as the evaluation indicators, but for different projects, the dam height is different, the inflow is different, the discharge channel is different, and the ratio of reservoir capacity to dam height is different, so it is impossible to compare the horizontal drawdown capacity based on the above three parameters. Even for the same project, only the above three parameters cannot qualitatively express the actual engineering significance of the drawdown effect on project safety, maintenance, and other aspects. Therefore, it is an important judgment basis for a safe drawdown of high dams and large reservoirs to explore the evaluation of the drawdown effect in many aspects.

(5) Risk prevention and control technology of cascade basin: On the premise that the drawdown safety of a single project can be guaranteed, the drawdown process should pay special attention to the impact of the cascade basin. Therefore, the multi-step emergency dispatching and stepwise drawdown mode of the catastrophic flood should be studied, and the stepwise drawdown time and flow technical requirements considering the impact of the cascade basin should be put forward. With the protection object of the basin as the target, a multi-objective collaborative emergency drawdown mode for cascade reservoirs is established. From the basin system level, it is an important requirement to study the risk prevention and control technology of engineering safety and coastal public safety in the process of multi-objective collaborative drawdown of cascade reservoirs and to build the risk prevention and control technology of emergency venting of high dams and large reservoirs as well as cascade reservoirs to adapt to the development of cascade basins.

7. Key Technologies to Be Solved Urgently

In the emptying facility, let the head of the pressure center of the gate be h_0, the cross-sectional area of the emptying facility be S, and the bulk density of water be γ. Then, the water pressure on the gate is given by means of $F = \gamma h_0 S$. Ignoring the traveling head of the reservoir and considering the effects of energy loss, the constant coefficient $\sqrt{2g}$, and structural shape coefficient as k, the discharge flow capacity is $Q= kSh_0{}^{1/2} = kF\gamma^{-1}h_0{}^{-1/2}$ [20]. F is proportional to the emptying depth H (the head of the gate pressure center is h_0) and the gate orifice area S. As the emptying depth increases, the gate's water pressure F tends to F_{max}, the gate orifice area S decreases, and the discharge Q decreases accordingly. When the discharge flow is lower than the incoming flow of the reservoir, the water level does not decrease but increases and the reservoir cannot be emptied. Emergency emptying capability is expressed in terms of emptying time, discharge volume, and discharge orifice area. The orifice area is the crucial factor: the larger the orifice area, the larger the discharge capacity, and the shorter the corresponding emptying time. The effectiveness of the emergency emptying response is therefore largely determined by the size of the discharge orifice. However, as we have seen, until now, discharge structures have been limited by the pressure-bearing capacity of the gate and the capacity of the hoist: the floor elevation cannot be deeper and the orifice cannot be too large. This results in limited emptying depth and discharge volume, which cannot meet any period of time. To meet the requirement of full emptying, this key technological problem urgently needs to be solved.

8. Conclusions and Outlook

The high dam and large reservoir projects have large storage capacities, high-water heads, period ring benefits, and high technical requirements for safety assurance purposes. From the perspective of long-term operation requirements, comprehensive inspection, maintenance, and reinforcement require a deeper lowering of the reservoir water level. In response to emergencies, it is necessary to have the capability of the emergency emptying reservoir water at any time. Based on a thorough investigation of the current situation of venting facilities for high dams and large reservoirs with a height of 200 m or more in the world, this paper finds that drawdown facilities currently face difficulties such as insufficient drawdown capacity, poor safety and stability of high head structures, extremely high lift hoisting equipment, and high difficulty in high head water seal technology. It is pointed out that the key technologies that need to be urgently addressed for a deep drawdown of high dams and large reservoirs are the pressure-bearing capacity of gates and the capacity limit of hoists. As a result, the elevation of the bottom plate of the drawdown building cannot be arranged, the orifice is limited, and the drawdown depth and drawdown discharge capacities are limited.

The emptying technology of high dams and large reservoirs needs to be further improved. First, it is necessary to conduct research into the formulation of relevant specifications and specify emptying capacity, emptying depth, emptying storage capacity, and emptying time. Second, limited to the particularity of the current emptying tunnel, it is difficult to design the structure under the conditions of the high-water head and large water level variation. The principle of an inverted siphon can be considered. The height of the inlet is higher than that of the outlet, and the gate position is relatively high. The water's self-weight and the inverted siphon principle are used for emptying dams. We used the hierarchical emptying concept for arranging emptying facilities and multiple emptying tunnels. Each emptying tunnel is equipped with multi-level gates. The downstream of the gate is filled with water to reduce the total water pressure of the gate, increase the safety and stability of the structure, and reduce the difficulty of structural design. In the future, we can also break through the limits of thinking and find new ways to solve the problem of safety and emergency emptying, e.g., drawing lessons from the working principle of deep-water submarines, designing high-pressure inflatable gates, using hollow inflatable gates to resist the high-water head upstream, reducing the total water pressure of the gates, and using composite nanomaterials to make gates, increase the strength of the gates, and achieve the purpose of deep-hole water blocking and emptying.

Author Contributions: Conceptualization and methodology, X.Z. and J.Y.; data collection, X.Z. and S.D.; writing—original draft preparation, X.Z. and L.Z.; writing—review and editing, S.W.; supervision and funding acquisition, S.W. and S.L. All authors have read and agreed to the published version of the manuscript.

Funding: This research was funded by the National Key Research and Development Program of China (Grant No. 2021YFC3090105).

Data Availability Statement: Not applicable.

Acknowledgments: The authors sincerely thank the reviewers who contributed their expertise and time to review this manuscript.

Conflicts of Interest: The authors declare no conflict of interest.

References

1. Zheng, S. Discussion on the construction and operation safety of high dams in China. In *Dam Construction and Management, Technical Progress in China Dam Association 2012 Annual Meeting of the Academic Papers;* Yellow River Water Conservancy Press: Zhengzhou, China, 2012; pp. 12–20.
2. Wang, L.; Jiang, L.; Ma, B.; Cai, W.; Chen, B.; Hu, Z. Key technologies of manned submersible in the field of high dam reservoir. *China Water Transp. Second. Half* **2019**, *19*, 17–18, 110.
3. Cai, Y.; Sheng, J.; Xiang, Y. Dam deepwater detection, emergency monitoring, early warning and emergency response of major water conservancy projects. *China's Environ. Manag.* **2017**, *9*, 105–106.
4. Chen, M. Emergency operation experience and Enlightenment of Baige Barrier Lake in Jinsha River. *People's Yangtze River* **2019**, *50*, 14–18.
5. Zhou, X.; Du, X.; Yao, Y. Dam break flood analysis of Baige barrier lake in Jinsha River. *Hydropower* **2019**, *45*, 8–12, 32.
6. Nie, G.; Xing, L. Necessity of setting reducing or empting reservoir facilities in hydropower station junction projects. *Dam Saf.* **2004**, *5*, 30–34.
7. Gao, J.; Guo, J. Some problems about setting the emptying hole in concrete-faced rockfill dam. *J. Hydraul. Power Gener.* **2013**, *32*, 179–183.
8. Zhou, X.; Zhou, J.; Du, X. Analysis and Enlightenment of spillway accident of Oroville dam in the United States. *J. Water Conserv.* **2019**, *50*, 650–660.
9. Savinov, O.A.; Ivanov, P.L.; Krasnikov, N.D. Concerning the problem of the seismic stability of the Nurek dam. *Hydrotech. Constr.* **1986**, *20*, 278–285. [CrossRef]
10. Bronshtein, V.I.; Lomov, I.E.; Menabde, A.Y. Arch dam of the Inguri hydroelectric station. *Hydrotech. Constr.* **1979**, *13*, 1162–1173. [CrossRef]
11. Borovoi, A.A.; Moiseev, I.S. Construction of the Chicoasen dam. *Hydrotech. Constr.* **1977**, *11*, 534–537. [CrossRef]
12. Aleksandrovskaya, É.K. Stress-strain state of the Sayano-Shushenskoe dam during filling of the reservoir. *Hydrotech. Constr.* **1986**, *20*, 123–129. [CrossRef]
13. Larrahondo, J.; Ross, W.; Ruiz, W.; Marulanda, C. Long-Term Behavior of a Rockfill Dam: La Esmeralda Dam, Colombia. In Proceedings of the Geo-Congress 2014, Atlanta, GA, USA, 23–26 February 2014.
14. McDonald, C. Oroville Dam highlights infrastructure risks. *Risk Manag.* **2017**, *64*, 6–9.
15. Alavez, A.C.; Santillan, D.Y.; Filloy, J.E. Properties of Compaction Reached in the Rockfills of the El Cajón Dam, Modern Rockfill Dams. China Water Resources and Hydropower Publishing House: Beijing, China, 2009.
16. Bandpey, H.K.; Ahangari, K.; Shirvani, M.H. Displacements Prediction in Double-Arch Dam Rock Abutment Using SPSS Software Based on Extensometer Readings Case study: Karun 4 Concrete Dam, Iran. *Res. J. Appl. Sci. Eng. Technol.* **2012**, *4*, 4607–4616.
17. Uppal, H.L.; Gulati, T.D.; Sharma, B.A.D. A Study of Causes of Damage to the Central Training Wall Bhakra Dam Spillway. *J. Hydraul. Res.* **1967**, *5*, 209–224. [CrossRef]
18. Giroux, R.P. Building Hoover Dam (Men, Machines, and Methods). In Proceedings of the Hoover Dam Anniversary History Symposium, Las Vegas, NV, USA, 21–22 October 2010.
19. Deng, G.; Li, Y.; Li, W. The Structure, Layout and Use Method of the Emptying Tunnel of the Super-High Dam Reservoir above 200 Meters. ZL201710547785.8, 6 July 2017.
20. Yang, J.; Du, S.; Zhan, Z.; Li, X. Ch on rapid and deep venting technology of high dam and large reservoir based on combined water retaining of multi-stage gate. *J. Water Conserv.* **2020**, *51*, 906–914.
21. *NB/T10966-2022*; Technical Guideline for Reservoir Drawdown. National Energy Administration: Beijing, China, 2022.
22. Bao, W.; Du, S.; Zheng, X. Study on calculation of shaft lining structure under high head sluice gate. *Sci. Technol. Innov.* **2020**, *2*, 24–28.
23. Zhu, K.; Xiao, M.; Chen, J.; Wang, A. Analysis of load-bearing characteristics of reinforced concrete lining for hydraulic tunnel with high head. *China Rural. Water Hydropower* **2019**, *12*, 124–128, 134.

MDPI AG
Grosspeteranlage 5
4052 Basel
Switzerland
Tel.: +41 61 683 77 34

Water Editorial Office
E-mail: water@mdpi.com
www.mdpi.com/journal/water

www.ingramcontent.com/pod-product-compliance
Lightning Source LLC
Chambersburg PA
CBHW040815250926
51850CB00007BA/1172
* 9 7 8 3 7 2 5 8 6 0 7 3 9 *